INTELLIGENT ADAPTIVE CONTROL

Industrial Applications

Edited by
Lakhmi C. Jain
Clarence W. de Silva

CRC Press

Boca Raton   London   New York   Washington, D.C.

Library of Congress Cataloging-in-Publication Data

Catalog record is available from the Library of Congress.

PREFACE

Fuzzy logic, neural networks, and evolutionary computing have provided important tools and techniques for system control. Specifically, the field of Intelligent Control is fertile with techniques of computational intelligence; particularly, soft computing, that are, by and large, based on a fuzzy-neural-evolutionary framework. Intelligent control seeks to achieve good performance in machines, industrial processes, consumer products, and other systems, by using control approaches that, in a loose sense, tend to mimic direct control by experienced humans. Many of these techniques can learn, adapt to compensate for parameter changes and disturbances, and are able to provide satisfactory control even in incompletely-known and unfamiliar situations. With this backdrop, the main purpose of this book is to present some important techniques, developments, and applications of computational intelligence in system control.

Intelligent control is rapidly becoming an established field of education and research, not simply because of its pedagogical importance but also in view of its tremendous success in industrial applications. Most of the books in this field, however, are written by university educators and researchers, in the form of textbooks, research monographs, or collections of research papers. People who implement the techniques of intelligent control in industrial systems and products are often reluctant to publish their work due to proprietary restrictions, and the lack of direct professional benefit and opportunity. Of course, there are some university-industry collaborative activities where intelligent control techniques have been developed, designed, implemented, and evaluated, with the end result of successful commercial products and systems. Such collaborative work has a greater probability of entering into archival literature than those that are exclusively industrial developments. Facing this situation, we have made a serious attempt, in this book, to report truly industrial developments and applications of intelligent control.

This book consists of fourteen chapters. Chapter 1 provides an introduction to the fundamentals of neural networks, fuzzy logic, and evolutionary computing. This material forms the foundation upon which the remaining chapters are based. Chapters 2 through 7 fall into the category of theory and applications. In particular, these chapters will provide a more rigorous treatment of intelligent control and also will demonstrate some applications of the associated techniques. The remaining seven chapters primarily present industrial applications of intelligent control and soft computing. Applications presented in these chapters come from a variety of industries including transportation, petroleum, motor drive, and fish processing, The emphasis of the book is on the application of fuzzy logic, neural networks, and evolutionary computing. It is these three general approaches that are utilized either separately or cooperatively, in most applications presented in the book. Other knowledge-based techniques are also used in some applications, albeit in a lir manner; specifically, in Chapters 13 and 14.

The book provides a state-of-the art treatment in practical applicatior computational intelligence in system control. Introductory and advanced th design, practical implementation, and industrial use are covered in a coh manner. The book will be useful to engineers, scientists, researchers, and gra students, as a practical guide to theory and implementation of intelligent co particularly employing fuzzy logic, neural networks, and evolutionary computir

■ ACKNOWLEDGMENTS

We are very grateful to the contributors of this book for their aptitude, rigor professionalism. The significance of their individual contributions is reflected i overall utility of the book. We wish to thank Jan Vogel, Pieter Grimmerink Jerry Papke for their assistance in the preparation of the book. Our gra students and research colleagues have made important contributions to the co and development of the book, both directly and indirectly. In this context, we to particularly mention here, with thanks, Min-Fan Ricky Lee of National Res Council and the University of British Columbia, Scott Gu of Precix, Inc., Kurnianto of Tantus Electronics, Corp., Fakhri Karray of University of Wate and Jonathan Wu of National Research Council.

Lakhmi Jain
University of South Australia, Adelaide

Clarence de Silva
University of British Columbia, Vancouver.

THE EDITORS

Lakhmi C. Jain is the Founding Director of the Knowledge-Based Intelligent Engineering Systems (KES) Centre, Faculty of Information Technology, University of South Australia, Adelaide, Australia. His Ph.D. degree is in Electronic Engineering. A Fellow of the Institution of Engineers, Australia, he has received a number of awards for his papers including the Sir Thomas Ward Memorial Medal in 1996, and the Best Paper Award from the Electronics Association of South Australia, in 1995. Jain holds a patent (with Udina) for Skylight Light Intensity Data Logger. He is Founding Editor-in-Chief of the *International Journal of Knowledge-Based Intelligent Engineering Systems* and serves as an Associate Editor of the *IEEE Transactions on Industrial Electronics*. He has organized several international conferences and was appointed as Vice-President (International Liaison) by the Electronics Association of South Australia An expert in knowledge-based systems, artificial neural networks, fuzzy systems, and genetic algorithms with application to practical engineering problems, he has authored or co-authored over 100 books and research papers in his field. He is the Editor-in-Chief of the International Series in Computational Intelligence of CRC Press.

Clarence W. de Silva, Fellow ASME and Fellow IEEE, is Professor of Mechanical Engineering at the University of British Columbia, Vancouver, Canada, and occupies the NSERC Chair Professorship in Industrial Automation since 1988. He obtained his first Ph.D. from the Massachusetts Institute of Technology in 1978 and, twenty years later, another doctorate from the University of Cambridge, England. De Silva has served as a consultant to several companies including IBM and Westinghouse in the U.S.A., and has led the development of many industrial machines. He is a recipient of the Meritorious Achievement Award of the Association of Professional Engineers of BC, the Outstanding Contribution Award of the IEEE Systems, Man, and Cybernetics Society, and the Outstanding Large Chapter Award of the IEEE Industry Applications Society. He has authored 13 technical books, 10 edited volumes, over 100 journal papers, and a similar number of conference papers and book chapters. He has served on the editorial boards of twelve international journals and is the Senior Technical Editor of *Measurements and Control*, and Regional Editor, North America, of the *International Journal of Intelligent Real-Time Automation*. He has been a Lilly Fellow, Senior Fulbright Fellow to Cambridge University, ASI Fellow, and a Killam Fellow.

CONTENTS

1 | INTELLIGENT CONTROL TECHNIQUES

A. Filippidis and L.C. Jain
Knowledge-Based Intelligent Engineering Systems Centre
University of South Australia
Adelaide, The Levels, SA 5095, Australia

C.W. de Silva
Industrial Automation Laboratory
Department of Mechanical Engineering
The University of British Columbia
Vancouver, Canada

Fuzzy logic is useful in representing human knowledge in a specific domain of application, and in reasoning with that knowledge to make useful inferences or actions. Artificial neural networks are massively connected networks that can be trained to represent complex nonlinear functions at a high level of accuracy. They are analogous to the neuron structure in a human brain. Genetic algorithms are optimization techniques that can evolve through procedures analogous to human evolution, where natural selection, crossover, and mutation are central. Computational procedures associated with these techniques fall into the area of soft computing. This chapter will introduce some fundamental techniques of fuzzy logic, neural networks, and genetic algorithms. The main focus here will be their use in intelligent control. The basic treatment given in this chapter will lay the groundwork for a more rigorous analysis and application presented in the subsequent chapters of the book.

1 Introduction

This introductory chapter will outline the core techniques that form the basis of the developments and applications presented in the remaining chapters of the book. The presentation is not intended to be exhaustive, in view of the purpose of the chapter. A more rigorous treatment of some of the topics introduced here may be found in the subsequent chapters.

The term "Intelligent Control" may be loosely used to denote a control technique that can be carried out using the "intelligence" of a human who is knowledgeable in the particular domain of control [1]. In this definition, constraints pertaining to limitations of sensory and actuation capabilities and information processing speeds of humans are not considered. It follows that if a human in the control loop can properly control a plant, then that system would be a good candidate for intelligent control. Information abstraction and knowledge-based decision making that incorporates abstracted information, are considered important in intelligent control. Unlike conventional control, intelligent control techniques possess capabilities of effectively dealing with incomplete information concerning the plant and its environment, and unexpected or unfamiliar conditions. The term "Adaptive Control" is used to denote a class of control techniques where the parameters of the controller are changed (adapted) during control, utilizing observations on the plant (i.e., with sensory feedback), to compensate for parameter changes, other disturbances, and unknown factors of the plant. Combining these two terms, one may view "Intelligent Adaptive Control" as those techniques that rely on intelligent control for proper operation of a plant, particularly in the presence of parameter changes and unknown disturbances.

Computational procedures of Fuzzy Logic (FL), Neural Networks (NNs), and Genetic Algorithms (GAs) fall into the class of "soft computing" techniques, which can be directly utilized in intelligent control, either separately or synergistically. In particular, fuzzy logic may be employed to represent, as a set of "fuzzy rules," the knowledge of a human controlling a plant. This is the process of knowledge representation. Then, a rule of inference in fuzzy logic may be used according to this "fuzzy" knowledge base, to make control decisions for a given set of plant observations. This task concerns "knowledge processing" [1]. In this sense, fuzzy logic in intelligent control serves to represent and process the control knowledge of a human in a given plant.

Artificial neural networks represent a massively connected network of computational "neurons." By adjusting a set of weighting parameters of an NN, it may be "trained" to approximate an arbitrary nonlinear function to a required degree of accuracy. Biological analogy here is the neuronal architecture of a human brain. Since intelligent control is a special class of highly nonlinear control, neural networks may be appropriately employed there, either separately or in conjunction with other techniques such as fuzzy control. Fuzzy-neural techniques are applicable in intelligent adaptive control, in particular, when parameter changes and unknown disturbances have to be compensated.

Genetic algorithms belong to the area of evolutionary computing. They represent an optimization approach where a search is made to "evolve" a solution algorithm that will retain the "most fit" components, in a procedure that is analogous to biological evolution through natural selection, crossover, and mutation. It follows that GAs are applicable in intelligent control, particularly when optimization is an objective. Summarizing, the biological analogies of fuzzy, neural, and genetic approaches are: fuzzy techniques attempt to approximate human knowledge and the associated reasoning process; neural networks are a simplified representation of the neuron structure of a human brain; and genetic algorithms follow procedures that are crudely similar to the process of evolution in biological species.

Modern industrial plants and technological products are often required to perform complex tasks with high accuracy, under ill-defined conditions. Conventional control techniques may not be quite effective in these systems whereas intelligent control has a tremendous potential. The emphasis of the present book is on practical applications of intelligent control, primarily using FL, NN, and GA techniques. The remainder of this chapter will give an introduction to some fundamental techniques of fuzzy logic, neural networks, and genetic algorithms, within the context of intelligent control [1].

2 Knowledge-Based Systems

The expert systems are offshoots of artificial intelligence (AI) which is concerned with using computers to simulate human intelligence in a limited way. Artificial intelligence can be defined as the science of making machines do things that would require intelligence if done by humans. An expert system is defined as a software system (Figure 1) with a high symbolic and descriptive information content, which can simulate the performance of a human expert in a specific field or domain [2,3]. Expert systems acquire knowledge mostly from human experts. These systems deal with complex knowledge for which real expertise is required. Knowledge-Based Systems (KBS) are a general form of expert systems in view of the fact that they are quite general in application and not limited to mimicking the role of a human expert. Furthermore, very often they acquire knowledge from non-human information sources.

The major logical components of an expert system are a Knowledge Base, an Inference Engine, an Interface between the system and the external environment, an Explanation Facility, and a Knowledge Acquisition Facility.

A "knowledge engineer" gathers the expertise about a particular domain from one or more experts, and organizes that knowledge into the form required by the particular expert system tool that is to be used. Forms of knowledge representation may include logic, production system (rules), semantic scripts, semantic primitives, and frames and scripts. A typical expert system uses a data structure as the basis of the particular representation technique it implements. Consequently, the knowledge engineer needs to know the general form in which the knowledge is to be represented and the particular form of that representation required by the expert system itself. The knowledge that is "engineered" in this manner is the Knowledge Base [4].

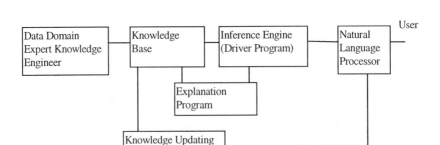

Figure 1: Simplified block diagram of a typical expert system.

The Inference Engine is the "driver" program. It traverses the Knowledge Base, in response to observations and other inputs provided to it from the external world, and possibly previous inferences and results from the KBS itself, and will identify one or more possible outcomes or conclusions [5]. This task of making inferences for arriving at solutions will involve "processing" of knowledge. It follows that representation and processing of knowledge are central to the functioning of a KBS [1]. The data structure selected for the specific form of knowledge representation determines the nature of the program created as an Inference Engine.

Keyboards, screen displays, sensors, transducers, and even output from other computer programs including expert systems, usually provide the Interface between an expert system and the external world.

In the case of production (rule-based) systems, the Knowledge Base consists of a set of rules written as an ASCII file. Therefore, the Knowledge Acquisition Facility that is required for these systems is often merely an editor [3]. In systems based on other forms of representation, the Knowledge Acquisition Facility will generally be an integral part of the expert system and can be used only with that system.

Typically, what is commercially available for developing expert systems is an expert system "shell." It will consist of the software programs required, but will not contain a knowledge base. It is the responsibility of the user, then, to organize the creation of the required knowledge base, which should satisfy the system's requirements with respect to the form of knowledge representation that is used and the structure of the knowledge base [2].

An expert system that is used to supervise the control system of a plant is called Control Expert System. Such expert systems are directly applicable at a high level, for process monitoring, supervision, and diagnosis, in intelligent control.

3 Neural Networks

3.1 Introduction

The successful operation of an autonomous machine depends on its ability to cope with a variety of unexpected and possibly unfamiliar events in its operating environment, perhaps relying on incomplete information [1]. Such an autonomous machine would only need to be presented a goal; it would achieve its objective through continuous interaction with its environment and automatic feedback about its response. In fact, this is an essential part of "learning." By enabling machines to possess such a level of autonomy, they would be able to learn higher-level cognitive tasks that are not easily handled by existing machines [6]. In designing a machine to emulate the capabilities found in biological controls, which depend on "intelligence." some experience on the structural, functional, and behavioral biological neural systems would be valuable. Neural networks have a great potential in the realm of nonlinear control problems [7]. A significant characteristic of neural networks is their ability to approximate arbitrary nonlinear functions. This ability of neural networks has made them useful in modeling nonlinear systems. A neuro-controller (neural network-based control system), in general, performs a specific task for adaptive control, with the controller taking the form of a multilayer network, and adaptable parameters being defined as the adjustable weights. In general, neural networks represent parallel-distributed processing structures, which make them prime candidates for use in multi-variable control systems. The neural-network approach defines the problem of control as the mapping of measured signals of system "change" into calculated "control actions," as shown in Figure 2.

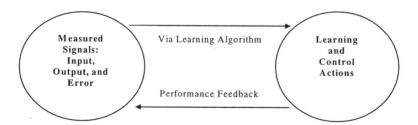

Figure 2: Representation of learning and control actions in a neural network approach. Mapping of the measured signals onto the learning and control space [8].

3.2 Biological Neuronal Morphology

The human brain consists of approximately 10 billion individual nerve cells called neurons. Each neuron is interconnected to many other neurons, forming a densely connected network called neural network. These massive interconnections provide an exceptionally large computing power and memory. The basic building block of the central nervous system is the neuron, the cell that processes and communicates information to and from various parts of the body [8]. From an information-processing point of view, an individual neuron consists of the following three parts, each associated with a particular mathematical function:

(1) The dendrites are a receiving area for information from other neurons.
(2) The cell body, called soma, collects and combines incoming information from other neurons.
(3) A neuron transmits information to other neurons through a single fiber called an axon. The axon is a tubular structure bounded by a typical cell membrane.

The junction of an axon with a dendrite of another neuron is called a synapse. Synapses provide memory for accumulating experience or knowledge. A single axon may be involved with hundreds of other synaptic connections. A schematic diagram of the biological neuron is shown in Figure 3(a). From a system-theoretic point of view, the neuron can be considered a multiple-input—single-output (MISO) system, as depicted in Figure 3(b).

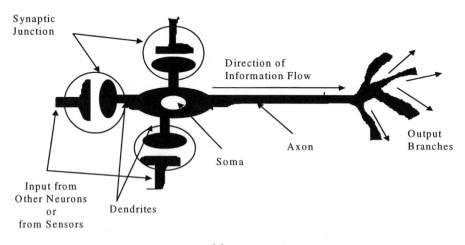

(a)

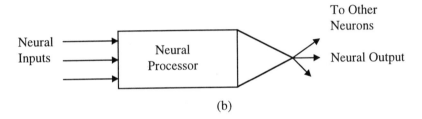

(b)

Figure 3: (a) A schematic view of the biological neuron.
(The soma of each neuron receives parallel inputs through its synapses and dendrites, and transmits a common output via the axon to other neurons.)
(b) Model representation of a biological neuron, with multiple inputs and a single output [8, 9].

Each neuron acts as a parallel processor because it receives pulses in parallel from neighboring neurons and then transmits pulses in parallel to all neighboring synapses [10]. The processing of information within the biological neuron involves two distinct operations [11,12]:

(1) Synaptic operation: This provides a weight to the neural inputs. Thus, the synaptic operation assigns a relative weight (significance) to each incoming signal according to the past experience (knowledge or memory) stored in the synapse.
(2) Somatic operation: This provides aggregation, thresholding, and nonlinear activation to the dendritic inputs. If the weighted aggregation of the neural inputs exceeds a certain threshold, the soma will produce an output signal.

3.3 Static Neural Networks

A neuron receives inputs from a number of other neurons or from the external world. A weighted sum of these inputs constitutes the argument of a nonlinear activation function as shown in Figure 4 [8]. The neuron is said to have been fired if the weighted sum of its inputs exceeds a certain threshold, w_0. Mathematically, the function of a neuron can be modeled as

$$y(t) = \Psi\left[\sum_{i=1}^{n} w_i x_i - w_0\right] \qquad (1)$$

where $[x_1,...,x_n]$ represent neuron inputs, $[w_1,...,w_n]$ are the synaptic weights, $y(t)$ is the neural output, and $\Psi[.]$ is some nonlinear activation function with threshold w_0. Using this model, many neural morphologies, usually referred to as feedforward neural networks, have been reported in literature [8,11]. These feedforward networks respond instantaneously to inputs because they possess no dynamic elements in their structure. Therefore, these are called static neural networks. A schematic representation of a static neural network is shown in Figure 4(b).

The neural network attributes, such as learning from examples, generalization, redundancy, and fault tolerance, provide strong incentives for choosing neural networks as an appropriate approach for modeling biological systems. The potential benefits of such a network can be summarized as follows [8,11]:

(1) The neural network models have many neurons (the computational units) linked via adaptive (synaptic) weights, arranged in a massive parallel structure. Because of its high parallelism, failure of a few neurons does not significantly affect the overall performance (known as fault tolerance).
(2) The ability to adapt and learn from the environment means that the neural network models can deal with imprecise data and ill-defined situations. A suitably trained network can generalize; i.e., it can accurately deal with inputs that do not appear in the trained data.
(3) The ability to approximate any nonlinear continuous function to a desired degree of accuracy.
(4) Neural networks have many inputs and many outputs; hence, they are easily applicable to multivariable systems.
(5) With advances in hardware technology, many vendors have introduced dedicated VLSI hardware implementations of neural networks. This brings additional speed to neural computing.

3.4 Common Types of Artificial Neural Networks

The commonly used neural network structures [9,11] include feedback and feedforward networks.

The feedback networks are neural networks that have connections between network output and some or all other neuron units (see Figure 4(c)). Certain unit outputs in the figure are used as activated inputs to the network, and other unit outputs are used as network outputs. Due to the feedback, there is no guarantee that the networks become stable. To guarantee stability, constraints on synaptic weights are introduced so that the dynamics of the feedback network is expressed by a Lypunov function. Concretely, a constraint of equivalent mutual connection weights of two units is implemented. The Hopfield network is one such neural network.

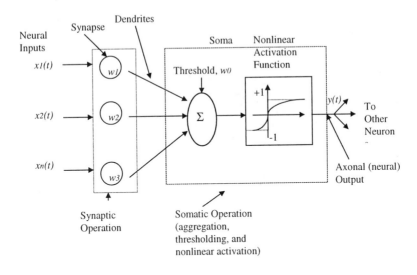

(a)

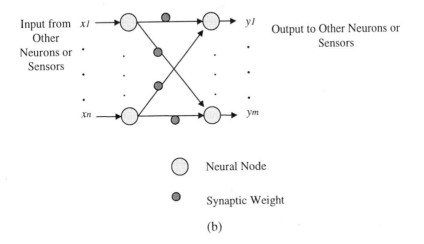

(b)

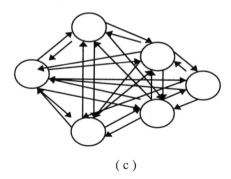

(c)

Figure 4: (a) A static neural node.([$x_1(t)$,..., $x_n(t)$] represent neural inputs,
[w_1 ,..., w_n] are the static synaptic weights, w_0 is the threshold, and
$y(t)$ is the axonal (neural) output).
(b) A static (feedforward) neural network with n-inputs and m-outputs.
(c) A feedback neural network [9,11].

Two aspects of a Hopfield network are as follows:

(1) Synaptic weights are determined by analytically solving the constraints, not by performing an iterative learning process. The weights are fixed during operation of the Hopfield network.
(2) Final network outputs are obtained by operating feedback networks, for the solution of an application task.

Another type of neural network which should be compared with the feedback network is the feedforward type as shown in Figure 4(b). The feedforward network is a filter whose output is the processed input signal. An algorithm will determine the synaptic weights to make the outputs match the desired result. These learning algorithms are categorized into supervised and unsupervised learning.

In a supervised learning algorithm, the synaptic weights are adjusted using input-output data, whereby input-output characteristics of the network are modified to obtain a desired output. In reinforcement learning, only a qualitative indicator of whether the desired output is achieved would be available. A popular algorithm of supervised learning, the backpropagation, will be described in detail in the next section.

In an unsupervised learning algorithm, the synaptic weights are adjusted according to the input values of the network, and not according to supervised output data of the network. Since the output characteristics are determined by the network itself, without the benefit of the desired output data, the unsupervised learning mechanism is called self-organization. Hebbian learning and competitive learning are representative of unsupervised learning algorithms.

A Hebbian learning algorithm increases the weight w_i between a neuron and an input x_i, if the neuron y is fired by that input (see Figure 5(a)). Specifically, we have

$$\Delta w_i = ayx_i \qquad (2)$$

where a is a learning rate. Weights are strengthened if units connected with the weights are activated. Weights are normalized to prevent an infinite increase in weights.

In a competitive learning algorithm, weights are modified to generate a unit with the greatest output. Some variations of the algorithm also modify other weights by lateral inhibition, to suppress the outputs of other units whose outputs are not the greatest. Only one unit can become active as the winner of the network. Kohonen's self-organizing feature map, a well-known competitive network, modifies the weight connected to the winner-take-all as
$$\Delta w_i = a(x_i - w_i), \qquad (3)$$
where the sum of input vectors is normalized as 1 (see Figure 5(b)).

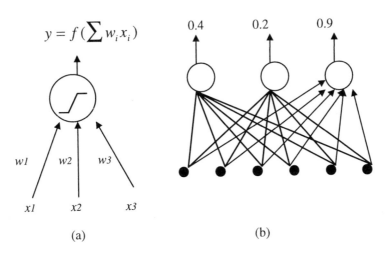

(a)   (b)

Figure 5: (a) Hebbian learning algorithm strengthens weight w_i when
input x_i activates neuron y.
(b) Competitive learning algorithm strengthens only the weights
connected to the unit whose output is the largest.

3.5 Backpropagation Learning Algorithm

One of the most popular learning algorithms which iteratively determines the weights of a feedforward network is the backpropagation algorithm. A simple learning algorithm that modifies the weights between the output and the hidden layers is called the delta rule. The backpropagation algorithm is an extension of the delta rule that can train the weights, not only between the output and hidden layers but also the hidden and input layers. Historically, several researchers proposed this idea independently [9,11]. Eventually, Rumelhart, in [13], provided a detailed mathematics basis for this algorithm.

Let E be the error between the network outputs, v^3, and supervised data, y. Since the network outputs are changed when synaptic weights are modified, E must be a function of the synaptic weight w:

$$E = \frac{1}{2}\sum_{j=1}^{N_k}(v_j^3 - y_j) \tag{4}$$

Since E is a function of w, the searching direction for the sharpest decrease in error is obtained by calculating partial derivatives with respect to w, which form the gradient. This technique is called the gradient method, or the steepest descent method, and is the basis of the backpropagation algorithm [9,11]. The searching direction is $g = -\partial E(w) / \partial w$, and the modification of weights is given by $\Delta w = \varepsilon g$. From this result we obtain the following general backpropagation algorithm (see Figure 6):

$$\Delta w_{i,j}^{k-1,k} = -\varepsilon d_j^k v_i^{k-1} \tag{5}$$

$$d_j^k = \begin{cases} (v_j^3 - y_j)\dfrac{\partial f(U_j^k)}{\partial U_j^k} & \text{for \quad output \quad layer} \\[4mm] \sum_{h=1}^{N_{K+1}} d_j^{k+1} w_{i,h}^{k,k+1} \dfrac{\partial f(U_j^k)}{\partial U_j^k} & \text{for \quad hidden \quad layer(s),} \end{cases}$$

$$\tag{6}$$

where $w_{i,j}^{k-1,k}$ is the connection weight between the i-th unit in the $(k-1)$-th layer and the j-th unit in the k-th layer, and U_j^k is the total amount of input to the j-th unit at the k-th layer. To calculate d_j^k, d_j^{k+1} must be previously calculated. Since the calculation must be conducted in the direction starting from the output layer to the input layer, in order, this algorithm is named the backpropagation algorithm. When a sigmoidal function is used as the characteristic function, $f()$, of the neuron units, the algorithm involves the following computations:

$$f(x) = \frac{1}{1+\exp^{-x+T}} \qquad (7)$$

$$\frac{\partial f(x)}{\partial x} = (1 - f(x))f(x) \qquad (8)$$

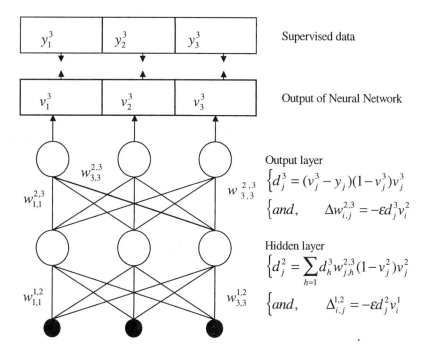

Figure 6: A schematic representation of the programming of backpropagation algorithm [9].

4 Fuzzy Logic

Fuzzy logic was first developed by Zadeh in the mid-1960s for representing some types of "approximate" knowledge that cannot be represented by conventional, crisp methods. In classical Boolean logic, truth is represented by the state 1 and falsity is by the state 0. Boolean algebra has no provision for approximate reasoning. Fuzzy logic is an extension of crisp Boolean logic in the sense that it provides a platform for handling approximate knowledge. Fuzzy logic uses fuzzy set theory, in which a fuzzy set is represented by a membership function, usually denoted by the Greek letter μ [1,14]. A particular variable value in the range of definition of the fuzzy set will have a grade of membership, which gives the degree to which the particular value belongs to the set. In this manner, it is possible for an element to belong to the set at some degree and simultaneously not belong to the set at the complementary degree, thereby allowing a non-crisp (or fuzzy) membership.

Fuzzy logic provides an approximate, yet effective, means of describing the behavior of systems that are too complex or ill-defined and difficult to tackle mathematically. As the complexity of a system increases, the ability to make precise and yet significant statements about its behavior diminishes until a threshold is reached beyond which precision and relevance become mutually exclusive characteristics [14].

A fuzzy set is a set without clear boundaries or without binary membership characteristics. Unlike an ordinary set where each object (or element) either belongs or does not belong to the set, a partial membership in a fuzzy set is possible [1,15]. An example of a fuzzy set could be "the set of long streets in Canberra." There are streets that clearly belong to the above set, and others that cannot be considered as long. But, if the concept of long is not exactly defined (for example, >1700m), there would be a "gray" zone where the judgment of the outcome is not obvious (e.g., somewhat long streets). In real life, one can effectively use the concept of long and short without taking physical measurements. Formally a fuzzy set is defined as follows:

Let X be a set that contains every set of interest in the context of a given class of problems. This is called the universe. Then, A is called a fuzzy set in X if A is a set of ordered pairs [1,15]:

$$A = \{(x, \mu_A(x)); x \in X, \mu_A(x) \in [0,1]\} \tag{9}$$

The function $\mu_A(x)$ is called the membership function of A, and it represents the grade of possibility (membership) that an element x belongs to set A. The closer the $\mu_A(x)$ is to 1, the more the x is considered to belong to A, and the closer it is to 0, the less it is taken as belonging to A.

Some basic set operations that can be defined on fuzzy sets are described next. Several methods are available to define the intersection and the union of fuzzy sets. The classical ones suggested by Zadeh (1965) are as follows [1]:

(1) The membership function of the complement $C = A^c$ of the fuzzy set A is defined by:

$$\mu_C(x) = 1 - \mu_A(x) \tag{10}$$

(2) The membership function of the intersection $D = A \cap B$ of two fuzzy sets A and B is defined by

$$\mu_D(x) = \min(\mu_A(x), \mu_B(x)) \tag{11}$$

(3) The membership function of the union $E = A \cup B$ of two fuzzy sets A and B is defined by

$$\mu_E(x) = \max(\mu_A(x), \mu_B(x)). \tag{12}$$

4.1 Fuzzy Systems and Rules

An important feature of fuzzy systems is its ability to represent a complex input-output relation as a set of simple input-output relations (or rules). The boundary of the rule areas is not sharp but "fuzzy." Another feature of fuzzy systems is the ability to separate logic and fuzziness. A fuzzy system is able to modify fuzzy rules when logic should be changed, and is able to modify membership functions when fuzziness of the rule variables should be changed [9].

Designing the antecedent part of a rule base involves deciding how to partition the input space. Most rule-based systems assume that all input variables are independent, and will partition the input space of each variable separately. This assumption makes it easy to not only partition the input space but also to interpret partitioned areas into linguistic rules. For example, the rule of "IF Temperature is $A1$ and Humidity is $A2$, THEN ..." is easy to understand, because the variables of temperature and humidity are separated. The crisp and fuzzy rule-based systems will differ in the manner in which their input space is partitioned [9,15]. The idea of fuzzy systems is based on the premise that in the real world, change is not catastrophic but would take place from one action rule to another. The degree of this overlapping is defined by membership functions. This gradual characteristic allows smooth fuzzy control.

Designing the consequent part of a rule base involves deciding the control (action) value of each rule. Fuzzy models are categorized into three types according to the expressions of the consequent part [9]:

(1) Mamdani model: $y = A$ (A is a fuzzy number)

(2) TSK model: $y = a_0 + \sum a_i x_i$ (a_i is a constant, and x_i is the input variable)

(3) Simplified model: $y = c$ (c is a constant).

The Mamdani type of fuzzy system has a fuzzy variable defined by a membership function in its consequents, such as $y = big$ or $y = negative\ small$. Although it is more difficult to analyze this type of fuzzy system than one whose consequents are numerically defined, it is easier for this fuzzy system to describe qualitative knowledge in the consequent parts. The Mamdani type of fuzzy system seems to be suitable for knowledge processing expert systems and high-level control systems, rather than for low-level direct control [1,15].

The consequents of a TSK (Takagi—Sugeno—Kang) model are expressed by a linear combination of weighted input variables. The TSK models are frequently used in direct fuzzy control and in simplified fuzzy models.

A simplified fuzzy model has fuzzy rules whose consequents are expressed by constant values. This model is a special case of both Mamdani type of fuzzy system and the TSK model. Even if each rule output is constant, the output of the whole fuzzy system is nonlinear, because the characteristics of membership functions are embedded into the system output [9]. The primary advantage of the simplified fuzzy models is that they are relatively easy to design.

4.2 Fuzzy Reasoning and Aggregation

The next stage of operation of a fuzzy system is to determine the final system output from the designed multiple fuzzy rules. There are two steps: (1) determination of rule strengths, and (2) aggregation of each rule output.

The first step of determining rule strengths will correspond to establishing how active or reliable each rule is. Antecedents include multiple input variables:

IF $x1 \in \mu1$ and $x2 \in \mu2$ and ... $x_k \in \mu_k$, THEN ...

In this case, one fuzzy rule has k membership values in the antecedent (condition) part: μ_i ($i = 1,...,k$). We need to determine how active a rule is, or its strength, from the k membership values. The class of fuzzy operators used for this purpose is known as a t-norm operator. There are many operators in the t-norm category. A frequently used t-norm operator is the algebraic product:

$$\text{Rule strength } w_i = \Pi_{j=1}^{k} \mu_j(x_j). \tag{13}$$

The "min" operator that Mamdani used in his first fuzzy controller is also described in fuzzy-logic textbooks [1]; where the rule strength is given by

$$w_i = \min(\mu_j(x_j)). \tag{14}$$

The overall system output, y^*, is determined by weighting each rule output with the obtained rule strength,

$$w_i: \quad y^* = \sum w_i y_i / \sum w_i. \tag{15}$$

Mamdani type of fuzzy controllers defuzzify the aggregated system output to determine the final non-fuzzy control value. The centroid approach, as given by Equation (15), is commonly used for this purpose.

4.3 Fuzzy Control

Existing fuzzy logic controllers commonly employ fuzzy reasoning that is based on the compositional rule of inference of Zadeh [1]. In particular, the "min-max-gravity method" [16] is used. A brief overview of this method is given below.

Consider the following fuzzy control problem [1, 17]:

Rule 1: *A1* and *B1* => *C1*
Rule 2: *A2* and *B2* => *C2*
... ...
Rule *n*: *An* and *Bn* => *Cn*
Facts (Context): x_0 and y_0

Control Inference: *C'*

Here Ai is a fuzzy set in the universe X; Bi in Y; and Ci in Z; and $x_0 \in X, y_0 \in Y$. Each fuzzy rule [Ai and $Bi => Ci$], for $i=1,\dots,n$, is defined as

$$\mu_{Ai} \text{ and } Bi => Ci^{(x,y,z)=\mu_{Ai}(x)\wedge\mu_{Bi}(y)\wedge\mu_{Ci}(z)} \tag{16}$$

where $\wedge$ denotes the "min" operation. An individual inference Ci' which is determined from the context (facts) x_0 and y_0, and the fuzzy rule [Ai and $Bi => Ci$] is given by

$$\mu_{Ci'}(z) = \mu_{Ai}(x_0) \wedge \mu_{Bi}(y_0) \wedge \mu_{Ci}(z) \tag{17}$$

The overall inference C' is the aggregate formed by taking the union (U) of the inferences from the individual rules, obtained as $C1', C2', \dots Cn'$ in Equation (17). Specifically,

$$C' = C1' \text{ U } C2' \text{ U } \dots \text{ U } Cn \tag{18}$$

or, in terms of membership functions,

$$\mu_{C'}(z) = \mu_{C1'}(z) \vee \quad \cdots \quad \vee \mu_{Cn'}(z) \tag{19}$$

where $\vee$ denotes the "max" operation. The representative crisp value z_0 of the resulting fuzzy inference set C' is obtained as the center of gravity (centroid) of C':

$$z_0 = \frac{\int z \cdot \mu_{C'}(z)dz}{\int \mu_{C'}(z)dz} \tag{20}$$

Application of this fuzzy reasoning method is graphically represented in Figure 7.

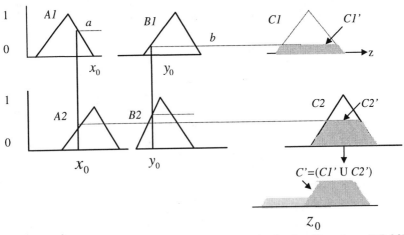

Figure 7: Application of the min-max-gravity method using Equations (17-20).

5 Evolutionary Computing

Searching or optimizing algorithms inspired by biological evolution fall into the category of evolutionary computing. The features of evolutionary computing [9,18,19] are that its search or optimization is conducted: (1) based on multiple searching points or solution candidates (population-based search), (2) using operations inspired by biological evolution, such as crossover and mutation, (3) based on probabilistic search and probabilistic operations, and (4) using little information on the search space. Typical paradigms that consist of evolutionary computing include genetic algorithms (GAs), evolution strategies (ES), evolutionary programming (EP), and genetic programming (GP).

Usually, GAs represent solutions for chromosomes with bit coding (genotype) and for searches for better solution candidates in the genotype space using GA operations of selection, crossover, and mutation. The crossover operation is the dominant operator.

ESs represent solutions as expressed by the chromosomes with real number coding (phenotype) and searches for better solutions in the phenotype space using ES operations of crossover and mutation. The mutation of a real number is realized by adding Gaussian noise, and ES controls the parameters of a Gaussian distribution allowing it to converge to a global optimum.

EP is similar to GA. The primary difference is that the mutation is an EP operator. EP uses real number coding, and the mutation sometimes changes the structure (length) of the EP code.

GP uses tree structure coding to represent a computer program or create new structures of tasks. The crossover operation is not for a numerical value, but for a branch of the tree structure. The rest of this chapter will focus on the GA and will illustrate how a GA searches for solutions.

5.1 GA Searching Algorithm

The basic principles of GAs were first laid down by Holland [20]. GAs simulate those processes in natural populations that are essential to evolution. Exactly which biological processes are essential for evolution, and which processes have little or no role to play is still a matter of research.

In nature, individuals of a population compete with each other for resources such as food, water, and shelter. Also, members of the same species often compete to attract a mate. Those individuals who are more successful in surviving and attracting mates will often have a relatively larger number of offspring [9]. Poorly performing individuals will often produce few, or even no offspring. This means that the genes from the highly adapted or "fit" individuals will spread to an increasing number of individuals in each successive generation. The combination of good characteristics from different ancestors can sometimes produce "super-fit" offspring, whose fitness is greater than that of either parent. In this way, species evolve to become more and more well suited for their environment [18].

GAs use a direct analogy of natural behavior (see Figure 8). They work with a population of individuals, each representing a possible solution to a given problem. Each individual is assigned a fitness score according to how good its solution to the problem is. The highly fit individuals are given opportunities to reproduce, by cross breeding with other individuals in the population. This produces new individuals as offspring, who share some features taken from each parent. The least fit members of the population are less likely to get selected for reproduction, and will eventually die out.

A whole new population of possible solutions is thus produced by selecting the best individuals from the current generation, and mating them to produce a new set of individuals. This new generation will contain a higher proportion of the characteristics possessed by the "good" members of the previous generation. In this way, over many generations, good characteristics are spread throughout the population, being mixed and exchanged in the process, with other good characteristics. By favoring the mating of the individuals who are more fit, the most promising areas of the search space would be explored.

Some common definitions of the technical terms used are described below [9,19]:

- **Chromosome** is a vector of parameters which represents the solution of an application task; for example, the dimensions of the beams in a bridge design. These parameters, known as genes, are joined together to form a string of values called chromosomes.
- **Gene** is a solution which will combine to form a chromosome.
- **Selection** is the process of choosing parents or offspring chromosome for the next generation.
- **Individuals** are the solution vectors of chromosome.
- **Population** is the collection of individuals.
- **Population size** is the number of chromosome in a population.
- **Fitness Function**   is the function which evaluates how each solution is suitable for a given task.
- **Phenotype** defines the expression type of solution values in the task world; for example, "red," "blue," "80kg," etc.
- **Genotype** are the binary (bit) expression type of solution values used in the GA search space; for example, "011," "000111011," etc.

Some advantages of GAs are

(1) Fast convergence to near-global optimum
(2) Superior global searching capability in a space that has a complex searching surface, and
(3) Applicability to a searching space where one cannot use gradient information of the space.

A GA determines the next set of searching points using the fitness values of the current searching points, which are widely distributed throughout the searching space. It uses the mutation operator to escape from a local minimum. A key disadvantage of GAs is that their convergence speed near the global optimum can be quite slow.

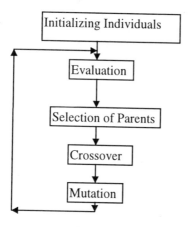

Figure 8: The flow of a GA process.

5.2 GA Selection

Selection is an operation that will choose parent solutions. New solution vectors in the next generation are calculated from them. Since it is expected that better parents generate better offspring, parent solution vectors that have higher fitness values will have a higher probability to be selected. There are several selection methods. The roulette wheel selection is a typical selection method [18]. In this case, the probability of winning is proportional to the area rate of a chosen number on a roulette wheel. According to this analogy, the roulette wheel selection assigns a selection probability to individuals in proportion to their fitness values.

The elitist strategy is an approach that copies the best n parents into the next generation. The fitness value of an offspring does not always become better than those of its parents [19]. The elitists strategy prevents the best fitness value of the offspring generation to become worse than that in the present generation, by copying the best parents to the offspring.

5.3 GA Reproduction

During the reproductive phase of a GA, individuals are selected from the population and recombined, producing offspring which, in turn, will comprise the next generation. Parents are selected randomly from the population using a scheme that favors the more fit individuals. Having selected two parents, their chromosomes are recombined using the mechanism of crossover and mutation [21, 22].

Crossover takes two individuals and cuts their chromosome strings at some randomly chosen position, to produce two "head" segments and two "tail" segments. The tail segments are then swapped over to produce two new full-length chromosomes (see Figure 9). Each of the two offspring will inherit some genes from each parent. This is known as a single-point crossover. Crossover is not usually applied to all pairs of individuals that are chosen for mating. A random choice is made, where the likelihood of crossover being applied is typically between 0.6 and 1.0.

Mutation is applied to each child individually, after crossover. It randomly alters each gene with a small probability (typically 0.001). Figure 10 shows the fifth gene of the chromosome being mutated. The traditional view is that crossover is the more important of the two techniques for rapidly exploring a search space. Mutation provides a small amount of random search, and helps ensure that no point in the search space has zero probability of being examined.

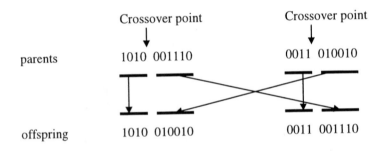

Figure 9: Single-point crossover.

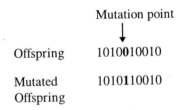

Figure 10: A single mutation.

6 Summary

This chapter introduced the analytical foundation on which the developments and applications that will be presented in the remaining chapters are based. Specifically, the fundamental concepts of neural networks, fuzzy logic, and genetic algorithms were outlined, in the context of intelligent control, and analogies to humans and human intelligence were indicated. The underlying numerical techniques fall into the general category of soft computing. The chapter started with a general introduction to knowledge-based systems, indicating the primary structure of such a system. Next the basic architecture of a neural network was described. Common types of artificial neural networks were mentioned, and the associated computational operations were given. The notions of fuzzy logic and fuzzy sets were introduced in a subsequent section. Representation of, and decision making with, fuzzy logic were discussed, with a focus on fuzzy-logic control. The main operations in fuzzy logic control were indicated. The chapter ended with an outline of evolutionary computing. Significance and application of evolutionary computing in intelligent control were pointed out. The nature of a genetic algorithm was outlined, and the underlying operations were described.

References

[1]   De Silva, C.W., *INTELLIGENT CONTROL: Fuzzy Logic Applications*, CRC Press, Boca Raton, FL, 1995.

[2]   Bernold, T. (Editor), *Expert Systems and Knowledge Engineering*, North-Holland, Amsterdam, 1986.

[3]   Bielawaski, L. and Lewand, R., *Expert Systems Development*, QED Information Sciences, Inc., Wellesley, MA, 1988.

[4]   Hart, A., *Knowledge Acquisition for Expert Systems*, McGraw-Hill, NY, 1986.

[5]   Forsyth, R. (Editor), *Expert Systems,* Chapman & Hall, NY, 1984.

[6]   De Silva, C.W. and Lee, T.H., "Knowledge-Based Intelligent Control", *Measurements and Control*, Vol. 28(2), pp. 102-113, April 1994.

[7]   Lee, T.H., Yue, P.K., and de Silva, C.W., "Neural Networks Improve Control," *Measurements and Control*, Vol. 28(4), pp. 148-153, September 1994.

[8]   Gupta, M. and Rao, H., *Neural-Control Systems Theory and Applications*, New York Press, NY, 1994.

[9]   Takagi, H., "Introduction to Fuzzy Systems, Neural Networks and Genetic Algorithms," Tutorial Notes, *Int. Conf. on Knowledge-Based Intelligent Electronic Systems* (KES'97), Adelaide, Australia, May 1997.

[10] Fu, K.S., "Learning Control Systems -- Review and Outlook," *IEEE Trans. Automatic Control*, Vol. 15, pp. 210-221, April 1970.

[11] Lippman, R.P., "An Introduction to Computing with Neural Nets," *IEEE ASSP Magazine*, Vol. 4, No. 2, pp. 4-22, April 1987.

[12] Stevens, F., "Synaptic Physiology," *Proc. IEEE*, Vol. 79, No. 9, pp. 916-930, June 1968.

[13] Rumelhart, D.E., Hinton, G.E., and Williams, R.J., "Learning Internal Representations by Error Propagation," *Parallel Distributed Processing*, D.E.

Rumelhart and J.J. McClelland (Editors), Chapter 8, pp. 318-360, MIT Press, Cambridge, MA, 1986.

[14] Jain, L.C. (Editor), *Soft Computing Techniques in Knowledge-Based Intelligent Engineering Systems*, Springer-Verlag, Berlin, 1997.

[15] Bardossy A. and Duckstien, L. *Fuzzy Rule-Based Modeling with Application to Geophysical, Biological and Engineering Systems*, CRC Press, Boca Raton, FL, 1995.

[16] Mamdani, E.H., "Application of Fuzzy Algorithms for Control of a Simple Dynamic Plant," *Proc. IEEE*, Vol. 121, pp. 1585-1588, 1974.

[17] Kadel, A. and Langholz, G., *Fuzzy Control Systems*, CRC Press, Boca Raton, FL, 1994.

[18] Davis, L., *Handbook on Genetic Algorithms*, Van Nostrand-Rienhold, NY, 1991.

[19] Grefenstette, J.J., "Optimization of Control Parameters for Genetic Algorithms," *IEEE Trans. SMC*, Vol. 16, pp. 122-128, 1986.

[20] Holland, J.H., *Adaptation in Natural and Artificial Systems*, MIT Press, Cambridge, MA, 1975.

[21] Vonk, E., Jain, L.C., and Johnson, R.P., *Automatic Generation of Neural Network Architecture using Evolutionary Computing*, World Scientific Publishing Co., Singapore, 1997.

[22] Van Rooij, A., Jain, L.C., and Johnson, R.P., *Neural Network Training using Genetic Algorithms*, World Scientific Publishing Co., Singapore, 1997.

2 | LEARNING AND ADAPTATION IN COMPLEX DYNAMIC STSTEMS

Fakhreddine O. Karray
University of Waterloo
Waterloo, Ontario
Canada

Clarence W. de Silva
University of British Columbia
Vancouver, British Columbia
Canada

A growing number of complex dynamic systems with learning capability have been recently studied and designed on the premises they are able to considerably improve their control performance through experiential knowledge. We first provide an overview on the major conventional approaches used for identification and adaptive control. Deficiencies of these techniques are then highlighted when applied to systems with complex dynamics and operating in partially unknown environments. Possible alternatives, based on the new generation of computational tools involving learning, are also presented as potential approaches for more effectively dealing with such class of systems. Pertinent aspects of symbolic and numerical learning techniques are then described with emphasis on the recently emerging tools of connectionist models and their abilities to tackle a growing range of complex real world problems.

1 Introduction

Conventional control techniques are usually implemented under the assumption of a good understanding of the process dynamics and its operating environment. These techniques fail, however, to provide satisfactory results when applied to poorly modeled processes, and/or processes functioning in an ill-defined environment. Even when a suitable analytical model is available, model parameters might be incomplete. This is the case when dealing with complex systems for which the physical processes are not fully understood, hence preventing the derivation of an adequate model for the system.

System identification techniques, based on experimentally determined input-output data, are known to provide powerful alternatives for dealing with modeling difficulties and can be used as valuable tools for adequate description of the system dynamics. This can be very useful for purposes of control system adaptation. In this respect, adaptive control techniques can be regarded among very powerful and effective control schemes that make the best use of system identification tools. At first, a model is derived experimentally through system identification. The identified model is then used as a basis for designing a controller with adjustable gains, hence capable of sustaining large changes in the system parameters, while providing desired responses. Adaptive control schemes are generally intended for plants with partially unknown dynamics and/or slowly time-varying parameters.

In response to the growing demand for the design of sophisticated and highly reliable processes, designers from many fields have developed systems characterized by a high level of component integration and quality precision. This has led, however, to an increase in the system complexities, and has created a new class of challenging problems concerning modeling and control. Such complex processes are usually characterized by highly nonlinear and time-varying behavior, extensive dynamic coupling, hierarchies, multiple time scales, and high dimensional decision space. As a result, their control may no longer be satisfied by the existing control theory including adaptive control and other known robust control techniques such as H-infinity (H_∞) and linear-quadratic-Gaussian (LQG) based methods [1,2]. Even experimental model identification may reveal only some facets of the process, because it may happen that the controlled variables of the process are not accessible, or the system is not completely detectable.

Research carried out recently in the area of Knowledge-Based Systems (KBS) has shown that it is possible to circumvent some of these difficulties through the design of so-called *expert* or *intelligent* systems. A major feature of these systems is their capabilities of autonomous identification and learning. This is achieved by making them learn, by themselves, the system behavior from a set of example situations. Such capabilities make them highly adaptable even to significant, unanticipated changes in process dynamics. Symbolic learning and numerical learning are commonly recognized as the two main learning approaches.

The well-known area of Artificial Neural Networks (ANN) represents an important class of numerical-learning-based tools. ANN techniques have been implemented with success in a large range of applications, including system identification. As such, they can be used to model any nonlinear mapping between input and output variables. This is indeed a very powerful identification tool that can circumvent special modeling problems. Such problems arise when the controller does not have sufficient capability to acquire knowledge in the conventional way, especially when the system dynamics are highly nonlinear and are subject to structured and/or unstructured uncertainties. This is usually the case when the systems are so complex that describing them with conventional mathematical models such as differential or difference equations becomes virtually impossible. Figure 1 represents a family of various types of dynamic systems and suitable identification and learning tools for controlling them.

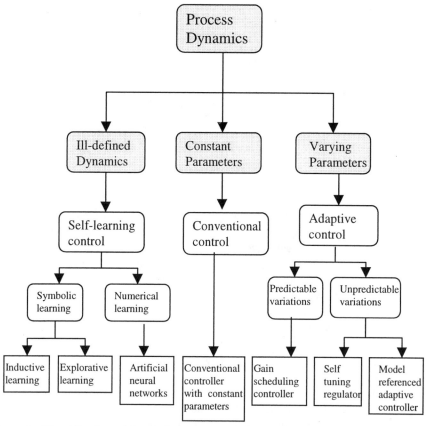

Figure 1: Classification of dynamic systems and appropriate controllers.

2   Systems Identification and Adaptive Control

In broad terms, a system can be defined as a set of interconnected objects, with a clearly defined boundary, in which a variety of variables interact to produce observable outputs (Figure 2). A system is termed a "dynamic system" when the rates of changes of the system-response variables are not negligible. In this chapter, we are primarily concerned with such systems. A system may be subjected to external signals, which can be classified into two categories: those that are manipulated by the observer or environment (known as inputs) and those that are not (known as disturbances).

To explain the behavior of a given system or to synthesize solutions for potential problems related to it, it is necessary to provide a working model that enables the scientist or the designer to analyze or even react to system outputs through adequate control laws. System models can be either analytical or experimental. The analytical models are based on the application of well-known, basic physical laws, continuity relations, and compatibility conditions [3], while experimental models are derived from the application of system identification tools [4,5].

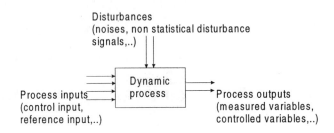

Figure 2: Block diagram of a dynamic system.

In some cases, particularly when the system structure is so complex that it prevents derivation of a straightforward and tractable model from analytical laws, experimental modeling techniques are sought to learn about the dynamic behavior of the system. This can also serve as a platform for designing adaptive controllers that are capable of providing the desired behavior, even under structural parametric variations of the system. This area of control is known as adaptive control [5]. Experimental modeling aspects and adaptive control procedures are the subject of the next two subsections.

2.1   Techniques for Systems Identification

The main goal of system identification is to experimentally establish an adequate dynamic model of the system such that the parameters which provide the best fit between the measured input-output data and the model-generated values are estimated. Model structure, as well as the model parameters, may be identified. The two well-known techniques for system identification are the nonparametric and the parametric estimation methods.

2.1.1   Nonparametric methods

Nonparametric system identification is often used to derive an input-output description of a system model, for which the dynamics are a *priori* known to be linear and time invariant. The method is called nonparametric since it does not involve a parameter vector for use in the search for the best description of system dynamics. A linear-time-invariant model can be described either by its impulse response (time domain method) or by its frequency response (frequency domain method).

The nonparametric, time-domain method is useful if the designer has access to the step or impulse response of the system, known also as the transient response data. These data are usually quick and relatively easy to obtain. The main drawback of the method is that the transient response should be highly noticeable to have a high level of signal-to-noise ratio. This is difficult to achieve in real world applications, and further special tests may have to be conducted as part of the overall identification process.

The other aspect of nonparametric system identification pertains to the frequency-domain method. This approach is based on the collection of frequency response data when the system is excited by input signals of different frequencies. The method has the advantage of being simple in terms of data collection. However, it requires more time to apply than the time domain method. This is particularly true for systems with long time constants, which is the case, for example, in chemical processes.

2.1.2   Parametric methods

Another approach for system identification is the one pertaining to the parametric estimation method. This method is particularly effective in providing on-line information on process parameters, which is very useful in designing controllers that can adapt to process-dynamic changes while providing a desired response. The very well-known technique of least squares parameter estimation represents the basic approach here. It is based on the least squares principle, which states that the unknown parameters of a mathematical model should be chosen to minimize the sum of the squares of the differences between the actually observed and computed values. Let us presume that the estimated output $\hat{y}(T)$ of a given system at a given discrete time T, is given as a linear function of n unknown parameters a_i, each multiplied by a known coefficient γ_i, called regressor variable, which is provided through measurements and depends on input-output data; i.e.,

$$\hat{y}(T) = \sum_{i=1}^{n} \gamma_i(T) a_i \tag{1}$$

Suppose that several pairs of data $(y(t_k), \gamma(t_k))$ are collected through a series of experiments, with $\gamma(t_k)$ being the regressor vector given by

$$\gamma(t_k) = [\gamma_1(t_k), \gamma_2(t_k), \cdots, \gamma_n(t_k)]^T,$$

and $k = 1, 2, \bullet\bullet\bullet\bullet, m$ with $t_m = T$. Let us also denote by $\underline{a}$ as the unknown vector given by $\underline{a} = [a_1, a_2, \bullet\bullet\bullet\bullet, a_n]^T$ and $\underline{y}$ as being the measurement vector given by. $\underline{a} = [y(t_1), y(t_2), \bullet\bullet\bullet\bullet, y(T)]^T$ Now, the problem can be stated as one of finding the parameters a_1, which minimize the least square errors between the measured output vector $\underline{y}$ and its estimated counterparts y. In other words, it is required to determine the parameters a_i that minimize the error δ given by

$$\delta = (1/2) \sum_{k=1}^{k=m} (y(t_k) - \hat{y}(t_k))^2.$$ (2)

This is a standard optimization problem, for which there exists a unique solution given by

$$\underline{a} = [a_1, a_2, \cdots, a_n]^T = (\Gamma^T \Gamma)^{-1} \Gamma^T \underline{y}$$ (3)

The solution (3) exists provided that the matrix inverse $(\Gamma^T \Gamma)^{-1}$ exists, with Γ defined as

$$\Gamma = [\gamma(t_1), \gamma(t_2), \cdots, \gamma(T)]^T.$$

For more detailed analysis on this and other identification techniques, the reader may wish to consult [4], in which the theory is extensively covered and [5–7], where some applications are provided.

2.2   Adaptive Control

An adaptive control system uses a control scheme that is capable of modifying its behavior in response to changes in process dynamics. Adaptive controllers have been extensively used in several industries including chemical, aerospace, automotive, and pulp and paper.

The rapid growth in the design of integrated and powerful information processors has made the use of adaptive controllers even more versatile. There are three well-known adaptive control schemes: gain scheduling, model-referenced adaptive control, and self-tuning regulators. A description of the main features of these techniques follows.

2.2.1   Gain scheduling

This type of adaptive control system is based on the adjustment of controller parameters in response to the operating conditions of a process [5]. This control scheme is particularly useful when the variations of the process dynamics are predictable. In fact, for a class of dynamic systems, it is possible that an explicit model of the system can be accurately described every time the operating conditions of the system take new values. Gain scheduling can be regarded as a mapping from the process parameters to the controller parameters. In practice, a gain scheduler can be implemented as a look-up table.

A block diagram of this adaptive scheme is shown in Figure 3. The two main drawbacks of this method are related to its *open loop* behavior and to the discrete assignment of controller gains according to a look-up data table. Indeed, for intermediate operating conditions, no explicit control gains are assigned to the system, and the designer must apply interpolation techniques to avoid instabilities.

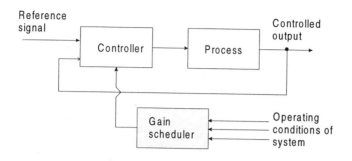

Figure 3: Block diagram of gain scheduling controller.

2.2.2 Model-referenced adaptive control

The Model-referenced adaptive control (MRAC) is an adaptive control scheme capable of handling processes with unpredictable changes [5]. This control procedure is based on the design of an adaptive scheme whose objective is to drive the error signal between the response of the process and that of the reference model to zero. The overall block diagram (Figure 4) consists of two main loops with different time constants. The inner loop, which is the faster one, is used for regulation of the process, while the outer loop is designed for adjustment of the parameters of the inner loop regulator to drive the error signal to zero. Algorithms for designing the adaptation scheme for the adjustment mechanism of the outer loop are discussed in [5] and [7]. Some instability problems for applying the MRAC procedure have been observed, and remedies have been proposed as well.

2.2.3 Self-tuning regulators

Self-tuning regulation is another adaptive control scheme characterized by its ability to handle dynamic processes that may be subjected to unpredictable changes in system parameters [5]. A self-tuning regulator (STR) uses the outputs of a recursive identification scheme for plant parameters (outer loop) to adjust, through a suitable adaptation algorithm, the parameters of a controller located in the regulation loop of the system (inner loop), as shown in Figure 5. One can easily notice some similarities in terms of inner and outer loop structuring between STR and MRAC. The main difference between the two schemes is, however, that while the STR design is based on an explicit separation between identification and control, the MRAC design uses a direct update of controller parameters to achieve asymptotic decay of the error signal to zero. In view of this fundamental difference in design, MRAC is referred to as a direct adaptive control scheme, while STR is known as an indirect adaptive control scheme.

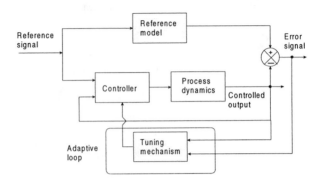

Figure 4: Block diagram of a model-referenced adaptive controller.

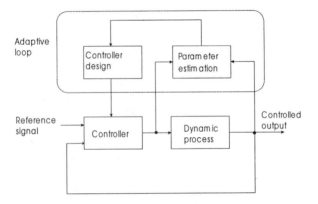

Figure 5: Block diagram of the self-tuning regulator.

3   Learning Techniques

The adaptive control techniques described in Section 2.2 assume the availability of an explicit model of the system dynamics (as in the gain scheduling technique) or, at least, an experimental model determined through identification (as in STR and MRAC). This may not be the case for a class of complex systems characterized by poorly known dynamics and time varying parameters, that may operate in an ill-defined environment. Besides, adaptive control techniques lack the important feature of learning. This means that an adaptive control scheme cannot use the knowledge it has acquired in the past to tackle similar situations in the present or the future. In other words, while adaptive control techniques can be used effectively for controlling systems with slowly time-varying parameters, they, nevertheless, lack the ability of learning, and as a result the same adaptation operation must be repeated every time the system is confronted with similar operating situations.

To tackle the control problem of such systems, designers have devised new control approaches through which the system becomes implicitly modeled and, using well-defined learning algorithms, the system adapts its controller parameters to accommodate unpredictable changes in the internal dynamics of the system. This is a resurgent area of research which focuses on the design of *expert* or *intelligent* controllers [8], based primarily on learning techniques. Expert controllers are built in a manner that makes use of the knowledge gained by the system through learning procedures. This allows the system to infer or construct decision rules, thereby making it autonomously adjust to a large class of unpredictable dynamic variations.

Learning is accomplished by developing algorithms that allow the system to learn for itself from a set of input/output training data. These algorithms are designed in such a way that they efficiently combine the main features of a computing machine with those of human expertise to draw as many correct decisions as possible. Expert controllers have the very distinctive feature of learning and adjusting their parameters in response to unpredictable changes in the dynamics or operating environment of the system. This is done without the need for an explicit knowledge of a system model or rules that guide its behavior.

Intelligent controllers based on learning algorithms have the capabilities for modeling and control of a large class of systems which is characterized by highly nonlinear behavior, time varying parameters, and operating in an unpredictable environment. This class of controllers, which is based on self learning algorithms, should have the ability of learning from past experiences and use the stored ("memorized") knowledge to infer right decisions for similar future situations. Two well-known approaches for designing learning systems employ symbolic-learning and numerical-learning-based techniques.

3.1 Symbolic Learning

Symbolic learning denotes the ability of a system to formulate and alter its knowledge base, composed mainly of facts and rules, from a set of well-structured feedback data related to the performance of the system. Systems with symbolic learning capabilities can rearrange their operating set of rules in response to changes in the system dynamics or the operating environment. This depends to a large extent on the degree of supervision to which the system may be subjected. Several categories of symbolic learning have been proposed in the literature [9]. These include rote learning, explorational learning, inductive learning, and explanation-based learning. One major feature that distinguishes these learning algorithms from one another pertains primarily to the level of supervision. The inductive learning approach, in particular, has received more attention due to the relative ease of its implementation and the rich feedback information it generates.

An inductive-learning-based algorithm makes use of knowledge previously acquired by the system to construct and, if needed, to refine configuration rules capable of generating as many correct decisions as possible. Learning by induction involves the synthesis of inference rules from a set of specific example solutions. A typical inference rule is

If   (situation)   then   (decision)

It is assumed here that the system is fed with a set of example solutions and is taught to draw the right decision from each set. The algorithm keeps rearranging and modifying the rules to accommodate new situations. The mechanism is carried out by modifying the rules that do not match the positive set of example data and keeping unaltered those that do match. Self-organizing fuzzy logic systems represent a good example of inductive learning based tools.

3.2   Numerical Learning

Numerical-learning-based algorithms represent another class of very useful learning tools. They are primarily used for numerical modeling and control of those systems for which very little is known about their dynamics and operating environment. The learning mechanism is carried out through an optimization process, during which the system attempts to approximate the desired response vector with the output of the learning algorithm. This is done by solving an optimization problem involving the desired response, the input data, and a set of unknown optimization parameters.

To clarify the idea, let us presume that we have a set of n input/output data (x_i, y_i), for which the nonlinear mapping function f from the input set to the output set is unknown. The problem is to find a set of mapping functions given by $\mathbf{f(x,w)}$ that will best approximate the output of f for every set of inputs. Here $\mathbf{w}$ represents the vector of weights obtained through an optimization process usually aimed at minimizing the squared Euclidean distance δ between the vector $\mathbf{y}$ and $\mathbf{f(x,w)}$:

$$\delta = \left\| \mathbf{y} \quad \mathbf{f(x, w)} \right\|^2 \tag{4}$$

As more data become available, the learning system will continue to update its structure by automatically tuning its parameters to comply with the specific requirement of system performance and to adjust to new situations. Several techniques to solve this optimization problem have been proposed in the literature [9,10]. Some of the well-known approaches are: genetic algorithms, annealing algorithms, and gradient-descent algorithms. Details on the characteristics of these algorithms can be found in [9] and [10]. In recent years, a large body of research on numerical learning has focused on Artificial Neural Networks (ANN). This is an area of research that has seen a renewed interest recently, after a marginally enthusiastic start in the 1960s and the early 1970s. The following section outlines in more detail some of the main features of this powerful numerical learning tool, and its relatively wide range of applications [11].

4 Artificial Neural Networks

Artificial Neural Networks (ANN) represent an important class of numerical learning tools. An artificial neural network is typically composed of a set of parallel and distributed processing units, called nodes, usually ordered into a set of layers interconnected by means of unidirectional weighted signal channels, called connections (Figure 6) [10].

The internal architecture of ANN provides powerful computational capabilities, allowing for the simultaneous exploration of different competing hypotheses. Massive parallelism and computationally intensive learning through examples in ANN make them suitable for application in complex and nonlinear processes. The topological structure of an ANN is basically inspired from the biological architecture of neurons in humans. The recent resurgence of interest in ANN is mainly due to advances in design of new network topologies and learning algorithms capable of solving a wide range of problems such as nonlinear functional mapping, speech and pattern recognition, categorization, data compression, and many other problems related to different fields of applications.

This is made easier by new developments in learning algorithms and by advances in the design of powerful and parallel-processing computing systems, which fit well with the internal structuring of an ANN.

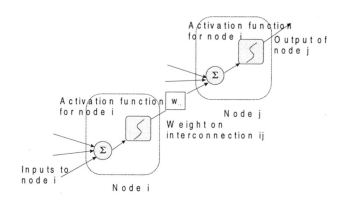

Figure 6: Topology of two adjacent nodes in an ANN.

A neural network is generally characterized by three important features.

(i)   Network topology, which corresponds to the ordering and organization of the nodes from the input layer to the output layer of the network.

(ii)   Network transfer functions or activation functions, which represent the nonlinear mapping at the nodes level. They can be of several types: sigmoid mappings, signum functions, or linear correspondences.

(iii)   Network learning algorithm, which updates the weighting parameters at theinterconnection level. This is done during the network training process. The learning algorithms can be one of two types: supervised or unsupervised. These notions are illustrated in more detail in Sections 4.1 and 4.2.

The way the nodes and interconnections are arranged determines the topology of a given ANN. The choice for using a given topology is mainly dictated by the type of problem being considered. Several network topologies and learning algorithms have been proposed in the literature. A summary on the main features of commonly used ANN is given in Table 1. For a comprehensive coverage of the different network topologies and their learning algorithms, one may consult [10,11].

To illustrate the notions of supervision and of self-organizations in ANN, the features of two widely used ANN, each belonging to a class of its own, are outlined next. These ANN are the Multilayer Perceptron and the Kohonen Self-Organizing Network.

Table 1: Classes of neural networks and their corresponding learning algorithms.

Networks Features	Multilayer Perceptron Network	Kohonen Self-Organizing Network	Hopfield Network	Adaptive Resonance Theory Net
Learning Algorithms	Backpropagation learning (supervised)	Competitive learning (unsupervised)	Recurrency based learning (supervised)	Adaptive learning (unsupervised)
Areas of Application	Functional approximation; System identification	Phonetic Typewriting; Texture Segmentation	Content addressable memory; Classification	Clustering; Data compression

4.1 Multilayer Perceptron

Multilayer perceptron (MLP) belongs to the class of feedforward networks, characterized by a forward propagation of the information signal flow (Figure 7). It is of the supervised type, which means that the system updates its weighting parameters in response to the feedback it gets on its performance as measured by the difference between the network output and the desired output provided by the network trainer (also known as the *teacher*).

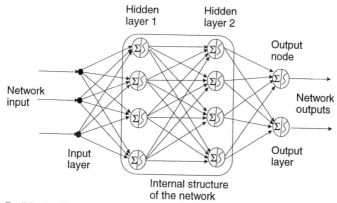

Figure 7: Block diagram of the multilayer perceptron with two hidden layers.

The activation function for each of the network nodes is of the sigmoid type. This is a differentiable mapping that permits the use of the widely known Backpropagation Learning (BPL) algorithm. This is a numerical learning algorithm based on the gradient descent optimization technique. Given their differentiability characteristics, MLP are very useful for functional approximation purposes. In fact, it is shown in [12] that with an appropriate number of hidden nodes, MLP can approximate any nonlinear mapping. Moreover, MLP have also been used successfully for classification purposes [13].

4.2   Kohonen Self-Organizing Network

Kohonen Self-Organizing Network (KSON) belongs to the class of unsupervised learning networks. This means that the network updates its weighting parameters without the need of a performance feedback from a *teacher* or a network trainer. One major feature of this network is that the nodes distribute themselves across the input space to recognize groups of similar input vectors, while the output nodes compete among themselves to be fired one at a time in response to a particular input vector. This process is known as competitive learning.

When suitably trained, the network will produce a low-dimension representation of the input space that preserves the ordering of the original structure of the network. This implies that two input vectors with similar characteristics excite two physically close layer nodes. In other words, the nodes of the KSON can recognize groups of similar input vectors. This generates a topographic mapping of the input vectors to the output layer and depends primarily on the feature characteristics of the input vectors. Given their self-organizing capabilities, KSON can be very useful for clustering applications such as in speech recognition, vector coding, and texture segmentation.

4.3   Neural Networks as Identification Tools

The ability of a class of ANN to approximate complex nonlinear mappings makes them very attractive tools for system identification and nonlinear modeling. MLP belongs to this class of neural networks, in which the activation functions in the hidden layer are of the sigmoid type. These are differentiable mappings that will retain a continuous behavior in the output signal.

With enough nodes in the hidden layer, it has been shown that any nonlinear function can be approximated by a three-layer MLP provided that the Backpropagation Technique is used as the learning algorithm. This makes MLPs excellent candidates for nonlinear input-output modeling. Other types of newly developed networks have provided similar results to those of MLP. Some examples are the Radial Basis Function Network and the Functional Link Network [10].

Recent advances made in the field of adaptive neural networks will certainly help in tackling more complex problems involving nonlinear modeling and real-time identification of special control systems. Such systems are generally characterized by large parameter variations and subjected to unpredictable disturbances and dynamic alterations. Designing stable learning algorithms for neural networks with adaptive topologies is a very challenging problem, the solution of which may lie within the next revolution in artificial neural network design.

4.4   Other Applications of ANN

In recent years, an increasing number of ANN with various learning algorithms and topologies have been tested and implemented into a wide range of applications. Several encouraging results have been obtained and have motivated designers, who have been trying to improve their training algorithms and network topologies to make ANN even more versatile and more reliable. Table 2 illustrates a sample of some recent applications of ANN in the fields of automotive technology, electronics, meteorology, robotics, and speech/pattern recognition.

5   Summary

As the need for more complex and highly integrated systems increases, designers are confronted with new and challenging problems related to modeling and control. Conventional techniques of identification and control are not always suited to deal with complex dynamic systems operating in an ill-defined environment and subjected to large parameter variations and external disturbances. The growing research activity witnessed recently in the area of numerical learning, such as in the field of artificial neural networks, has lead to several encouraging results in tackling these problems. In fact, recent improvements in the design of ANN with superior learning algorithms and more flexible topologies have led to more versatile and more reliable systems. Several previously unsolved control problems were approached recently through implementation of ANN, and very positive results were obtained. Successful tests and implementations of ANN continue to be carried out by designers from many fields [11].

Table 2: Areas of application of ANN as classified by the field.

Fields	Speech/ Pattern Recognition	Electronics/ Signal Processing	Aerospace/ Meteorological Applications	Robotics	Automotive Technology
Areas of Application	Optical character recognition; Speech recognition; Hand-written character recognition	Signal identification; Data compression; Nonlinear circuit mapping	Radar/optical image identification; Pattern recognition; Radar target detection	Robot position/ velocity control; Autonomous robotics; Computer vision	Car speed control; Break anti-lock system; Optimization of fuel injection

Despite their success in solving a number of complex problems, ANN suffer from their highly opaque character. This is an undesirable feature of ANN, which prevents the designer or the network trainer from knowing what may exactly happen inside the network. To ensure a continuous growth in the area of ANN, more transparency of the networks is required. This will enhance our understanding of the internal dynamics of a neural network, and may induce major improvements in the network operations.

References

[1] Maciejowski, J.M., *Multivariable Feedback Design*, Reading, MA: Addison Wesley, 1989 pp. 265-323.

[2] Lewis, F., *Applied Optimal Control and Estimation,* Englewood Cliffs, NJ: Prentice-Hall, 1992, pp. 2526-572.

[3] Shearer, J.L., Murphy, A.T., and Richardson, H.H., *Introduction to System Dynamics*, Reading, MA: Addison-Wesley, 1971.

[4] Ljung, L., *System Identification-Theory for the User*, Englewood Cliffs, NJ: Prentice-Hall, 1988, pp. 20-45.

[5] Astrom, K. and Wittenmark, B., *Adaptive Control,* 2nd ed., Reading, MA: Addison-Wesley, 1995, pp. 3-63.

[6] Franklin, G., Powell, D., and Emami-Naeini, A., *Feedback Control of Dynamic Systems*, 3rd ed., Reading: MA, Addison Wesley, 1995, pp. 144-150.

[7]   Sastry, S. and Bodson, M., *Adaptive Control: Stability Convergence and Robustness,* Englewood Cliffs, NJ: Prentice-Hall, 1989, pp. 10-45.

[8]   De Silva, C.W., *Intelligent Control: Fuzzy Logic Applications*, Boca Raton, FL: CRC Press, 1995.

[9]   Hopgood, A., *Knowledge-Based Systems for Engineers and Scientists*, Boca Raton, FL: CRC Press, 1993, pp. 159-185

[10] Haykin, S., *Neural Networks: A Comprehensive Foundation*, Englewood Cliffs, NJ: McMillan College Publishing Company, 1994.

[11] Harris, C., editor, *Advances in Intelligent Control,* London, England: Taylor and Francis, 1994.

[12] Masters, T., *Practical Neural Network Recipe in C++*, Academic Press Inc., 1993.

[13] Lippmann, R., An Introduction to Computing with Neural Networks, *IEEE Acoustics, Speech and Signal Processing Magazine*, vol. 4, no.2, pp. 4-22, 1987.

3 APPLICATIONS OF EVOLUTIONARY ALGORITHMS TO CONTROL AND DESIGN

Tatsuya Nomura
Evolutionary Systems Department
ATR Human Information Processing Research Laboratories
Kyoto, Japan

In this chapter, we introduce some recent applications of evolutionary algorithms (EA) to the areas of control and design. In Section 1, we introduce the concept and basic procedures of evolutionary algorithms. Next, Section 2 gives applications of them to control problems. Applications in design problems, including computer graphics and music are given in Section 3. Although there have been many applications of EA in these two areas, we will only discuss relatively recent contributions due to space limitation. Finally, we will discuss how the research in these areas should be developed.

1 Introduction

In this section, we will briefly explain the concept of evolutionary algorithms and will provide an overview of the recent trends.

1.1 Some Basics

Evolutionary algorithms (EA) are probabilistic optimization algorithms inspired by principles of biological evolution. Although the mechanism of evolution has not yet been completely clarified, the following properties are generally accepted:

- Evolution of an organism consists of processes of operations on systematic substances in which information on the structures of the organism is coded as symbols, called *chromosomes*.

- Through the process of *natural selection*, chromosomes corresponding to a structure that is more appropriate for survival of an organism are reproduced at higher probability.
- New chromosomes are produced through the processes of *mutation* and *recombination*.
- Through these processes, species gradually change and adapt to their environment.

Inspired by these properties, a variety of EAs and their application have been developed.

EA are used in optimization problems. In other words, given a stat space and a function in it, $f : S \rightarrow R$, EA search the minimum value of f and the corresponding points in S. A variety of optimization problems are found in scientific and engineering areas. EA is applicable in all areas where important problems can be formalized as optimization problems.

A variety of EAs have been proposed. The basic procedures common among them are:

1. Each state variable of the function to be optimized is coded into a chromosome represented by an array of symbols or numerical values.
2. A population is made of several individuals with chromosomes.
3. Each chromosome is assigned a fitness value for the environment, based on the function to be optimized. Here, chromosomes with higher fitness values are made to correspond to state variables with lower values of the original function.
4. Through alternation of generations by iteration of selection and reproduction (recombination and mutation), the individuals having chromosomes with higher fitness values increase in the population. In addition, a variety of individuals are produced by mutation and recombination (crossover).

Figure 1 shows the basic flow of an EA. As shown there, an EA basically does not use any information on the function to be optimized, such as its continuity, differentiability, multi-peakedness, and so on. It uses only the function value that assigns the fitness values to chromosomes. Moreover, mutation and recombination are basically executed in chromosomes as symbolic or numerical operations, without any information on what the corresponding state variables represent. That is, common procedures for selection, reproduction, mutation, and recombination are applicable for different problems, although the procedures for encoding/decoding between chromosomes and state variables as well as fitness assignment are different among problems. This property is called *"Blind Search"*. In real applications, however, there are many cases where these procedures are tuned based on the problem-dependent properties to get better performance.

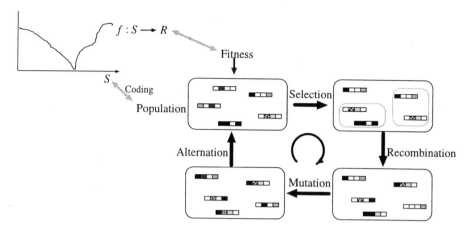

Figure 1: Basic flow of an EA.

1.2 Types of Evolutionary Algorithms

Evolutionary algorithm (EA) is a general term that is used for al algorithms that simulate the process of evolution. Several types of these algorithms have been independently developed. Table 1 summarizes the properties of each algorithm. Although the differentiation between these algorithms is generally ambiguous, some distinguishing properties may be identified, as follows [36]:

Table 1: Properties of evolutionary algorithms.

	GA	ES	EP	GP
Representation of Chromosomes	Discrete	Continuous	Continuous	LISP-like
Fitness Value	Scaled Version of Original Function Value	Original Function Value Itself	Scaled Version of Original Function Value	Evaluation of Behavior Expressed as a Program
Mutation	A Subsidiary Operation (Parameter Adjustment not Used)	Main Operation (Parameter Adjustment Used)	Unique Operation (Parameter Adjustment Used)	A Subsidiary Operation (Parameter Adjustment not Used)
Recombination	Main Operation	A Subsidiary Operation	Not Used	Main Operation
Selection	Stochastic	Deterministic	Stochastic	Stochastic

Genetic Algorithm (GA) is basically used for searching in a discrete space. In this algorithm, selection is stochastically executed based on fitness values (roulette selection); recombination by crossover is regarded as important; and mutation is randomly and uniformly done in each chromosome.

Evolutionary Strategy (ES) is basically used for searching in multi-dimensional, real-valued spaces. In this algorithm,   selection is deterministically executed based on the rank of individuals by   fitness value; only mutation is used; and it is based on normal distribution.

Evolutionary Programming (EP) is basically used for searching in multi-dimensional, real-valued spaces. Although this algorithm is similar to an ES, it is different in that selection is stochastically executed; the process of evolution that it simulates differs from that of ES in a strict sense, and so on.

Genetic Programming (GP) may be regarded as an extension of GA, and is basically used for automatic program generation. In this algorithm, chromosomes are represented by a tree structure corresponding to a LISP-like program, each fitness value is evaluated by the behavior of the corresponding   program, and mutation and recombination are defined as appropriate   for tree-structured chromosomes.

Of course, which algorithm should be used will depend on the specific application. A combination of algorithms may be considered as well, depending on their properties.

1.3 Types of Operations

The basic operations of EAs were introduced in Section 1.1. Many kinds of operations for each step have been proposed. We give several representative types of operations in this section.

1.3.1 Representation of Chromosomes

To apply an EA to an optimization problem, the elements in the state space of the function to be optimized must be coded with a chromosome; that is, an array of symbols to be operated by the EA.

Figure 2 shows chromosome representation. In optimization with discrete spaces, as in combinatorial optimization problems, the state variable is represented with a 0-1 bit string or integer-valued vector, and it is regarded as a chromosome representation.

In optimization with continuous-valued spaces (e.g., function approximation), the state variable is represented by a real-valued vector. Thus, the bit string obtained by digitizing each element of the vector. For digitization, the general binary coding or grey coding is used. Moreover, the vector itself is regarded as a chromosome in ES.

In optimization of programs like GP, the object is an equation represented by a programming language with the same structure as LISP. In this case, the equation is represented with a tree structure and the tree is regarded as a chromosome.

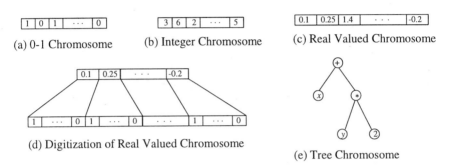

(a) 0-1 Chromosome   (b) Integer Chromosome   (c) Real Valued Chromosome

(d) Digitization of Real Valued Chromosome

(e) Tree Chromosome

Figure 2: Chromosome representation.

1.3.2 Fitness Value

The purpose of an optimization problem is to minimize or maximize a function related to the problem, defined in a state space. On the other hand, the purpose of EA is to increase the chromosomes having higher fitness values. Thus, the fitness value must be determined so that a chromosome has a high fitness value, just as the corresponding state variable has a suitable value for the function.

For minimization problems of a function f, the following values have been proposed:

Fitness(c) = (Maximum Function Value in the Population) - $f(x)$
   + Constant) × (Minimum Function Value in the Population)
Fitness(c) = A - $B f(x)$ (A, B : constant)
Fitness(c) = $1/f(x)$ (for $f(x) > 0$)

Here, c is a chromosome and x is the corresponding state variable. Typically, a fitness value is positive. Moreover, if the chromosome represents a picture or music, the fitness is manually assigned by the user who watches or listens to it (see Section 3.3).

1.3.3 Selection

The most basic selection method is roulette selection. In this method, the probability of survival in the next generation is assigned to each chromosome c_i in the current population { $c_1, \ldots \ldots, c_n$ } as

$$p_i = \frac{Fitness(c_i)}{Fitness(c_1) + \dots + Fitness(c_n)}$$

and it is stochastically determined based on the probability that each chromosome is left in the next population. Moreover, in rank-based selection, only the order of fitness values among the population is used in calculating the probability of survival. This is used to exclude the effect of distribution of fitness values among the population.

In deterministic selection, for each chromosome, $n \times pi$ copies of it are directly included in the next generation. In binary tournament selection, a pair of chromosomes in the population is randomly selected and the chromosome with the higher fitness value in the pair is kept in the next generation. In some cases like GP, an explicit fitness value is not used, and is determined based on comparison with the results from actual execution of programs [19].

In addition, an elite strategy is used as a combination of stochastic and deterministic selections, in which the chromosome with the best fitness value is deterministically retained in the next generation and the rest are stochastically selected.

1.3.4 Recombination and Mutation

Recombination in GA and GP is generally called crossover. Although this term is not frequently used in EA, the basic concept is to produce offspring chromosomes from more than two parent chromosomes. In general, recombination is done at a probability given in advance (this is termed recombination or crossover rate).

Figure 3 shows several recombination methods. For chromosomes represented by an array of symbols, one-point, multi-point, and uniform crossovers may be used. In one-point crossover, a crossover point on two parent chromosomes is randomly selected, and then offsprings are produced by replacement of the elements on the parents, based on the point. Multi-point crossover is done by replacement based on more than two randomly selected points. In uniform crossover, offsprings are produced by individual replacement of elements in the parents for each place.

Although these crossovers can also be applied to chromosomes represented by a real-valued vector, recombinations that use real values themselves have been proposed. In other words, offspring chromosomes are produced as average or unfair average vectors of the parent vectors [7, 26]. For chromosomes represented by a tree, a crossover is done by replacement of subtrees under randomly selected nodes on the parent trees.

Mutation of chromosomes represented by an array of symbols or a real-valued vector is done by replacement of the value in a randomly selected position on the chromosome by a random value (in the case of bit strings, bit flip is done). In particular, mutation in ES is done by a perturbation based on Gaussian distribution.

Mutation of chromosomes represented by a tree is done by replacement of the subtree under a randomly selected node by a random subtree. Mutation may be done at a probability given in advance (this is termed mutation rate).

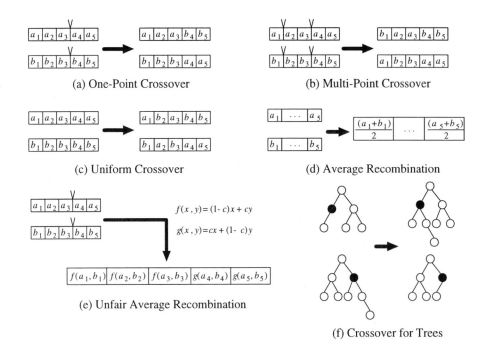

(a) One-Point Crossover

(b) Multi-Point Crossover

(c) Uniform Crossover

(d) Average Recombination

$$f(x,y) = (1-c)x + cy$$

$$g(x,y) = cx + (1-c)y$$

(e) Unfair Average Recombination

(f) Crossover for Trees

Figure 3: Recombination methods.

1.4 Applications

As mentioned previously, if the main problem of a given application can be formulated as an optimization problem, EA may be applicable. In fact, EAs have been applied to a variety of problems like scheduling, management, circuit design, image/signal processing, and modeling of social systems [7, 19]. This shows the generality of EA.

If there are conventional algorithms available that may be applied to answer a problem, however, EAs may not necessarily be superior to them. EAs are applicable when the function to be optimized is quite complex (multipeak, not differentiable, and so on) and cannot be solved using conventional algorithms, or when there are no conventional algorithms that are appropriate. Some control problems we deal with in this chapter belong to these categories.

EAs are also useful when the function itself cannot be explicitly formulated, as in the case of human intention or preference. In situations of this type, for example, graphics and music, system user directly determines the survival of chromosomes instead of explicit fitness values.

2 Applications in Control

In this section, we explain how evolutionary algorithms (EA) are used in control problems. Some concrete examples of applications in the control of robots and communication systems will be presented.

2.1 Robot Control

In this section, we introduce two kinds of applications of EA in robot control: the planning of joint configuration of robot manipulators and the acquisition of behavior rules of autonomous mobile robots.

2.1.1 Planning of Joint Configuration of Manipulators

A robot manipulator consists of several joints with links between them, and has an end-effector to accomplish tasks. Given a joint configuration that indicates the position of the joints, the corresponding position of the manipulator end-effector may be determined according to its forward kinematics. The inverse kinematics problem may not have a unique solution, however. That is, many joint configurations correspond to one position of the end-effector, but only one joint configuration should be selected for a given end-effector position. Figure 4 shows an application of EA for this problem.

For a given initial joint configuration and a desired final end-effector position, Aydin and Kocaoglan [2] used GA to find the most efficient join configuration among those that achieve the final end-effector position. In their method, it is assumed that the manipulator has four joints and a chromosome is represented by a 64 bit binary string (because any joint configuration is represented by a four-dimensional real-valued vector and each value is digitized as a 16-bit binary string). For the chromosomes, selection, mutation, and recombination are executed in this method.

The method evaluates the fitness value with a penalty function. The penalty function consists of a geometrical property of the space of possible configurations and the total displacement from the initial configuration to the final one. The selection is executed by deleting the chromosome with the highest penalty value. The mutation is bit-flip and the recombination is one-point crossover.

Davidor [7] dealt with the generation of sequences of joint configurations through which the end-effector is made to follow a desired trajectory. The main characteristic here is that chromosomes of variable lengths are used, while the crossover is modified based on the fact that one chromosome must correspond to one trajectory of the end-effector. This is because trajectories with variable numbers of joint configurations must be considered for the desired trajectory of the end-effector, and the conventional crossover methods for fixed length chromosomes are hard to apply to variable-length chromosomes. For this purpose, the crossover points are selected based on the nearness between points on a trajectory.

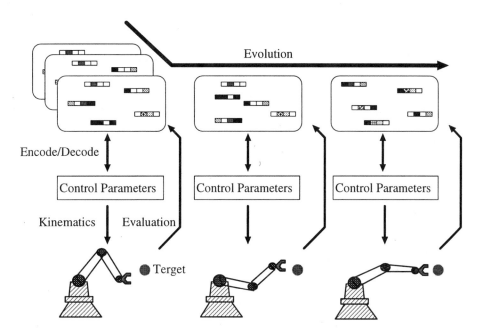

Figure 4: Application of EA for planning of robot joint configuration.

2.1.2 Acquisition of Behavior Rules of Autonomous Mobile Robots

A main problem with behavior-based autonomous robots is to design a control strategy to accomplish a task, that is, mapping the outputs of robotic sensors into the actuators. However, design of associated control strategies is difficult, because the desired behaviors are fuzzy and difficult to explicitly define, and furthermore, not all useful behaviors of autonomous robots can be determined *a priori* or recognized by humans. For this problem, many EA methods have been proposed that will automatically make robots learn control strategies. Figure 5 shows an application of EA to this problem.

The basic procedures in these methods are as follows:

1. Control strategies (sensory input-action relations) are represented with chromosomes.
2. A population of chromosomes is prepared, and the robot operates based on the strategy corresponding to each chromosome in the environment of a given task.
3. Based on the results of task accomplishment, the chromosomes appropriate for the task are found through selection, recombination, and mutation.

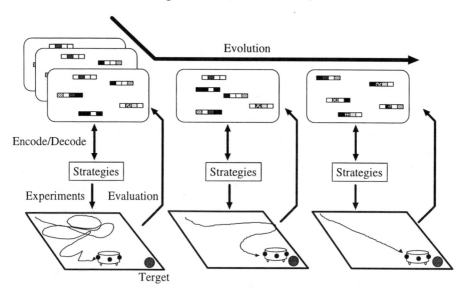

Figure 5: Application of EA for acquisition of behavior rules of robots.

Here, we give three cases where the control systems in robots are combinatorial logic systems. The first and second cases use a small physical robot and the third uses only simulations.

In the case of Naito et al. [21], the type of each logic element, the existence of a connection between each logic element and sensor, and one between each logic element and other elements are represented as chromosomes. Because the task to be achieved in their example is obstacle avoidance, the robot's behavior is evaluated based on the total mileage, the degree of straight movement, and the distance from any obstacle in each trial. In their case, one-point crossover, bit-flip mutation, and selection with elite strategy are used as operations of EA.

In the case of Louis and Li [17], the control system consists of three controllers, two primitive activators, and one arbitrator. Their example is a modified box-pushing task, which is to push a box toward a goal position indicated by a light source. This task is broken into two sub-tasks, box-pushing and box-side circling, and one of the activators corresponds to each of these sub-tasks. The arbitrator activates one of the activators at each step. Each controller consists of sensory inputs, logic components, and comparators. A controller is represented with a tree chromosome, where the leaves represent sensor responses or a threshold, the internodes represent logic components or comparators, and a general GP is applied to these tree chromosomes. The evolution of the control strategy is executed through three steps:

1. The activator corresponding to box-pushing is evolved based on the fitness value calculated from the degrees of orientation to the box, straight movement, and forward speed.
2. The activator corresponding to box-side-circling is evolved based on the fitness value calculated from the degrees of keeping a distance from the box and forward speed.
3. The arbitrator is evolved based on the fitness value calculated from the total distance between the box and goal.

In the case of Lee et al. [16], the control circuit that will map sensory inputs into binary control commands is assumed to be a gate array which must be encoded into a binary chromosome. Moreover, the GA they used differs from a standard GA in that not all of the chromosomes in the current generation are replaced with the offsprings, crossover between similar chromosomes is not allowed, two parents exchange exactly one half of their randomly selected nonmatching bits, and so on. The task is to approach and eat food, avoid obstacles, and follow walls in a bounded area such as an office. The basic concept is similar to that of Louis and Li. The evolution of the control strategy is executed through the following steps:

1. The strategy for approaching food is individually evolved in an environment with only food.
2. The strategy for obstacle avoidance is individually evolved in an environment with food and several obstacles.
3. The strategy for wall following is individually evolved in an environment with food and walls.
4. Chromosomes with the best fitness value for the above three basic strategies are selected as the chromosomes in the initial population of a subsequent GA, and the strategy for a more complex environment, including all food, obstacles, and walls, is evolved (this process is inspired by Case Based Reasoning (CBR)).

2.2 Communication System Control

In communication networks, an important problem is how to control the traffic, to maximize performance while preventing it from a breakdown.

In telephone circuit communication networks, Cox et al. [7] applied EA to the scheduling of calls. Given the information on the network (nodes, links between nodes, capacities of nodes, and recent traffic) and regular calls, they used GA to find a probabilistic strategy in which paths and calls are allocated to minimize the mean loss of the calls per unit time.
The method consists of the following two steps:

1. A master allocation strategy is determined, based on recent statistical traffic pattern and the loss value of each call type, to minimize mean loss of the calls per unit time (this problem itself is a nonlinear scheduling problem).
2. When a new call reaches the network, an appropriate path is allocated to it based on the above master strategy. If no appropriate path exists, reallocation for the waiting calls is done by GA, based on the recent network situation.

In this method, a chromosome is a permutation in the collection of waiting calls, and its fitness value is evaluated by the sum of lost calls that are not allocated in the corresponding permutation based on the recent network situation. Selection and recombination are repeated until a condition for termination is satisfied, and the best permutation found within this time is given as the output of GA.

Moreover, Lee et al. [15] applied EA to load balancing in a distributed computer network. In their case, many processors are connected to the network, and each processor independently decides to which processors the task should be transferred when its own load becomes heavy. GA is used for dynamic load balancing to determine a destination processor to receive a task.

In the method, each processor has its own population of chromosomes. Each chromosome is represented by a bit string $< v0, v1, \ldots \ldots, vn-1>$ that indicates a set of processors to which the request messages are sent. Here, n is the total number of processors and $vi=1$ if the message is transferred to the i-th processor, and $vi=0$ otherwise. The fitness of each chromosome is the inverse of the weighted sum of total message processing time, total message transfer time, and total task processing time for the destination processors indicated in the corresponding bit string.

The flow of the method consists of the following steps:

1. Each processor randomly generates a population of strings without duplication.
2. Each processor observes its own load. If the load becomes heavy, the following step is performed.
3. Selection and recombination are performed and a new set of destination processors is generated.

Here, the selection is stochastically executed and the recombination is one-point crossover.

A common property in the above cases is that EA is applied to dynamic environments, that is, situations where functions to be optimized are changed through the processes of the systems.

3 Applications in Design

In this section, we explain how EA is used for design problems by introducing concrete examples of applications for design of circuits, graphics, and music.

3.1 Circuit Design

Many problems in circuit design can be formalized as optimization problems and in the past several years many optimization methods for this purpose have been proposed. Of course, many methods using Eas have also been proposed, and we give only a few cases here. Figure 6 shows an application of EA to this problem.

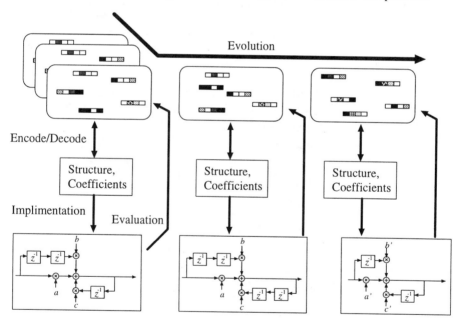

Figure 6: Application of EA in circuit design.

In the design of VLSI circuits or printed circuits, the problem is how to arrange cells or circuit elements to satisfy given constraints such as size, the number of connections, and so on. Moon and Kim [20] applied a GA to a VLSI ratio-cut partitioning problem whose objective is to determine the cell set partitioning that has optimal ratio-cut, which represents a balance between the sizes of two disjoint subsets of the set of cells.

$$\frac{\text{Number of connections between two subsets}}{(\text{Size of one subset}) \times (\text{Size of the other})}$$

For this kind of problem, the success of GA depends on the method of coding chromosomes. Although the chromosomes in their case are represented by bit strings, the coding depends on the numbering of the cells. Therefore, the chromosome that indicates the coding itself is independently prepared, and a crossover method including the coding chromosomes and original chromosomes is proposed.

Additionally, there are applications to the determination of coefficients in filters. Neubauer [23] represented the coefficient vector of an IIR filter with both a bit string by digitization and a real-valued vector, and applied both GA and ES. Fitness is evaluated by the difference between the response characteristics of the corresponding filter and the desired ones for a special frequency area such as a cut-off frequency. Chellapilla et al. [5] used EP to determine coefficients of an IIR filter. The fitness is evaluated by mean squared error for desired input-output signals.

Moreover, EAs have been directly applied to circuit design. Koza et al. [13, 14] apply a GP to the design of analog electrical circuits. In their case, the circuit-constructing program is represented with a tree chromosome and fitness is evaluated by execution of the compiled function in a SPICE simulator.

In another application to circuit design, Smith et al. [34] applied EA to generate test programs for verification of the design of a modern microprocessor. In their case, a GA generates variable lengths of sequences of instructions that function as stress test programs.

3.2 Graphics Design

The design of computer graphics requires more labor as the complexity of the graphics increases. One solution to this problem is the use of EAs based on interactive methods.

3.2.1 Interactive EA Approach

Figure 7 shows an application of interactive EA approach. The basic concept of the interactive EA approach is as follows:

1. The parameters of the graphic object to be designed are coded as a chromosome and a population of chromosomes is prepared.

2. The graphics corresponding to the chromosomes in the population are displayed to the user, and an observer selects the best graphics based on his intention or preference and leaves them in the next generation.
3. As new chromosomes are produced through recombination and mutation, the above process is repeated.

Through these processes, the observer is able to design graphics without understanding the details of the graphic parameters. The main characteristic of this approach is that a specific optimization goal, that is, an explicit function to be optimized, does not exist and the sense of the observer is directly reflected in the evolution. As we show in Section 3.3, this approach is also applied to music design.

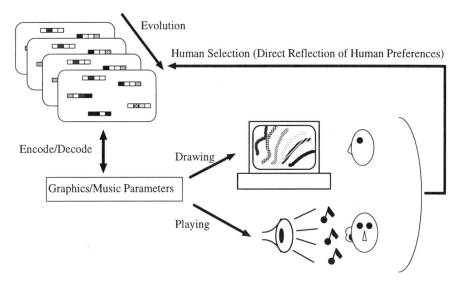

Figure 7: Application of the interactive EA approach.

Sims [33] used the above interactive EA approach to generate 2D textures, 3D parametric objects, and 2D animations. In his case, the functions that generate the graphics, such as mapping 2D pixels into color values, are represented by a program like LISP, and a GP is applied to them.

Aoki and Takagi [1] used GA to determine the parameters of lighting in 3D CG. In their case, the type of light, on/off condition, brightness, latitude, and so on are digitized and coded into a chromosome represented by a bit string. In the GA used for this application, two populations are prepared and different mutation rates are given to them.

3.2.2 Fitness Estimation in Interactive Approach

A main problem in using the interactive EA approach is the burden on the user due to the selection procedure. EA requires a large number of chromosomes and generation alternations. It repeatedly forces the user to select the best graphics among many samples and is thus a drawback for the development of design support systems that are easier to use. One solution to this problem is the improvement of the user interface in the interactive system, but this is outside of the scope of EA. Another solution is to generate a fitness function of EA for graphic parameter evolution by extracting the user's preference using another EA. Masui [18] applied GP to generate a function that evaluates graph layout. In his case, the function represented by LISP is evolved based on the examples of graph layouts that a user evaluates. Then, GA is applied to the chromosomes that represent graph layout based on the obtained evaluation function

Nakanishi [22] also used GP to generate a function that evaluates dress design. In his dress design aid system, dress graphics are drawn with spline curves, whose parameters are discretized as chromosomes within a real-valued area, and new dress graphics are stochastically produced from the selected designs. Between the production of new designs and human selection, an evaluation function, that reflects another designer's sense through GP, is used to reduce the number of candidate graphics.

3.3 Music Design

As with graphic design, the objects of computer music such as physical instrument models and phrase can be evolved by EA. Here, we give examples of sound synthesis and algorithmic composition.

3.3.1 Sound Synthesis

It is not easy to determine the parameters of a physical model emulating natural sounds; this generally requires trial and error. EAs, including an interactive approach, have been applied to this problem.

Horner et al. [10] use GA to find optimum parameters for reconstruction through FM synthesis of a sound having harmonic particles. In their problem, the following procedures are executed to reconstruct a given original sound:

1. The parameters of the original sound in the frequency domain (the amplitude for each harmonic frequency) are extracted through short-time Fourier analysis.
2. The chromosomes corresponding to the parameters of FM synthesis are evolved through GA based on a degree of matching between the parameters of synthesized sound and the original sound in the frequency domain.

The parameters of FM synthesis are carrier-to-modulator frequency ratio, modulation index, and time-varying carrier amplitudes. The carrier-to-modulator ratio and modulation index are coded into a bit string chromosome by digitization. The time-varying carrier amplitudes are calculated in a straightforward manner with squared error from the amplitudes of the original sound obtained by Fourier analysis.

Vuori and Valimaki [37] applied ES to estimate the parameters of a non-linear physical model emulating the sound of a flute by using the procedures similar to those of the above problem. In their work, the chromosome is directly represented by a vector of real-valued parameters.

Horner et al. also [11] applied an interactive EA approach to generate a creative timbre. Filtering and time warping are applied to a source sound to generate new sounds for the user to select.

Takala et al. [35] also applied an interactive approach in sound synthesis synchronized to animation. In their work, the sound and motion are synthesized with functions expressed in LISP, and a GP is applied to the tree chromosomes.

3.3.2 Algorithmic Composition

Algorithmic composition is a computer-aided music composition, and EAs have been applied to find novel music.

Biles [3] applied GA to generate jazz solos. His system, called "GenJam", is given, in advance, some information of the temporal and rhythmic style, the number of solo choruses, the chord progression, and MIDI sequences for piano, bass, and drum. Then, it improvises on the tune by building choruses of MIDI events decoded from the chromosomes. GenJam has two kinds of chromosomes, phrase chromosomes and measure chromosomes. The measure chromosome indicates a series of notes (an array of integers which indicate MIDI event numbers), and the phrase chromosome indicates a series of measures (an array of the numbers of individuals in the population of measure chromosomes). Thus, the phrase chromosomes are correlated with the measure chromosomes. The chromosomes are represented by bit strings because each integer of the chromosomes is digitized. The user of GenJam listens to the generated solos and assigns fitness value to them by key typing like a handclap, and new solos are generated through one-point crossover and mutation.

Jacob [12] applied GAs to larger-scale problems including composing and arranging. His system, called "Variations", consists of three modules, composer, ear, and arranger, and GA is applied to each module. The composer module has chromosomes corresponding to phrases, the ear module has those corresponding to data filters which indicate valid chord transitions, and the arranger module has those corresponding to the arrangement and orchestration of the phrases. The basic procedures of Variations for composition are as follows:

1. Before composition begins, the chromosomes in the ear module are evolved based on the interactive GA, that is, human selection.
2. The composer module produces phrases based on the GA operation, and the ear module selects better phrases based on a data filter corresponding to one of its chromosomes.
3. The phrases accepted through the ear module are collected, and the arranger module creates a set of chromosomes from them.
4. Better progressions are evolved by the interactive GA.

The main characteristic of this system is that several GA modules are used and one of them, the ear module, is used to evaluate the fitness of another module, the composer module. In this sense, the existence of the ear module is related to the concept of fitness estimation described in Section 3.2.2.

Moreover, Polito *et al.* [27] used GP to construct a system that composes counterpoints based on the theory of 16th-century music, called "Gpmuse".

4 Other Applications

Other applications of EA to control and design problems include the following:

- Control of an air-conditioning system [24].
- Design of gear trains and pressure vessels [4].
- Design of a real plant system [6].
- Generation of training exercises by battlefield simulations [28].
- Design of airplanes [7].

5 Summary

In this chapter, we introduced the concept of Evolutionary Algorithms (EA) and their applications to problems of control and design. In fact, there are many more applications than those introduced here.

In cases where a conventional application problem has been formalized and dealt with in operations research, such as circuit design in Section 3.1, EA has been successfully applied. In such cases, the conventional algorithms may be superior to EA in the context of optimization. If the function to be optimized is explicitly represented, then appropriate conventional algorithms should be used first.

As mentioned in Section 1.4, when such a conventional algorithm does not exist, or when the implementation is hard even if an algorithm exists, or when the function cannot be represented with any explicit form, EA might be a suitable solution to the problem because of the ease of implementation.

Moreover, recently it has been noted that EAs should not be used in static environments, but rather in dynamic environments. In this sense, interactive EA approach is compatible with supporting areas for human thinking or sensitivity, such as design of graphics and music as mentioned in Sections 3.2 and 3.3.

At any rate, further development of EA applications needs progress in the theory of EA. Although many researches on analysis of the behaviors of EA have recently been developed [8,9,25,29,30,31,32], only very few results are useful in real applications.

Extensive development of EAs, both the theory and applications, would be possible in the future.

References

[1]   Aoki, K. and Takagi, H. (1997), 3-D CG Lighting with an Interactive GA, In *Proc. First International Conference on Knowledge-Based Intelligent Electronic Systems (KES'97)*, Vol. 1, pp. 296-301, Adelaide.

[2]   Aydin, K. K. and Kocaoglan, E. (1995), Genetic Algorithm Based Redundancy Resolution of Robot Manipulators, In *Proc. ISUMA-NAFIPS'95*, pp.322-327, Maryland.

[3]   Biles, J. A. (1994), Genjam: A Genetic Algorithm for Generating Jazz Solos, In *Proc. International Computer Music Conference (ICMC'94)*, pp. 131-137, Aarhus.

[4]   Cao, Y. J. and Wu, Q. H. (1997), Mechanical Design Optimization by Mixed-Variable Evolutionary Programming, In *Proc. IEEE International Conference on Evolutionary Computation (ICEC'97)*, pp. 443-446, Indianapolis.

[5]   Chellapilla, K., Fogel, D. B., and Rao, S. S. (1997), Gaining Insight into Evolutionary Programming through Landscape Visualization: An Investigation into IIR Filtering, In Angeline, P. J. et al., Editor, *Evolutionary Programming VI (EP'97)*, Lecture Notes in Computer Science 1213, pp. 407-417, Indiana. Springer.

[6]   Chen, K., Parmee, I. C., and Gane, C. R. (1997), Dual Mutation Strategies for Mixed-Integer Optimization in Power Station Design, In *Proc. IEEE International Conference on Evolutionary Computation (ICEC'97)*, pp. 385-390, Indianapolis.

[7]   Davis, L. (1990), *HANDBOOK OF GENETIC ALGORITHMS*, Van Nostrand Reinhold.

[8]   Davis, T. E. and Principe, J. C. (1993), A Markov Chain Framework for the Simple Genetic Algorithm, *Evolutionary Computation*, Vol. 1, No. 3, pp. 269-288.

[9]   Dawid, H. (1994), A Markov Chain Analysis of Genetic Algorithms with a State Dependent Fitness Function, Complex *Systems*, Vol. 8, pp. 407-417.

[10] Horner, A., Beauchamp, J., and Haken, L. (1993), Machine Tongues XVI: Genetic Algorithms and their Application to FM Matching Synthesis, *Computer Music Journal*, Vol. 17, No. 4, pp. 17-29.

[11] Horner, A., Beauchamp, J., and Packard, N. (1993), Timbre Breading. In *Proc. International Computer Music Conference (ICMC'93)*, pp. 396-398, Tokyo.

[12] Jacob, B. L. (1995), Composing with Genetic Algorithms, In *Proc. International Computer Music Conference (ICMC'95)*, pp. 452-455, Banff Alberta.

[13] Koza, J. R., Andre, D., Bennett III, F. H., and Keane, M. A. (1997), Design of a High-Gain Operational Amplifier and Other Circuits by Means of Genetic Programming, In Angeline, P. J. et al., Editor, *Evolutionary Programming VI (EP'97)*, Lecture Notes in Computer Science 1213, pp. 125-135, Indiana. Springer.

[14] Koza, J. R., Bennett III, F. H., Lohn, J., Dunlap, F., Keane, M. A., and Andre, D. (1997), Automated Synthesis of Computational Circuits Using Genetic Programming, In *Proc. IEEE International Conference on Evolutionary Computation (ICEC'97)*, pp. 447-452, Indianapolis.

[15] Lee, S. H., Kang, T. W., Ko, M. S., and Chung, G. S. (1997), A Genetic Algorithm Method for Sensor-Based Dynamic Load Balancing Algorithm in Distributed Systems, In *Proc. First International Conference on Knowledge-Based Intelligent Electronic Systems (KES'97)*, Vol. 1, pp. 302-307, Adelaide.

[16] Lee, W. P., Hallam, J., and Lund, H. H. (1997), Applying Genetic Programming to Evolve Behavior Primitives and Arbitrators for Mobile Robots, In *Proc. IEEE International Conference on Evolutionary Computation (ICEC'97)*, pp. 501-506, Indianapolis.

[17] Louis, S. J. and Li, G. (1997), Combining Robot Control Strategies Using Genetic Algorithms with Memory, In Angeline, P. J. et al., Editors, *Evolutionary Programming VI (EP'97)*, Lecture Notes in Computer Science 1213, pp. 431-441, Indiana. Springer.

[18] Masui, T. (1994), Evolutionary Learning of Graph Layout Constraints from Examples, In *Proc. ACM Symposium of User Interface Software and Technology*, pp. 103-108.

[19] Mitchell, M. and Forrest, S. (1994), Genetic Algorithms and Artificial Life, *Artificial Life*, Vol. 1, pp. 267-289.

[20] Moon, B. R. and Kim, C. K. (1997), Genetic VLSI Circuit Partitioning with Dynamic Embedding, In *Proc. First International Conference on Knowledge-Based Intelligent Electronic Systems (KES'97)*, Vol. 2, pp. 461-469, Adelaide.

[21] Naito, T., Odagiri, R., Matsunaga, Y., Tanifuji, M., and Murase, K. (1996), Genetic Evolution of a Logic Circuit Which Controls an Autonomous Mobile Robot, In *Proc. ALIFE V Poster Presentations*, pp. 117-119, Nara.

[22] Nakanishi, Y. (1996), Applying Evolutionary Systems to Design Aid System, In *Proc. ALIFE V Poster Presentations*, pp. 147-154, Nara.

[23] Neubauer, A. (1997), Genetic Design of Analog IIR Filters with Variable Time Delays for Optically Controlled Microwave Signal Processors, In *Proc. IEEE International Conference on Evolutionary Computation (ICEC'97)*, pp. 437-442, Indianapolis.

[24] Nishimura, T., Sakamoto, S., Hiei, T., Itoh, H., and Shimadzu, T. (1996), Advanced Control for Air Conditioner Using Artificial Life Model, In *Proc. ALIFE V Poster Presentations*, pp. 80-86, Nara.

[25] Nix, A. E. and Vose, M. D. (1992), Modeling genetic algorithms with Markov chains, *Annals of Mathematics and Artificial Intelligence*, Vol. 5, pp. 79-88.

[26] Nomura, T. and Miyoshi, T. (1995), Numerical Coding and Unfair Average Crossover in GA for Fuzzy Clustering and their Applications for Automatic Fuzzy Rule Extraction, In *Proc. IEEE/Nagoya University WWW'95*, pp. 13-21, Nagoya. Nagoya University.

[27] Polito, J., Daida, J. M., and Bersano-Begey, T. F. (1997), Musica ex Machina: Composing 16th-Century Counterpoint with Genetic Programming and Symbiosis, In Angeline, P. J. et al., Editors, *Evolutionary Programming VI (EP'97)*, Lecture Notes in Computer Science 1213, pp. 113-123, Indiana. Springer.

[28] Porto, V. W. and Fogel, L. J. (1997), Evolution of Intelligently Interactive Behaviors for Simulated Forces, In Angeline, P. J. et al., Editors, *Evolutionary Programming VI (EP'97)*, Lecture Notes in Computer Science 1213, pp. 419-429, Indiana. Springer.

[29] Qi, X. and Palmieri, F. (1994), Theoretical Analysis of Evolutionary Algorithms With an Infinite Population Size in Continuous Space part I: Basic Properties of Selection and Mutation, *IEEE Transactions on Neural Networks*, Vol. 5, No. 1, pp. 102-118.

[30] Qi, X. and Palmieri, F. (1994), Theoretical Analysis of Evolutionary Algorithms With an Infinite Population Size in Continuous Space part II: Analysis of the Diversification Role of Crossover, *IEEE Transactions on Neural Networks*, Vol. 5, No. 1, pp. 120-129.

[31] Rudolph, G. (1994), Convergence Analysis of Canonical Genetic Algorithms, *IEEE Transactions on Neural Networks*, Vol. 5, No. 1, pp. 96-101.

[32] Rudolph, G. (1996), Convergence of Evolutionary Algorithms in General Search Spaces, In *Proc. IEEE International Conference on Evolutionary Computation (ICEC'96)*, pp. 50-54, Nagoya, May.

[33] Sims, K. (1993), Interactive evolution of equations for procedural models, *The Visual Computer*, Vol. 9, pp. 466-476.

[34] Smith, J. E., Bartley, M., and Fogarty, T. C. (1997), Microprocessor Design Verification by Two Phase Evolution of Variable Length Tests, In *Proc. IEEE International Conference on Evolutionary Computation (ICEC'97)*, pp. 453-458, Indianapolis.

[35] Takala, T., Hahn, J., Gritz, L., Geigel, J., and Lee, J. W. (1993), Using Physically-Based Models and Genetic Algorithms for Functional Composition of Sound Signals, Synchronized to Animated Motion, In *Proc. International Computer Music Conference (ICMC'93)*, pp. 180-183, Tokyo.

[36] Tamaki, H., Kita, H., and Iwamoto, T. (1996), Genetic Algorithms – VI – Recent Trends in Evolutionary Computation, *SYSTEMS, CONTROL AND INFORMATION*, Vol. 40, No. 4, pp. 170-177 (in Japanese).

[37] Vuori, J. and Valimaki, V. (1993), Parameter Estimation of Non-Linear Physical Models by Simulated Evolution – Application to the Flute Model, In *Proc. International Computer Music Conference (ICMC'93)*, pp. 402-404, Tokyo.

4 | NEURAL CONTROL SYSTEMS AND APPLICATIONS

Q.M.J. Wu and **K. Stanley**
National Research Council Canada
Ottawa, Canada

C.W. de Silva
Department of Mechanical Engineering
The University of British Columbia
Vancouver, Canada

The technology of neural control has grown very rapidly in recent years. Neural control systems of many shapes and forms, often designed on the basis of heuristic considerations, have been proposed and widely used in a range of technical applications. The challenge the control community faces is to find the best ways of fully utilizing this powerful new tool in control system applications. This chapter provides an introductory overview of neural control systems. The chapter is intended to describe a selection of useful neural control techniques, and also to survey neural control research and applications. The primary aim in this context is to provide useful guidelines and references that are useful for research and application of neural control systems.

A modular neural-visual servo control system is presented in the second part of the chapter. One of the key problems of feature based visual servo is calculating the inverse Jacobian, relating change in features to change in robot position. Neural networks can learn to approximate the inverse feature Jacobian. However, the neural network approach can approximate only the feature Jacobian for a small workspace. In order to overcome this problem, we propose using a modular approach, where several networks are trained over a small area. Furthermore, we use a neural-fuzzy counterpropagation network to decide which subspace the robot is currently occupying. Preliminary results of the system's operation are also presented.

1 Introduction

The technology of neural control has grown very rapidly in recent years. This growth has been fuelled by the renewed attention to neural networks and its application in areas such as pattern recognition and intelligent control. Fuzzy control is the other control area that has attracted similar attention in the last decade. The rationale for using neural control or fuzzy logic control systems is related to difficulties, which are commonly experienced by control engineers in real-world applications. It is generally difficult to accurately represent a complex process by a mathematical model or by a simple computer model. Even when the model itself is tractable, control of the process using a "hard" (non-soft or crisp) control algorithm might not provide satisfactory performance. Furthermore, it is commonly known that the performance of some industrial processes can be considerably improved through the high-level control actions made by an experienced and skilled operator, and these actions normally cannot be formulated by crisp control algorithms [18].

From the control theory point of view, if the process model itself is inaccurate, model-based control cannot provide satisfactory results. Even when an accurate model is known, if the parameter values are partially known, ambiguous, or vague, then appropriate estimates have to be made. Crisp control algorithms based on incomplete information usually will not give satisfactory results. A primary purpose of classical feedback control is to increase robustness of the control system; i.e., to increase the performance of the system when there is uncertainty such as model errors, unknown disturbances, and noise. Adaptive and robust control techniques are designed to cope with uncertainties due to large variations in parameter values, environmental conditions, and signal inputs. However, the region of operability of the control system will be restricted, although it will be considerably large compared to nonrobust classical control systems [7][73]. In complex processes in practice, the range of uncertainty may be substantially larger than can be tolerated by crisp algorithms of adaptive and robust control. What are known as "intelligent" control techniques [18] are useful here, one example being neuro-control.

To understand why neural networks can be very attractive in control applications, some important properties of neural networks are summarized below [13]:

- Massive parallelism: Neural networks are highly parallel and can easily be implemented using parallel hardware.

- Inherent nonlinearity: Neural networks have the ability to model any piecewise continuous nonlinear mapping to an arbitrary degree of accuracy, by properly selecting the size and parameters of the network.

- Learning capability: Neural networks have a powerful learning capability. They can learn from examples.

- Capability for generalization: Most neural networks exhibit some structural capability for generalization. In particular, the network will cover many more situations than the examples used to train it. Therefore, they have the ability to deal with difficulties arising from uncertainty, imprecision, and noise, in a wide range of problems.

- Guarantied stability: Recent theoretical results prove that certain neural network control structures are guaranteed to be stable for certain nonlinear control problems.

Clearly, these characteristics are important as the need increases to deal with increasingly complex systems, and to accomplish highly demanding design requirements with less precise prior knowledge of a process plant and its environment. It has been argued by Werbos [90] that if a task can be done equally well using either a conventional method or the neural-network approach, then there are several advantages for using the latter. He indicated in 1991 that the computer chip company, Intel, produced a neural-net chip which had a more effective throughput than all of the Cray supercomputers of the world put together. He further stated that it is certainly much easier to develop algorithms to make use of such chips than incorporating Crays into every workstation. The challenge that the control community faces is to find convenient and efficient ways to fully utilize this powerful new tool in control system applications.

The chapter is organized into seven sections. Section 1 gives an introduction to the chapter. Section 2 provides a brief overview of various neural network architectures and learning algorithms, with particular reference to their relevance in neural control. Section 3 reviews modeling and identification of dynamic systems using neural networks. Useful neural control techniques are presented and example implementations are surveyed in Section 4. In that section, we will also present a survey of methods for design and evaluation of more complex systems for learning; i.e., specifically hybrid systems. There we will consider systems that use a hybrid combination of fixed controllers and learning elements, and also a hybrid combination of fuzzy systems and neural networks. Section 5 surveys various software languages, software packages, and hardware components for implementation of neural control systems. Section 6 presents a modular neural-visual servo control system which we developed for robotic applications. Conclusions are given in Section 7.

2   Artificial Neural Networks (ANN)

2.1   The Backpropagation Network (BPN)

Neural networks consist of a group of nodes and weighted connections, usually organized into layers. For a backpropagation network (the most common), the input is carried through the network by series of multiplications over the connections and summations at the nodes. The most common architecture is based on the multilayer perceptron shown in Figure 1.

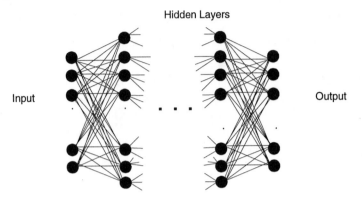

Figure 1: Multilayer perceptron.

The multilayer perceptron node is shown below in Figure 2. For each new input, the string of multiplications and additions is carried out. The result is a large number of nonlinear equations acting as a global estimator.

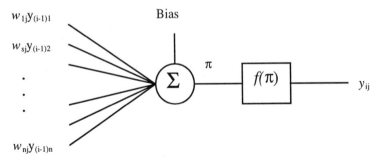

Figure 2: Perceptron node.

The connection weights are held between -1 and 1 and the activation function is usually sigmoidal. The multilayer perceptron node sums the outputs of all connected nodes and a local bias which it feeds into the activation function and then passes to the next layer. Therefore, the output can be written as [33]

$$\zeta_i = \sum_i \Theta\left(xw_{ij} + bias\right)$$
$$y_k = \sum_j \zeta_j w_{jk}$$

(1)

where $\Theta(\pi)$ is usually of the form

$$\Theta(\pi) = \frac{1}{1 + e^{-\pi}} \tag{2}$$

Initially a sigmoidal nonlinearity was chosen because it approximates the nonlinearity in human neurons. However, from a mathematical perspective, the use of a sigmoidal nonlinearity in Equation (2) gives a three-layer perceptron arbitrarily accurate nonlinear estimation. Unfortunately there are no guidelines for the size of the hidden layer or the number of iterations required to converge.

Kuntapreeda and Fullmer have recently proposed a variation on the traditional backpropagation learning algorithm which guaranties stability for control applications [43]. The proof for stability is given by proving the existence of a Lyapunov function. They use an additional training step to ensure that the weight vectors always satisfy the stability criterion.

2.2   Kohonen Networks

Another type of neural network is the Kohonen network. Based on work done by Kohonen in the early eighties, this net is an example of a Self Organizing Feature Map (SOFM) [33]. The Kohonen network does not require error terms to be fed back; it captures patterns at the input and acts as a classifier. The more patterns the net is exposed to, the greater the chance the network is able to correctly map the input into a smaller dimensional space. A two-dimension SOFM is shown in Figure 3.

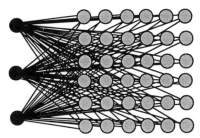

Figure 3: Two-dimensional SOFM network.

The Kohonen network consists of a single layer of input nodes and a single layer of output nodes. The input nodes usually capture a multidimensional time varying signal; in this case, the image of the object of interest. The input and output layers are fully connected, but the meaning of the weights is different from the BPN. The output of a Kohonen network is determined on a winner-take-all basis. The node in the Kohonen layer with weights most closely resembling the current input is the winner, and its output is set to 1. All other nodes in the network have their output set to zero. Kohonen used this network to classify the phonemes of the Finnish language, then ran the net as a phoneme recognizer, a step toward speech recognition [33].

The learning algorithm of the Kohonen network rewards the node closest to the input pattern using a simple learning algorithm, shown in Equation (3) [33].

$$\Delta w_i = \eta \left[x\ (n) - w_i(n) \right] \qquad (3)$$

The Kohonen network can be adapted like the forward propagation network to change the learning rate over time giving faster and more reliable convergence. The algorithm can also be changed from a winner-takes-all to a region-based approach, rewarding all neurons in the region of the winning neuron. This is known as a topology conserving feature map. Generally a topology conserving SOFM takes longer to converge than a regular SOFM but has some desirable qualities for image recognition and fuzzy-neural systems.

The Kohonen network and SOFM algorithm can be viewed as a neural network implementation of adaptive pattern classification or vector quantization, both well understood applications of statistics and systems theory. By proving the equivalence of the algorithms, scientists have provided a theoretical framework for analyzing the behavior of neural networks.

2.3   Counterpropagation Networks

The counterpropagation network (CPN) is based on a combination of Kohonen and backpropagation topologies. The counterpropagation network uses a self-organizing Kohonen layer and a linear output outstar layer. The input layer is a linear buffer where operations such as scaling and normalization take place. The Kohonen layer is useful for fuzzy neural applications because of its topology preserving properties. Each node can be viewed as the center of a fuzzy membership function, and nodes in each neighborhood have closely overlapping membership functions. The counterpropagation network can be used as a fuzzy decision maker and ,by extension, as a fuzzy controller and nonlinear approximator. The counterpropagation network is shown in Figure 4 [65].

Nie and Linkens proposed a fuzzy-neural CPN, but allowed only multi-input–single-output (MISO) systems and fuzzy singletons as output membership functions. By allowing multiple winners, the system can become a multi–single-output (MIMO) controller with arbitrary output membership functions [65]. The forward operation of the network maps input vectors onto output vectors following a fuzzy IF THEN pattern; that is, IF *x* THEN *y*, where *x* and *y* are crisp, but the IF and THEN are fuzzy.

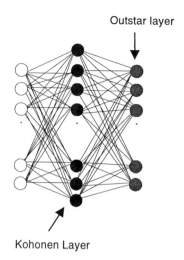

Outstar layer

Kohonen Layer

Figure 4: Counterpropagation network.

2.4   Hebbian Networks

Perhaps the most famous self-learning paradigm is based on the work of Hebb in the late 1940s. Hebb's learning rule was based on observations of the functioning of the brain [33]. His work can be described qualitatively as

1. If two neurons on either side of a synapse (connection) are activated simultaneously (i.e., synchronously), then the strength of that synapse is selectively increased.

2. If two neurons on either side of a synapse are activated asynchronously, then that synapse is selectively weakened or eliminated.

The above statement emulates the synaptic learning algorithm in the brain. Numerically the learning rule is expressed as follows. First the output of each neuron is calculated using Equation (4), then the weights are updated using Equation (5).

$$y(n) = \sum_{i=0}^{p-1} w_{ji}(n)\, x_i(n) \tag{4}$$

$$\Delta w_{ji}(n) = \eta\left[y_j(n)\, x_i - y_j(n)\sum_{k=0}^{j} w_k(n)\, y_k(n) \right] \tag{5}$$

where p is the number of nodes in the Hebbian (output) layer, and j is the number of nodes in the input layer. The second term of Equation (5) is added to keep the learning rule from growing without bound.

Hebbian networks have a two-layer topology very similar to the Kohonen. Links from the input layer to the Hebbian layer are feed forward and fully connected. For feature extraction, Hebbian layers are often cascaded and connected randomly between Hebbian layers. Hebbian networks are also used in principal component analysis. In principal component analysis the network learns to extract the p-1 most influential eigenvalues, that is, the eigenvalues that contain the most information about the image or signal. The Hebbian network compresses the image or signal, while still preserving statistically important information.

2.5   Radial Basis Function Networks

The radial basis function (RBF) network is based on the simple intuitive idea that an arbitrary function $y(x)$ can be approximated as the linear superposition of a set of localized basis functions $\bullet_j(x)$ [12]. The radial basis function networks look much like the common feed-forward architecture used with backpropagation training, as shown in Figure5. It has three layers; the input, hidden, and output layers. The input layer is simply connected to the hidden layer via unweighted links. Therefore, inputs x_1, x_2, ..., x_n are directly applied to all neurons in the hidden layer. Each hidden layer neuron computes the following exponential function [87]:

$$h_i = e^{\frac{-|x-u_i|}{2\sigma_i}} \tag{6}$$

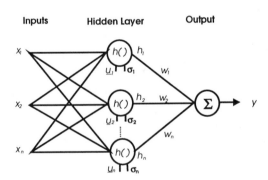

Figure 5: Architecture of RBF network.

where x is an input vector, h_i is the output of the ith neuron of the hidden layer, u_i is a vector representing the center of the ith basis function, and σ_i is the width of its Gaussian.

The output of the network is the weighted sum of the outputs from hidden layer and is given by

$$y = \sum_{1}^{n} h_i w_i \qquad (7)$$

where, w_i, $i = 1, 2, ..., n$, are the weights of the connections between the output layer and the hidden layer, and n is the number of RBF neurons. A more detailed explanation of the structure and operation of RBF networks can be found in [87].

The radial basis function networks are attracting a great deal of interest because they are capable of rapid training, generality, and simplicity; however, radial basis functions suffer from the curse of dimensionality. The number of nodes required to perform complex tasks can be very large. Fabri and Kadirkamanathan proposed a new network growing scheme designed for control applications [22].

2.6   Hopfield Networks

A Hopfield network consists of two layers: an input and a Hopfield layer, as shown in Figure 6. Each node in the input layer is directly connected to only one node in the Hopfield layer. Nodes in the latter layer are neuron models as described earlier. Different threshold functions such as hard-limiting or sigmoidal functions can be used [28]. The net has N nodes each with a binary input x_i. Here it is considered that x_i can be either 1 or 0. The output of each node is fed back to all remaining nodes via active delay and weights denoted by t_{ij}. The N discrete binary x_i values constitute a pattern that can also be considered to be the state vector of the system.

This net can be used as a memory to store a number of different patterns (state vectors). If any one of these stored patterns is presented to the memory, the memory will remain in that state. However, if a modified version of any of the stored patterns is presented, the memory will evolve until it reaches a stable state which will be nearest to that modified state. When the memory is functioning properly, it can regenerate the original prototype (the stable state in terms of which the modified input had been recognized). So, in this case it is considered to be working as an auto-associative memory. The network can also be used as a classifier to distinguish among M specific classes. In this case, the output after convergence is compared with the M exemplars to determine if it matches an exemplar directly. If so, the output will be that class whose exemplar matches the output pattern. If not, a "no match" result will occur [47].

This net can be used as a memory to store a number of different patterns (state vectors). If any one of these stored patterns is presented to the memory, the memory will remain in that state. However, if a modified version of any of the stored patterns is presented, the memory will evolve until it reaches a stable state which will be nearest to that modified state. When the memory is functioning properly, it can regenerate the original prototype (the stable state in terms of which the modified input had been recognized). So, in this case it is considered to be working as an auto-associative memory. The network can also be used as a classifier to distinguish among M specific classes. In this case, the output after convergence is compared with the M exemplars to determine if it matches an exemplar directly. If so, the output will be that class whose exemplar matches the output pattern. If not, a "no match" result will occur [47].

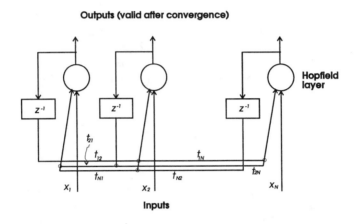

Figure 6: Hopfield network.

The memory learns by modifying the values of t_{ij}, the internode weights, which can be thought of as long term memory. Moreover, it can record and retain any pattern for a while by modifying x_i, which can be thought of as short-term memory [67]. Hopfield network has been found very useful for pattern classification. The feedback connections, with a dynamic delay, of the Hopfield network make it attractive for use in modeling and control applications of dynamic systems.

3   Neural Modeling and Identification

System modeling and identification is an important issue in many control applications. Several approaches have been proposed during the last two decades [20]. However, many difficulties still exist in applying the existing methods to complex systems with partially known, nonlinear, and time-varying characteristics. Recently, a number of researchers used neural networks for modeling and identification of dynamic systems [16], [60], [94], [13]. In [16], neural networks were considered to be an alternative form of functional representation of nonlinear systems as opposed to the nonlinear autoregressive moving average models with exogeneous (NARMAX). A comprehensive treatment of the topic in identification and control applications is given by Narendra and Parthasarathy [60] where several models for identifying (or emulating) an unknown, nonlinear, single-input, single-output (SISO) plant by using multilayer neural networks are proposed. A detailed treatment of their techniques is given in [60], [61], [62]. Very recently Narendra's team used a Taylor expansion to create a more accurate approximation of their NARMAX model for a given system [63].

Many other approaches have also been proposed. Yamada and Yabuta [98] proposed a practical neural network design method for the identification of both direct and inverse transfer functions of an object plant. They show that for a second-order plant, identification can be satisfactorily achieved and that neural network identifiers can represent nonlinear plant characteristics quite well. Brown, Ruchti, and Feng [13] presented a method for incorporating *a priori* information about an uncertain nonlinear system into the structure of a multilayer feedforward artificial neural network. They used the technique in the identification of a dynamic system and compared it with conventional feedforward artificial neural network identifier. Their results exhibit an improvement in the quality of the identification model and an increase in the rate of convergence. In an industrial example, Bhat et al. [10], [11] used neural networks with backpropagation to model chemical processes including an industrial distillation tower with seven inputs and four outputs. Wu and Stanley used a combination of backpropagation and compression networks to create a complete neural robot vision system [96], [97].

Recently, approaches have been proposed to combine fuzzy systems and neural networks for system identification. Fuzzy identification has been found to be effective in forming relationships between input and output data, especially in situations where vagueness, subjectivity, and uncertainties are involved. Pedrycz [69] proposed an approach based on resolving fuzzy relational equations. Takagi and Sugeno [81] have suggested a fuzzy identification method based on multidimensional fuzzy reasoning. More recently, Wang and Vachtsevanos [86] presented an approach based on fuzzy associative memory (FAM) for system identification. Wu and de Silva [95] presented an improved method for identifying the structure of a fuzzy model and determination of a fuzzy rule-set which maps input data to output data of a dynamic system. In this method, identification of the model is based on the evaluation of 'conflicting rules' in the fuzzy associative memory (FAM) cells, and the primary rule-set is obtained by using the adaptive FAM method. A neural-network-based error compensator has been developed to improve the accuracy of the fuzzy model.

4 Neurocontroller-Design Methods

Neural control has existed for many years and there is a large body of empirical evidence of its viability in nonlinear control applications. Many different control structures have been proposed ranging from fuzzy/intelligent control to pure model approximation. More recently, theoretical results have guaranteed the stability of neural controllers for problems that can be composed as a nonlinear autoregressive moving average model. [63], [82]. These results give weight to the neural control argument, removing some of the mathematical uncertainties inherent in using neural networks.

A number of neurocontrol structures have been proposed. They can be loosely categorized into six classes [88], [5], [75], [80]:

- Supervised control.

- Direct inverse control.

- Neural adaptive control.

- Backpropagation through time.

- Reinforcement learning control.

- Hybrid control.

4.1 Supervised Control

In this method, a neural network is trained to perform the control task similar to a desired controller or human operator. Therefore, training data must be collected either in advance or on-line from an existing physical system. Figure 7(a and b) show the on-line training phase and on-line control phase, respectively. This approach is very useful for transferring human knowledge or experience into an automatic controller [88]. Typically, a backpropagation algorithm is used for training the neural networks, although any learning algorithms may be used [88]. The objective here is to find a mapping which will map the sensor inputs to desired actions as accurately as possible.

This approach has been employed in many control applications. Windrow's pole balancing controller [92] is one of the earliest reported in the literature. A more sophisticated, updated version was later provided in [30]. Rjorgensen and Schley [74] used neural network based controllers for landing aircraft. For training the neural controller, information from both existing controller and human expert knowledge has been used. Other studies include those of Asada [3] and Liu and Bekey [50].

Asada and Liu [2] developed a method to transfer human skills to neural net robot controllers. Their focus is to determine the conditions under which the skill transfer method using neural net is applicable. They noted that, while a neural net controller resulted from supervised learning does not require explicitly stated control objectives, the training data must be studied and examined to avoid inconsistent, contradictory, and insufficient data, since such data may incur difficulties in training neural nets (e.g., lack of convergence) and may even cause instability and erratic behavior in robots.

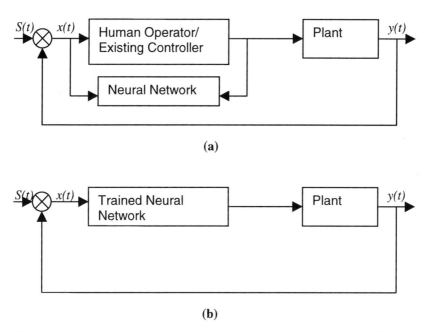

(a)

(b)

Figure 7: Supervisory control. (a) Network training phase. (b) On-line operating
phase.

4.2 Direct Inverse Control

In inverse control, a neural network is trained to learn the inverse dynamics of a
system. Inverse dynamics provide the input that will generate a particular output.
This approach has been found useful in control of systems such as robotic
manipulators.   By moving the manipulator around, the system inputs (motor
signals) and the manipulator position (actual coordinates) can be recorded and used
to train a neural network so that given the desired positions of a trajectory, the
neural controller will generate the desired inputs to the joint motors. Figure 8(a
and b) illustrate two learning architectures for learning the inverse function [71],
[87]. Psaltis and his colleagues [71] refer the configuration shown in Figure 8(a) as
generalized   learning architecture. Such an architecture provides a method for
training a neural controller that minimizes the overall error. The training sequence
can be described as follows:

1.   Select a plant input u and apply it to the plant.
2.   Obtain the corresponding output y from the plant.
3.   Train the neural network to reproduce u when y is given as the input
     network.

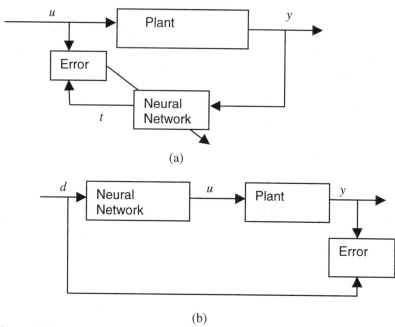

(a)

(b)

Figure 8: Direct inverse control. (a) Generalized learning architecture.
(b) Specialized learning architecture.

Once the neural network is trained, during the on-line operation phase, it will reproduce input u to the plant for a desired response d. This way, the output of the plant should approach the desired response d. Psaltis and his colleague indicated that this architecture by itself can be very difficult to use and may not provide satisfactory performance in real applications, because it is difficult to know in advance the region of interest in which the plant may operate. To overcome this problem, they proposed a specialized learning architecture which is shown in Figure 8(b). This architecture has been widely used in the literature [83]. It uses the difference between the actual and desired outputs of the plant to modify the weights of the inverse neural model. This approach allows the network training to take place exactly within the operational range of the plant, and, therefore, is usually more accurate than generalized learning. To make use of the advantages of both methods, they also proposed a method that combines the two approaches [71].

As indicated in [40], learning an inverse model is currently among more viable techniques in the application of neural networks for control. In work such as [39], [36], and [57], CMAC networks have been used to learn the inverse dynamics of a plant such as a robot manipulator. Khalid and Omatu [40] investigated the use of appropriately trained back-propagation neural networks as physical controllers similar to conventional feedforward controllers in real-time control systems. Their results show that the neural network controllers perform well. Sofge and White [78] successfully applied direct inverse neurocontrol to the manufacture of thermoplastic composite structures. More examples are found in [58].

It may be noted that problems can arise when the plant inverse is not causal or well-defined, as indicated in [75] and [88]. More detailed discussions on this can be found in [37] and [39]. It has also been pointed out by Khalid [40] that much effort is required in order to apply the approach to real plants, which typically exhibit some nonlinearity and variations in their parameters due to noise, disturbances, and other environmental factors.

4.3 Neural Adaptive Control

In neural adaptive control, a neural network is used to identify the parameters of the system and tune the controller in a conventional, adaptive-control structure such as model-referenced adaptive control (MRAC) or self-tuning regulator (STR) [60]. A number of papers have been published over the past few years to report the progress made in neural adaptive control. The work reported by Narendra's group [58] is especially noteworthy. Their studies have formed a solid foundation for the application of both gradient methods and stability methods for control. Figure 9 shows the model-referenced control structures which they have proposed for nonlinear systems utilizing neural network models. The control structure for the identification model as well as the controller are derived based on the linear case [60]. However, nonlinear neural networks are used here in place of gains in linear control. Their neural adaptive control scheme includes two major steps:

- The input-output behavior of the plant using one of the identification models proposed in [58] which contains neural networks and linear dynamical elements as subsystems.

- The parameters of the controller using the identification model that resulted in the previous step.

- Other examples in this class can be found in [72], [77].

Self-tuning regulator is another type of neural adaptive control structure that has been implemented [79], [84]. In [84], Tokita developed a self-tuning regulator for force control of a robotic manipulator. The neural network was used to identify system dynamics and to predict the future process outputs. Recently adaptive, variable-structure, model following control has been implemented by Karakasoglu, Sudharsanan, and Sundareshan [38] using a recurrent neural network for control of multijointed robotic manipulators. Sbarbaro and Hunt [76] pursued the receding horizon approach to optimal control. They used a control structure which consists of two networks: one models the plant and provides prediction data for optimization, and the other is trained as an approximate inverse model of the plant. Studies have also been conducted to compare conventional control methods and neural network-based controllers [41], [42], [66].

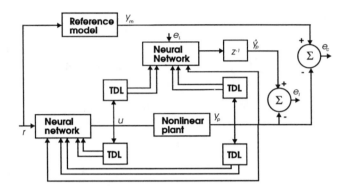

Figure 9: Indirect adaptive control.

4.4   Backpropagation Through Time

Backpropagation through time is a popular learning algorithm for recurrent neural networks [90]. Suppose that a neural network is used to describe the dynamics of the system to be controlled. The model may predict that

$$r\ (t) = f\ (\ r\ (t\text{-}1),\ u\ (t))$$

where $r(t)$ is a state vector representing the current state, u is a vector of controls, and f is a vector-valued function implemented by the neural network. Using backpropagation through time, the derivative of some predefined function $U(r)$ is usually calculated with respect to $u\ (t)$ at all times (or with respect to the weights of a network generating $u(t)$). By adapting $u(t)$ in response to these derivatives, the optimal actions of weights can be found. For example, Nguyen and Widrow [64] used the backpropagation-through-time algorithm to back up a truck. The motion of the truck is broken down into steps, each representing an instant. Backpropagation goes backward, from the ending position to the starting position, adjusting controller weights at each instant. Plumer [70] successfully used multilayer neural networks, trained by the backpropagation-through-time algorithm, as state-feedback controllers for nonlinear terminal control problems. The technique is useful wherever a sequence of actions leads from a starting state toward a desired state [86].

Park, Choi, and Lee extended the BTT algorithm to provide optimal tracking for a neural controller [68]. Their system is cast in the form of a general nonlinear autoregressive moving average filter (NARMA). In addition they use a quadratic cost function to evaluate the tracking error of the system. Their control structure uses a feed forward neural controller to adjust the system relative to the set point and a BTT feedback controller to stabilize tracking error dynamics when following an arbitrary reference output [68].

4.5   Reinforcement Learning Control

In supervised learning, information for training the controller is assumed to be readily available or can be conveniently obtained. However, in many practical situations, the information available cannot be directly used as training data. Instead, it can provide information, usually in an approximate manner, on the performance of the system. In other words, such information or data can be used only to evaluate the state of the system. This class of methods is often referred to as reinforcement learning [5].

One of the earliest reinforcement learning control systems was implemented by Waltz and Fu [85]. They used reinforcement learning to learn bang-bang control. The most notable work in this area is that of Barto, Sutton, and Anderson [6]. The system they developed consists of a single associative search element (ASE) and a single adaptive critic element (ACE), as shown in Figure 10. The task was to learn to balance a pole on a wheeled cart. The cart moved in one direction with the movement caused by the application of either a positive or negative fixed force on the cart along the direction of motion. They assumed that the dynamics of the process are not known. The feedback that evaluates the performance was based on a failure signal that occurs when the pole fell or the cart reached a stop at either extreme of its range of travel. One major improvement of this system over that by Michie and Chambers [54] is that an adaptive critic element is introduced which can compute a prediction of reinforcement so that, if the punishment is less than its expected level, it acts as a reward.

Fabri and Kadirkamanathan's dynamic RBF network uses a reinforcement learning scheme to control simulated dynamic systems [22]. Their network dynamically scales as it receives training data using selective node activation. They show that they can optimize the network size while maintaining guaranteed stability of the controller.

Anderson extended such an adaptive critic method to the case of multilayer networks [1]. Berenji [9], Lee [45], Lee and Berenji [44], and Berenji and Khedkar [8] combined the adaptive heuristic critic model [6] with fuzzy logic to design the membership functions of conclusions of control rules. They also extended the adaptive heuristic critic algorithm of Barto, Sutton, and Anderson to include prior control knowledge of human operators. They demonstrated significant improvements in terms of the speed of learning and robustness to changes in parameters of the dynamic system over pure neural network-based schemes. In their system, two networks, referred to as the evaluation network and the action network, replace the two neuron-like elements. These networks are multilayered networks based on the backpropagation learning algorithm.

Some interesting results have been reported by Grant and Zhang [29]. They derived some reinforcement learning equations for modifying neural network back-propagation weights. They have shown that a significantly improved level of some reinforcement learning equations for modifying neural network back-propagation weights. They have shown that a significantly improved level of learning can be achieved using their algorithm, and have indicated that the algorithms could generalize and accommodate changes in the control environment, without a need for further training.

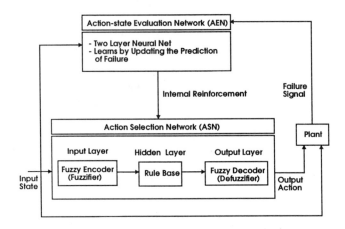

Figure 10: Adaptive learning control.

Recently, comparative studies [31] have been conducted to compare supervised learning control with reinforcement learning control. The comparison shows that the reinforcement learning method performed better for tasks involving learning to control unstable, and non-minimum phase dynamic systems such as cart-pole balancing. Indeed, reinforcement learning control systems have at least five attractive features which have been summarized in a paper by Zomaya, et al. [99]; specifically,

- Mathematical model of the system is not necessary for the development of a control law. The controller learns to develop by association of input and output signals.

- Wide class of performance measures can be optimized.

- System to be controlled can be time-varying and/or nonlinear.

- A non-uniform sampling rate can be used.

- Algorithms are naturally adaptive. Hence, they can be used to directly control the system or to optimize the performance of an existing control system.

It may be noted, however, that learning times can be quite long for this type of learning control scheme. Usually many failures occur before a successful control is realized [75].

4.6   Hybrid Control

The need to meet demanding control requirements in increasingly complex systems makes hybrid control very attractive. These hybrid techniques may be of the following forms:

- Conventional control and learning control methods.

- Knowledge-based control and neural network techniques.

- Fusion of fuzzy logic and neural networks methods.

Conventional Control + Learning Control Methods

A typical example in this category is the refinement learning control method presented by Franklin [24], [23]. It is a hybrid combination of reinforcement learning control and a conventional controller. They designed a conventional controller using pole placement technique based on a linear estimation of the dynamics. Then they developed a secondary learning controller using a reinforcement learning-based neural network to refine the performance of the control system.

Other techniques in this category include those presented by Miller et al. [55], [58] and Baird and Baker [4]. Miller et al. used a hybrid combination of a fixed gain PID feedback controller and learning elements. The CMAC-based neural networks were used to learn a model of robot manipulator dynamics. Supervised learning techniques were employed. Baird and Baker [4] proposed a hybrid controller which combines adaptive control with learning control techniques. Conventional adaptive control is used to respond quickly to local plant perturbations, and the learning controller learns to control the resulting adaptive plant on a more global level.

Expert Control + Neural Networks

Artificial neural networks and expert systems offer very different capabilities concerning control system design, implementation, and performance. Knowledge-based systems provide a control system designer with a convenient mechanism for automated complex decision making. Task-specific knowledge may be defined explicitly. Neural networks, on the other hand, encode knowledge implicitly, adjusting internal weights so that their input/output relationships remain consistent with observed training data. An appropriate combination of these techniques could be formed that would exploit the advantages of both worlds and generate more efficient and reliable control systems. Successful use of this technique has been demonstrated by Handelman et al. in [32] where they have integrated knowledge based systems with CMAC-type of neural networks for the purpose of robotic control. The task was to teach a two-link manipulator how to make a specific type of swing. The rule-based system was used to determine how to make a successful swing using rules alone. By observing the rule-based execution, the CMAC network would learn how to accomplish the task and subsequently take over the operation.

Their simulation studies show that this approach is promising, although many issues need to be addressed.

Fuzzy Control + Neural Networks

The development of hybrid control systems by the fusion of fuzzy logic and neural networks is now becoming popular and holds much promise for the control of nonlinear, uncertain, and complex systems. Fuzzy logic control which is based on the concepts of fuzzy logic, first developed by Lotfi Zadeh, is a class of knowledge-based control that incorporates control knowledge in the form of a set of linguistic rules. This allows the application of heuristic information to be used in the control of complex systems [18].

Like neural network based control, fuzzy logic control has attracted much attention in the field of intelligent control in the last few years because of its ability in dealing with difficulties arising from uncertainty, imprecision, and noise. Deriving fuzzy control is, however, often difficult and time-consuming. Currently there is no systematic way of designing fuzzy logic controllers. A typical approach is to establish a rule base by studying and extracting the required information from an existing or human-operated controller. It has been noted that a fuzzy controller based directly on operator-specified or common sense rules often exhibits poor performance characteristics [52], because the initial rules are invariably rather crude and, although qualitatively correct, have to be refined to achieve better performance.

In the current designs of many fuzzy logic control systems, trial-and-error approaches are used to tune the rule base and the associated membership functions of the fuzzy variables until satisfactory performance is achieved. A more effective class of fuzzy control systems constitutes the adaptive fuzzy control systems which are capable of learning from process data as well as incorporating linguistic data. An adaptive fuzzy controller could automatically generate a set of fuzzy control rules and improve on them as the control process evolves [21]. Much work has been done in this area. Surveys of such work have been presented in [21], [52]. Attempts have been made to combine the powerful learning and generalization capabilities of a neural network with the structure-rich characteristics of fuzzy logic control. The following is a brief summary of such cases:

- *Fuzzy rule generation using NNs:* Given a series of sample data vectors, neural networks can be used to adaptively infer the fuzzy rule base using an unsupervised learning algorithm (such as an adaptive vector quantizer (AVQ) with differential competitive learning (DCL)). This is useful when sample data can be relatively easily obtained. For example, in many industrial processes, a control system can be fine-tuned manually and the input-output data pairs can be recorded. When building the knowledge base, one can use these data vectors to infer the control surface using NNs. This can then be used either to replace the process of interviewing operators, or as a compliment to that process.

- *Use of NNs to tune fuzzy membership functions:* In fuzzy control applications, often a fuzzy control system is developed by choosing fuzzy sets for the condition and action variables and building the knowledge base by interviewing operators. In this process, the selection of fuzzy sets is very subjective and standard rules to follow do not exist. With NNs, the fuzzy membership functions can be 'optimized.' This can be achieved via supervised learning. Given the training input data and the desired output value, NNs can be used to adjust the parameters of the membership functions to minimize the error between the desired output and the real output. As a result, the membership functions would be optimized for this particular application.

- *Use fuzzy rules to pretrain NNs:* Once the fuzzy rules are known, they can be used to train a NN which can subsequently be used to replace a fuzzy tuner. This idea has been implemented in [18] where a NN trained by fuzzy rules is used to tune a servo-motor system. A similar idea may also be used to implement a fuzzy tuning algorithm for a hierarchical supervisory control system. Since a neural network can extrapolate and interpolate, reasonable tuning is maintained even when the conditions demanded by a rule are not exactly satisfied. Unlike the conventional approaches to fuzzy tuning, a NN-based fuzzy tuning method has a coupling effect on the input attributes; i.e., if two or more performance attributes are out of specification at one time, all of them will be taken into consideration in generating the output action attributes. Hence, less time would be required to tune the system.

A review paper in this area has been written by Buckley [14]. While excellent work has been done by many researchers, e.g., [19], [46], the work by Berenji and his colleagues [8], [9], [44], [45] deserves special mention. They proposed the Approximate Reasoning Based Intelligent Control (ARIC) architecture for integrating fuzzy logic controller with neural networks. Figure 11 shows a diagram of the ARIC architecture. The two main elements in this model are the action-state evaluation network (AEN) and the action selection network (ASN). AEN plays the role of an adaptive critic element and provides advice to the main controller. ASN includes a fuzzy controller. Both ASN and AEN are modeled by multilayer neural networks, consisting of an input layer, a hidden layer, and an output layer. A reinforcement learning scheme and a modified error backpropagation algorithm are used for the output layer and hidden layer, respectively.

Lin and Lin proposed a fuzzy neural controller based on the ART (active resonance theory) network. Their system used a reinforcement learning and an adaptive critic network. Both the critic and action network are created dynamically to represent a set of fuzzy functions representing the control algorithm. Their dynamic network creation algorithm creates nodes for each fuzzy set then tunes them during the parameter learning phase. They demonstrate several simulated control problems such as the pole balancing problem [48].

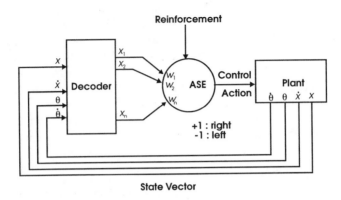

Figure 11: The ARIC architecture.

5 Hardware and Software for ANN

At present, most research and application projects on neural networks are conducted by using conventional software languages (e.g., C or C++) and by running on standard computer platforms (e.g., Sun Sparcstations or PCs) [12]. While this is adequate in many applications, it can be expected that more parallel software/hardware implementations of neural networks will be seen in future projects to make use of the massive inherent parallelism of the neural networks.

There are many software packages (available either in public domain or in commercial format) for neural computing. Some of the software packages provide excellent facilities for simulation and training of multiple nets with various architectures and learning algorithms. Access to such existing software can be a tremendous time saver for researchers and engineers in the field of neural control. On the other hand, a variety of industrial companies have produced various types of hardware platforms for neurocomputing. They could be very valuable for real-time implementations of neural network based systems including neural control systems. To date, the best source of information on these various types of software and hardware packages is seen in the FAQ posting to the Usenet newsgroup comp.ai..neural-net. Based   on   the   information   obtained   there,   the software/hardware can be categorized into three categories:

- Public domain software package for neural networks,

- Commercial software packages for neural networks, and

- Companies supplying neural network hardware.

The public domain software packages are free and available via anonymous ftp. Most of these packages are usually developed based on one or more research projects related to neural networks. Some of the packages are, in fact, very advanced. They provide a complete simulation environment for building, training, editing, and examining neural networks. They include a variety of neural network architectures and learning algorithms. No doubt, packages in this category can be extremely useful for researchers and professional scientists alike who are working in the neural networks field. Some of the packages are relatively simple. They can be useful for newcomers to the field or students who wish to get started with this new technological area. A key problem of almost all public domain software packages is that they are not guaranteed to be optimized or may not even be well-debugged, and, therefore, there is no assurance that they will consistently deliver what they claim. Besides, there is usually very little support available for this type of software. While these should not present serious problems to researchers, difficulties will arise if they are to be used in commercial projects.

Generally speaking, the commercial software packages for neural computing are much more user-friendly. Various user interfaces including graphical displays are provided. They also provide easy-to-use facilities or building blocks for developing neural systems. They usually allow users to build applications quickly, easily, and even interactively. Some packages will allow users to develop applications using block diagrams and subsequently generate source code in high level languages that can be subsequently downloaded to the target hardware.

When compared to hardware systems in other fields, the hardware development for neural networks is still in an early stage. A number of companies supplying hardware for neural networks can be found in the FAQ posting to the Usenet comp.ai.neural-net, and hardware systems for neural networks reported in the literature are listed as in [148] through [165]. As can be seen from the references, both analog and digital implementations have been reported. Analog artificial neural networks can offer high speed, low power, and less complexity of the circuitry, and they are suitable for applications where a small number of neurons are required. For very large, reconfigurable networks, which require a large number of neurons and high levels of connectivity, mixed analog/digital implementations are more suitable [78].

6 Modular Neural-Visual Servo Control System

This section presents a modular neural-visual servo control system which we developed for vision-guided robotic applications.

6.1 Introduction

The use of camera-manipulator systems has grown as progress in both fields has advanced sufficiently so using visual servo is a viable solution for industrial automation problems. These manipulator systems involve the integration of robotics, computer vision, and control knowledge to create automation systems that were previously unworkable.

Feature-based visual servo calculates the robot position based on features extracted from the object in the camera's field of view. This method requires the computation of the feature Jacobian, which relates change in robot position to change in image features. In general, this feature Jacobian is a complex, highly nonlinear transform which is unique to each visual servo application. To automate the derivation of the feature Jacobian, neural networks can be used to learn the nonlinear mapping directly from data generated by the system.

Many past visual servo systems have attempted to learn the mapping of features to position over the entire workspace. This is sufficient if the workspace is kept small or simple, but for an arbitrary workspace, this method is no longer sufficient. Modular neural network systems, on the other hand, more closely approximate the functioning of the brain [121], [33]. In general, modular solutions require fewer computational units and are trained faster than global networks [33].

In order to provide a robust general solution for neural-visual servo, we propose the use of a set of modular neural networks that divide the workspaces into several subspaces. Furthermore, switching between the separate networks is accomplished using a neural-fuzzy network. This modular structure finds solutions for any subspace, with the neural-fuzzy controller providing smooth switching between subspaces. In this chapter we will outline the derivation of the neural-fuzzy system, outline the operation of the system as a whole, and present preliminary results of the system operating on a 6 DOF robot.

6.2 Modular Neural-Control System

A block diagram of the system illustrating the execution flow, is shown in Figure 12. The control networks are composed of simple backpropagation networks, connected in a complete or sparse manner. The indexing and decision network are based on the neural-fuzzy model proposed by Nie and Linkens[65].

The overall operation of the system is based on an idea originally suggested by Hashimoto et al. [121]. However, they used a heuristic to advance the network, rather than the learned behavior from a switching network used in our system.

6.2.1 Control Networks

The control networks are based on simple backpropagation networks (see Figures 13 and 14). The networks are connected in two ways, completely or with internal modular connections. A completely connected network is identical to the traditional backpropagation network used by Chan and Lo to create their limited controller [122]. However, in order to get the control task to converge for more difficult subspaces, internal modularity was introduced.

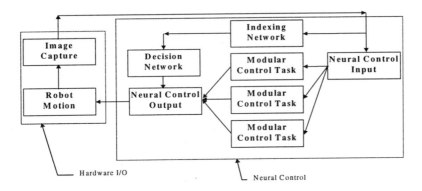

Figure 12: System structure.

We used simple three-layer backpropagation networks (input, output, and hidden) whose output is governed by the equations:

$$\zeta_i = \sum_i \Theta\left(xw_{ij} + bias\right)$$
$$y_k = \sum_j \zeta_j w_{jk} \tag{8}$$

The backpropagation algorithm performs a gradient descent over error space to minimize the mean squared error. The change in weight is given by the following equation:

$$\Delta \mathbf{w}_{ij} = \eta \delta_i \Theta'_{(net)} \mathbf{x}_j \tag{9}$$

where $\bullet'$ is the derivative of the activation function and δ_j is determined below:

$$\delta_j = \begin{cases} y_d - y_{net} & \textit{if output layer} \\ \sum_j w_{ij} \delta_j & \textit{otherwise} \end{cases} \tag{10}$$

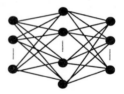

Figure 13: Standard backpropagation network

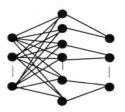

Figure 14: Sparse backpropagation network

In this manner, the error is propagated from the output layer to the input layer.

The sparse backpropagation network (Figure 14) is identical in function to a normal backpropagation network, except the connections from the hidden layer to the output layer are connected in an outstar fashion. The hidden layer is divided into sublayers. Each sublayer is connected to a single output node. This has the effect of decoupling the error terms for each joint from each other. This helps the network converge for motion in the z plane, where the position of 2 of the links are determined from the feature data with the third determined from the feature data and the position of the other links. Essentially, the network constructs different functions for each output instead of creating a global vector mapping for the entire system. Given that the operation of the network is expressed as a vector function, the differences between the two networks can be described as follows:

$$y = F(\mathbf{x}) \qquad \textit{fully connected}$$
$$y_i = F_i(\mathbf{x}) \qquad \textit{sparsely connected} \tag{11}$$

where the output y is determined by the global approximator F for the fully connected network and by the local approximator F_i for each value y_i in the output vector for the sparsely connected case.

The sparse backpropagation network uses the same forward and learning algorithms as the regular forward network, and still obeys Equations (8), (9), and (10). However, the limits on the summation are different for evaluating the step from the hidden to the output layers. Instead of being summed over all hidden nodes, the input is summed over the sublayer connections. Similarly, when evaluating

Equation (10) for the hidden layer, the error at the hidden node depends on a single output node, not the entire layer, so the nodes in that hidden layer only approximate the function dictated by the single error signal.

6.2.2 Decision Networks

There are two decision networks: one decides which subspace the robot currently occupies; the other decides which network(s) should be operating. The decision network itself takes the output of the indexing network and outputs a set of weights for each control network. Each decision network is based on a neural-fuzzy counterpropagation network.

The counterpropagation network (CPN) uses a self-organizing Kohonen layer and a linear output outstar layer. The input layer is a linear buffer where operations such as scaling and normalization take place. The Kohonen layer is useful for fuzzy neural applications because of its topology preserving properties. Each node can be viewed as the center of a fuzzy membership function, and nodes in each neighborhood have closely overlapping membership functions. The counter-propagation network can be used as a fuzzy decision maker, and by extension, as a fuzzy controller and nonlinear approximator.

Nie and Linkens originally proposed the fuzzy-neural CPN, but allowed only MISO systems and fuzzy singletons as output membership functions. By allowing multiple winners, the system can become a MIMO controller with arbitrary output membership functions [65]. The forward operation of the network maps input vectors onto output vectors following a fuzzy IF THEN pattern; that is, IF x THEN y, where x and y are crisp, but the IF and THEN are fuzzy.

To eliminate unnecessary computation of small contributions from nodes in the Kohonen layer with low activation values, a thresholding function was added to the normal Kohonen distance metric. The threshold is global and is not adjusted during the network learning process [33].

$$\textbf{input}_i = \Phi(x) = \begin{cases} \|x - w_i\| & if\ \|x - w_i\| < \delta \\ 0 & Otherwise \end{cases} \qquad (12)$$

This has the effect of finding all fuzzy subsets which are activated by the input, with the fuzzy subsets also defined as above. As the network is trained, the input space is partitioned into N (where N is the number of Kohonen nodes) fuzzy hypercubes with centers at w_i, much like the functioning of a radial basis function neural network. The above step also has the effect of selecting the minimum activation for each fuzzy subset.

In order to produce the output from the Kohonen layer, the output fuzzy membership functions are generated and max-min decomposition is used to find the result. For the output layer, the membership functions are not learned but defined as simple triangular functions centered at each node. The output functions are shown in Equation (13).

$$\Theta(\xi) = \begin{cases} \dfrac{(1-input_i)}{neighborhoodSize}(i-j) & if \ i < j \\ \dfrac{(1-input_i)}{neighborhoodSize}(j-i) & otherwise \end{cases} \tag{13}$$

where i is the index of the winning node, j is the index of the current node, and *neighborhoodSize* is a function of the Kohonen.

Therefore, the output of node j at the output layer is given by the maximum excitation of all output functions that cover that area, or the max-min composition of the original input vector x.

$$output_j = \max_k(\Theta(\Phi(x))) \tag{14}$$

where k represents all active nodes in the neighborhood of j and the minimum is implicitly calculated in $\bullet \ (x)$.

The output of the entire system is given by the outputs of the outstar layer. The output of each linear node is given by Equation (15).

$$y_l = \sum_j w_{jl} output_j \tag{15}$$

Using the weighted sum to find the output is similar to calculating the crisp output of a fuzzy function using the centroid method. However, the CPN does not approximate the output of a fuzzy function by a known heuristic. Rather, it finds the optimal fuzzy-to-crisp conversion by adjusting the weights w_{jl} to match the desired output.

Initially each CPN is trained in isolation similar to the control tasks. During this initial training, standard CPN training algorithms are used.

The neural-fuzzy CPN takes advantage of the topology conserving properties of the Kohonen layer to create fuzzy partitions of the input space. The neuron with weights closest to the input vector alters other neurons in its topological area, creating topology conserving clusters. These topology conserving clusters can be interpreted as fuzzy partitions as explained earlier. Kohonen's training algorithm is given by [123]

$$\Delta \mathbf{w}_j(n) = \begin{cases} \alpha_j(n) e^{-(i-j)} \left[\mathbf{x}\ (n) - w_j(n) \right] & if\ j \in N_i \\ 0 & otherwise \end{cases} \quad (16)$$

where i is the winning neuron, j is the current neuron, $\bullet_j \bullet n\bullet$ is the time dependent learning rate of node j, and N_j is the set of neurons in the neighborhood of I. The learning rate begins at 1 and descends inversely toward zero, according to Equation 17.

$$\alpha_j(n) = \frac{1}{n_{selected}} \quad (17)$$

where $n_{selected}$ is the number of times the jth neuron has been trained.

The exponential term in Equation (16) makes the change in weights of the neurons in the neighborhood of i descend exponentially with distance from i. This prevents a few training samples from dominating the neighborhood.

The output layer is trained according to the Grossman equation, which is really just the first step of the backpropagation algorithm [4].

$$\Delta w_j = \eta\ (y_d - y_{net}) \quad (18)$$

In the case of the fuzzy CPN, the output layer finds the optimum by minimizing the error at the output layer using Equation (17). Essentially, the Kohonen layer creates a fuzzy output corresponding to the desired system response. The output layer assigns the fuzzy output state of the Kohonen layer to the crisp desired output. By iteratively applying Equation (17), the output weights achieve the most accurate defuzzification procedure that represents the desired output.

6.3 Evaluation System

The development system is composed of three distinct hardware components: the robot and associated controller, the camera and associated frame grabber, and the computer. The robot is a CRS Robotics A465 6DOF manipulator, the camera is a Toshiba minicam, and the neural network software is running on a 100MHz Pentium PC. The object that the system is observing is a small woodchip used in the manufacture of particle board.

The control neural networks are responsible for calculating the inverse feature Jacobian in a specific workspace. Actuator control is performed by the analog controller that comes with the robot. The system functions as a path planner not a controller.

The evaluation system uses three control networks to divide the control task into three subspaces: FAR, MIDDLE, and NEAR. The control networks are trained using direct feedback of the change in robot position for a given change in error as described by Chan and Lo [122]. The control networks learn the feature Jacobian. The two control networks are trained from all data files recorded for training each of the control networks.

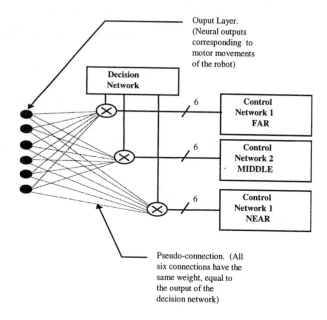

Figure 15: Pseudo-output layer.

The indexing network takes the robot position as input and has the workspace that the position came from as a training signal. There is a possibility that the subspaces could overlap in Cartesian space, but the subspaces are defined in joint space and are different enough to allow the indexing network to exactly determine the current workspace. If there is overlap, the indexing network would output the membership in each of the subspaces.

The decision network was trained using a simple heuristic. The network was to remain in the same subspace unless the image error was less than a predefined threshold. Instead of learning the hard-switching IF THEN rules, the decision network created a fuzzy IF THEN correlation between image error and subspace, allowing degrees of output in each subspace depending on the input vision error. As the error approached the threshold, the decision network began to "cheat" and allow combined motion in adjoining subspaces. This output was then used as the weights for the pseudolinks to the output layer as shown in Figure 15.

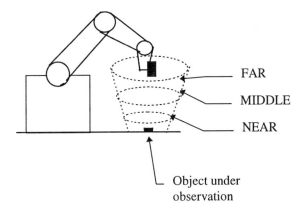

Figure 16: Evaluation system.

The NEAR and FAR networks were fully connected backpropagation networks. (see Figure 16) Each network had six input, four output, and twenty hidden nodes. The middle network was a sparsely connected backpropagation network, with nine input, four outputs, and sixty hidden nodes. The sixty hidden nodes were divided into four sublayers of fifteen nodes as described in Section 6.2. The sparse connection helped the network converge because the system is better represented as four independent approximations instead of a single global approximation.

The FAR network controls joints 1, 2, 3, and 6, that is, rotation, shoulder, elbow, and twist. The FAR workspace is defined as 800 encoder pulses from the starting point on joints 1 through 3, and 3000 encoder pulses on joint 6. The MIDDLE network is defined on joints 1, 2, 3, and 5. This allows the robot to move linearly in the z direction, approaching the object. The NEAR network, like the FAR network, is defined on joints 1, 2, 3, and 6; however, the NEAR network's workspace is half that of the FAR network.

Each network was trained over 30 iterations of a 1000 sample batch file. The NEAR and FAR networks converged to approximately 1.5% error and the MIDDLE network converged to approximately 4% error. Because the error in the input features is approximately 1%, the NEAR and FAR networks have excellent convergence properties. The larger error in the MIDDLE network is due to a few samples where the positions of joints 2, 3, and 5 could not be resolved. In an iterative system, landing at such a point delays the converges by a step, but does not prevent it.

The indexing network contains six input, three output, and 20 Kohonen nodes. The decision network uses 7 input, three output, and 100 Kohonen nodes. The indexing network learns to map subspaces to output to an extremely high accuracy (10^{-6}%) in approximately 10000 iterations. Iterations for CPN networks are given as the total number of iterations for Kohonen and outstar layers. The decision network achieves only 10% accuracy after 50000 iterations. The high error is actually an indication of the fuzziness of the output. While the training samples are crisply defined, the output cheats in the vicinity of the switching point between subspaces, which increases the error.

6.4 Preliminary Results

The performance of the system was evaluated using the system described above. At the time of writing, complete testing of the system had not occurred; however, preliminary results are available. The system was evaluated for two simple tasks, moving to a predefined position over the object and following the object along an arbitrary path.

The system was tested for the ability to move over a stationary object to evaluate its applicability for stationary tasks such as high precision drilling. The robot was placed in a ready position (the center of the FAR workspace) and the object was moved to an arbitrary position and orientation in the camera's field of view. The system was then run in the forward mode and the number of iterations and the final image error were recorded. The system was evaluated with different object starting positions. At each position, the system was run several times and the average error and number of iterations was recorded. The results are presented in Table 1.

The error is given in terms of the mean squared pixel error over the upper-leftmost and bottom-rightmost points on the object.

The error due to the frame grabber/feature extractor is approximately ±3 pixels. Therefore the expected error for the error quantity would be ±6 pixels. The mean error is slightly outside the input error range. However, because the neural network is an approximation, a small error is expected outside the input error. It may be noted that the frame grabber resolution is 640 by 480, so a pixel error of 7.5 is only a 1.5% shift in the image.

Table 1: System performance.

Trial	Iterations	Error
1	4.4	9.1
2	7	7.5
3	6.6	8.2
4	7.4	5.1
5	6.8	7.7
Net	**6.44**	**7.52**

The system was also tested at the extreme ranges of the FAR subspace. The object was placed at the extreme boundaries of the field of vision, and the experiment was repeated. The system did not fair well. It was possible to get the system "stuck" in an area near the transition point of the NEAR and MIDDLE networks, so the robot would move back and forth, but never forward. The system enters this state when the robot passes outside an area where the MIDDLE network's function approximation is valid. NEAR and FAR networks depend only on image data and are nearly global, but MIDDLE depends on joint positions and is defined on a finite subspace. Leaving that subspace results in incorrect operation of the system.

Finally, the robot was allowed to chase a moving object. The object was moved by hand between iterations, and the robot tried to follow it. The subspace partitioning worked well. The robot could switch back to the FAR network from the MIDDLE network if it had not progressed too far (approximately one iteration). When the object stopped, the robot would zero in on the target as before. If the object started moving again when the robot had entered the NEAR subspace, the robot could not switch back to the FAR subspace and could not "keep up." This would be simple enough to add algorithmically, but it is an interesting challenge to train the decision network to do so.

6.5 Conclusion

The modular neural-fuzzy system proved an adequate method of controlling a visually guided robot arm. The final pixel error was near the input error and the number of iterations was low. However, the true potential of this approach has not been tapped. Continued optimization of the neural networks to minimize error and iterations has not yet been implemented. We are currently looking at a simple adaptation of the backpropagation algorithm to minimize error. Because the inter-network topology is known, the credit assignment problem becomes much simpler and complex operators like genetic algorithms and the Jacobs Jordon method are not required.

7 Summary

This chapter has presented common neurocontrol techniques, overviewed hardware/ software implementations for neural networks, and surveyed their applications in various industries. Their nonlinear mapping capability, massive parallel hardware implementation potentials, and ability of learning and adapting under uncertain and noisy situations make them desirable for a wide class of control applications, such as cloning experts, tracking trajectories or setpoints, and optimization (e.g., approximate dynamic programming). It is clear that neural networks offer a great potential to devise new and innovative control strategies for complex systems. The chapter also showed that as systems are becoming increasingly complex with the advancement of technology, hybrid systems which can combine the capabilities of other technology such as fuzzy control techniques with the learning capability of neural networks will prove to be more powerful. These technologies can help future control systems become autonomous, flexible, adaptable, and self-learning in highly dynamic environments.

References

[1] Anderson, C., "Learning to control an inverted pendulum using neural networks," *IEEE Control Systems Magazine*, Vol. 9, No.3, pp. 31-37, 1989.

[2] Asada, H. and Liu, S. (1991), "Transfer of human skills to neural net robot controller," *Proc. R & A*, 1991, pp. 2442-2448.

[3] Asada, H., "Teaching and learning of compliance using neural nets," *Proc. 1990 IEEE Intl. Conf. on Robotics and Automation*, pp. 1237-1244, May 1990.

[4] Baird, L. and Baker, W., "A connectionist system for nonlinear control," *Technical Report CSDL-P-2958*, The Charles Stark Sraper Lab., Inc., Cambridge, MA, 1990.

[5] Barto, A.G. "Connectionist learning for control," *Neural Networks for Control*, T. Miller et al. (Eds.) MIT Press, Cambridge, MA, 1990.

[6] Barto, A.G., Sutton, R.S., and Anderson, C.W., "Neuronlike adaptive elements that can solve difficult learning control problems," *IEEE Trans. on Systems, Man, and Cybernetics*, Vol. 13, No. 5, pp. 834-846, 1983.

[7] Bavarian, B., "Introduction to neural networks for intelligent control," *IEEE Control Systems Magazine*, Vol. 8, Iss. 2, pp. 3-7, 1988.

[8] Berenji, H.R. and Khedkar, P., "Learning and tuning fuzzy logic controllers through reinforcements," *Technical Report FIA-92-02*, NASA Ames Research Center, 1992.

[9] Berenji, H.R., "An architecture for designing fuzzy controllers using neural networks," *Second Joint Technology Workshop on Neural Networks and Fuzzy Logic*, NASA Johnson Space Center, Houston, Texas, 1990.

[10] Bhat, N.V. and Mc Avoy, "Use of neural nets for dynamic modeling and control of chemical process systems," *Proc. 1989 American Control Conference*, Vol. 2, pp. 1342-1347, Pittsburg, 1989.

[11] Bhat, N.V., et al., "Modeling chemical process systems via neural computation," *IEEE Control Systems Magazine*, Vol. 10, Iss. 3, pp. 24-30, 1990.

[12] Bishop, C.M., "Neural networks and their applications," *Rev. of Sci. & Instrum.*, Vol. 65, No.6, pp. 1803-1832, 1994.

[13] Brown, R.H., Ruchti, T.L., and Feng, X., "Artificial neural network identification of partially known dynamic nonlinear systems," *Procs. of the 32nd IEEE Conference on Decision and Control*, Vol. 4, pp. 3694-9, 1992.

[14] Buckley, J.J. and Hayashi, Y., "Numerical relationships between neural networks, continuous functions, and fuzzy systems," *Fuzzy Sets and Systems*, Vol. 60, Iss. 1 pp. 1-8, 1993.

[15] de Silva, C.W., Wu, Q.M. and Barlev, S., "Enhancement of a Fuzzy Tuner with Neural Networks for a Servo Controller," *Intl. Conf. on Control and Robotics,* August 4-7, 1992, Vancouver, Canada.

[16] Chen, S., Billings, S.A., and Grant, P.M., "Nonlinear system identification using neural networks," *International Journal of Control,* Vol. 51, No. 6, pp. 1191-1214, 1990.

[17] Cybenco, G., "Approximations by superpositions of a sigmoidal function," *Mathematics of Control, Signals, and Systems,* Vol. 2, pp. 303-314, 1989.

[18] de Silva, C.W., *INTELLIGENT CONTROL: Fuzzy Logic Applications,* CRC Press, Boca Raton, FL, 1995.

[19] Dickerson, J., Kim, H.M., and Kosko, B., "Hybrid ellipsoidal learning and fuzzy control for platoons of smart cars," *IFIS '93. The Third Intl. Conf. on Industrial Fuzzy Control and Intelligent Systems,* pp. 60-5, Houston, TX, 1993.

[20] Eykhoff, P., *Trends and Progress in System Identification,* Oxford, England: Pergamon, 1981.

[21] Esogbue, A.O. and Murrel, J.A., "Advances in fuzzy adaptive control," *Computers & Mathematics with Applications,* Vol. 27, Iss. 9-10, pp. 29-35, 1994.

[22] Farbi, S., Kadirkamanathan, P., and Visakan, F., "Dynamic Structure Neural Networks for Stable Adaptive Control of Nonlinear Systems," *IEEE Trans. on Neural Networks,* Vol. 7, No. 5, pp. 1151-1167, 1996.

[23] Franklin, J.A., Sutton, R.S., and Anderson, C.W., "Application of connectionist learning methods to manufacturing process monitoring," *Procs. of IEEE International Symposium on Intelligent Control 1988,* pp. 709-12, Arlington, VA, 1989.

[24] Franklin, J.A., "Refinement of robot motor skills through reinforcement learning," *Procs. of the 27th IEEE Conf. on Decision and Control,* Austin, TX, 1988.

[25] Fukuda, T. and Shibata, T., "Neuromorphic control for robot manipulators: position, force and impact control" *Procs. of IEEE Int. Symp. Intell. Contr.,* pp. 310-315, 1990.

[26] Fukuda, T. and Shibata, T., "Theory and applications of neural networks for industrial control systems," *IEEE Trans. on Industrial Electronics,* Vol. 39, No.6, pp. 472-489, 1992.

[27] Glanz, F.H., Miller, T.W., and Kraft, L.G., "An overview of the CMAC neural network," *Neural Networks for Ocean Engineering,* pp. 301-308, 1991.

[28] Gomm, J.B., Page, G.F., and Williams, D., "Introduction to neural networks," *Application of Neural Networks to Modeling and Control,* Chapman & Hall, 1993.

[29] Grant, E. and Zhang, B., "Neural network based reinforcement learning," *Procs. of the 31st Conference on Decision and Control,* Tucson, Arizona, pp. 856-861, 1992.

[30] Guez, A. and Selinsky, J. (1988), "A trainable neuromorphic controller," *Journal of Robotic Systems,* Vol. 5, No. 4, pp. 24-32, 1988.

[31] Gullapalli, V., "A comparison of supervised and reinforcement learning methods on a reinforcement learning task," *Procs. of the 1991 IEEE Intl. Symposium on Intelligent Control*, pp. 394-399, August 1991, U.S.A.

[32] Handelman, D.A., Lane, S.H., and Gelfand, J.J., "Integrating neural networks and knowledge-based systems for intelligent control," *IEEE Control Systems Magazine*, pp. 77-86, April 1990.

[33] Haykin, S., *Neural Networks: A Comprehensive Foundation,* Macmillan College Publishing Company, New York, 1994.

[34] Huang, W.H. and Lippman, R.P., "Neural net and traditional classifiers," *NIPS Procs.*, 1987.

[35] Hunt et al., "Neural networks for control systems -- a survey," *Automatica*, Vol. 28, No.6, pp. 1083-1112, 1992.

[36] Hwang, H. and Choi, D.Y., "CMAC toward intelligent robot: motion and calibration," *Procs. IROS '90. IEEE Intl. Workshop on Intelligent Robots and Systems '90. Towards a New Frontier of Applications*, Vol. 1, pp. 517-26, 1990.

[37] Jordan, M.I., "Supervised learning and system with excess degree of freedom," COINS Tech. Rep., pp. 27-88, 1-41, 1988.

[38] Karakasoglu, A., Sudharsanan, S.I., and Sundareshan, M.K., "Identification and decentralized adaptive control using dynamical neural networks with application to robotic manipulators," *IEEE Transactions on Neural Networks*, Vol. 4, Iss. 6, pp. 919-30, 1993.

[39] Kawato, M. (1990), "Computational schemes and neural network models for formation and control of multijoint arm trajectory," *Neural Networks for Control*, T. Miller et al. (Eds.) MIT Press, Cambridge, MA, 1990.

[40] Khalid, M and Omatu, S., "A neural network controller for a temperature control system," *IEEE Control Systems Magazine*, Vol. 12, No.3, pp. 58-64, 1992.

[41] Kraft, L.G. and Campagna, D.A., "A summary comparison of CMAC neural network and traditional adaptive control systems," *Handbook of Intelligent Control: Neural, Fuzzy, and Adaptive Approaches*, D.A. White et al. (Eds.), New York: Van Nostrand Reinhold, 1992.

[42] Kraft, L.G. and Campagna, D.P., "Comparison of CMAC architectures for neural network based control," *Procs. of the 29th IEEE Conf. on Decision and Control*, Vol. 6, pp. 3267-9, Honolulu, HI, 1990.

[43] Kuntanapreeda, S. and Fullmer, R., "A Training Rule Which Guarantees Finite-Region Stability for a Class of Closed Loop Neural-Network Systems," *IEEE Trans. on Neural Networks*, Vol. 7, No. 3, pp. 745-751, 1996.

[44] Lee, C.C. and Berenji, H.R., "An intelligent controller based on approximate reasoning and reinforcement learning," *Proc. of IEEE International Symposium on Intelligent Control*, Albany, NY, 1989.

[45] Lee, C.C., "Self-learning rule-based controller employing approximate-reasoning and neural-net concepts, *International Journal of Intelligent Systems*, 1990.

[46] Lee, C.S.G. and Lin, C.T., "Supervised and unsupervised learning with fuzzy similarity for neural-network-based fuzzy logic control systems," *1992 IEEE Intl. Conf. on Systems, Man and Cybernetics*, Vol. 1, pp. 688-93, Chicago, IL, 1992.

[47] Leonard, L. and Kramer, M.A., "Improvement of the backpropagation algorithm for training neural networks," *Computers and Chem. Eng.*, Vol. 14, No.3, pp. 337-341, 1990.

[48] Lin, Cheng-Jian, and Lin, Chin-Teng, "Reinforcement Learning of an ART-Based Fuzzy Adaptive Learning Control Network," *IEEE Trans. on Neural Networks*, Vol. 7, No. 3, pp. 709-731, 1996.

[49] Lippmann, P., "An introduction to computing with neural nets," *IEEE Conf. on Acoustics, Speech and Signal Processing*, ASSP-35, p. 4-22, 1987.

[50] Liu, H. and Bekey, G.A., "Neural network architecture for robot hand control," *IEEE Control Systems Magazine*, pp. 38-43, 1989.

[51] Luebbers, P.G. and Pandya, A.S., "Neural networks for process control," *Intl. Conf. on Systems, Man and Cybernetics*, 1992.

[52] Mallampati, D. and Shenoi, S., "Self-organizing fuzzy logic control," *Knowledge-Based Systems and Neural Networks: Techniques and Applications*, Sharda, R. et al. (Eds.), Elsevier, Amsterdam, Netherlands, 1991.

[53] Messom et al., "Designing neural networks for manufacturing process control systems," *Procs. of the 1992 IEEE Intl. Symposium on Intelligent Control*, pp. 423-9, Glasgow, U.K., 1992.

[54] Michie, D. and Chambers, R., "BOXES: an experiment in adaptive control," *Machine Intelligence 2*, E. Dale and D. Michie (Eds.), Oliver and Boyd, Edinburgh, 1968.

[55] Miller, W., III, Glanz, F., and Kraft, L., III, "Application of a general learning algorithm to the control of robot manipulators," *The Intl. J. of Robotics Research*, Vol. 6, No.2, 1987.

[56] Miller, W., III, "Sensor-based control of robot manipulators using a general learning algorithm," *IEEE J. Robotics and Automation*, Vol. RA-2, N0.2, pp. 157-165. 1987.

[57] Miller, W.T. III and Arldrich, C.M., " Rapid learning using CMAC neural networks: real time control of an unstable system," *Procs. 5th IEEE Intl. Symposium on Intelligent Control 1990*, Vol. 1, pp. 465-70, Los Alamitos, CA, 1990.

[58] Miller et al. (Eds.), *Neural Networks for Control*, MIT Press, Cambridge, MA, 1990.

[59] Narendra, K.S. and Annaswamy, A.M., *Stable Adaptive Systems*, Englewood Cliffs, NJ: Prentice-Hall, 1989.

[60] Narendra, K.S. and Parthasarathy, K., "Identification and control of dynamic systems using neural networks," *IEEE Trans. on Neural Networks*, Vol. 1, pp. 4-27, 1990.

[61] Narendra, K.S. and Parthasarathy, M.,"Gradient methods for the optimization of dynamical systems containing neural networks," *IEEE Trans. on Neural Networks*, Vol. 2, No.1, pp. 252-262,

[62] Narendra, K.S., "Adaptive control of dynamical systems using neural networks," *Handbook of Intelligent Control: Neural, Fuzzy, and Adaptive Approaches*, D.A. White, et al. (Eds.), New York: Van Nostrand Reinhold, 1992.

[63] Narendra, K.S. and Mukhpadhyay, S., "Adaptive Control Using Neural Networks and Approximate Models," *IEEE Trans on Neural Networks,* Vol. 8, No. 3, pp. 475-485, 1997.

[64] Nguyen, P.H. and Widrow, M., "Neural network for self-tuning control system," *IEEE Control Systems Magazine*, Vol. 10, No.3, pp. 18-23, 1990.

[65] Nie, J. and Linkens, E., *Fuzzy Neural-Control, Principles, Algorithms and Applications* Prentice Hall, London, 1995.

[66] Nordgren, R.E. and Meckl, P.H., "An analytical comparison of a neural network and a model-based adaptive controller," *IEEE Transactions on Neural Networks*, Vol. 4 Iss:4, pp. 685-94, 1993.

[67] Pao, Y.H., *Adaptive pattern recognition and neural networks*, Addison - Wesley, 1989.

[68] Park, Young-Moon, Choi, Myeon-Song, and Lee, Kwang Y., "an Optimal Tracking Neural-Controller for Nonlinear Dynamic Systems," *IEEE Trans. on Neural Networks,* Vol. 7, No. 5, pp. 1099-1110, 1996.

[69] Pedrycz, W., "Numerical and Application Aspects of Fuzzy Relational Equations," *Fuzzy Sets and Systems*, Vol. 11, pp. 1-18, 1983.

[70] Plumer, E.S., "Time-optimal terminal control using neural networks," *1993 IEEE Intl. Conf. on Neural Networks*, Vol. 3, pp. 1926-31, San Francisco, CA, 1993.

[71] Psaltis, D., Sideris, A., and Yamamura, A.A., "A multi-layered neural network controller," *IEEE Control Systems Magazine*, pp. 17-20, April 1988.

[72] Puskorius, G.V. and Feldkamp, L.A., "Model reference adaptive control with recurrent networks trained by the dynamic DEKF algorithm," *IJCNN Intl. Joint Conf. on Neural Networks*, Vol. 2, pp. 106-13, Baltimore, MD, 1992.

[73] Rao, D.H., Gupta, M.M, and Wood, H.C, "Neural networks in control systems," *WESCANEX 93: Communications, Computers & Power in the Modern Environment*, pp. 282-290, 1993.

[74] Rjorgensen, C.C. and Schley, S. (1990), "A neural network baseline problem for control of aircraft flare and touching down," *Neural Networks for Control*, T. Miller et al. (Eds.) MIT Press, Cambridge, MA, 1990.

[75] Samad, T., "Neurocontrol: concepts and applications," *1992 IEEE Intl. Conf. on Systems, Man and Cybernetics*, Vol. 1, pp. 369-74, Chicago, IL, 1992.

[76] Sbarbaro, D. and Hunt, K.J., "Neural networks for nonlinear internal model control," *Procs. of the 30th IEEE Conf. on Decision and Control*, Vol. 1, pp. 172-3, Brighton, U.K., 1991.

[77] Shin, S. and Nishimura, H., "Neuro-adaptive control for general nonlinear discrete-time systems," *Procs. IECON '91. 1991 Intl. Conf. on Industrial Electronics, Control and Instrumentation*, Vol. 2, pp. 1454-8, Kobe, Japan, 1991.

[78] Sofge, D.A. and White, D.A. (1990), "Neural network based process optimization and control," *Proc. of the 29th Conference on Decision and Control*, 1990.

[79] Sudharsanan, S.I., Muhsin, I., and Sundareshan, M.K., "Self-tuning adaptive control of multi-input, multi-output nonlinear systems using multilayer recurrent neural networks with application to synchronous power generators," *1993 IEEE Intl. Conf. on Neural Networks*, Vol. 3, pp. 1307-12, San Francisco, CA, 1993.

[80] Tai, P., Ryaciotaki-Boussalis, H.A., and Hollaway, D., "Neural network implementation to control systems: a survey of algorithms and techniques," *Conference Record of the Twenty-Fifth Asilomar Conference on Signals, Systems and Computers*, Vol. 2, pp. 1123-7, Los Alamitos, CA, U.S.A., 1991.

[81] Takagi, T. and Sugeno, M., "Fuzzy Identification of Systems and its Applications to Modeling and Control," *IEEE Trans. on SMC*, Vol. SMC-15, No.1, pp. 116-132, 1985.

[82] Tanaka, Kazuo, "An Approach to Stability Criteria of Neural-Network Control Systems," *IEEE Trans. on Neural Networks*, Vol. 7, No. 3, pp. 629-643, 1996.

[83] Thibault, J. and Grandjean, B.P.A., "Neural networks in process control – a survey," *Advanced Control of Chemical Processes*, pp. 251-60, 1992.

[84] Tokita, M. et al., "Force control of robots by neural network models: control of one-dimensional manipulators," *J. Japan Society of Robotics Engineers*, Vol. 8-3, pp. 52-59, 1989.

[85] Waltz, M.D. and Fu, K.S., "A heuristic approach to reinforcement learning control systems," *IEEE Trans. on Automatic Control*, Vol. 10, No.4, 1965.

[86] Wang, B.H. and Vachtsevanos, G., "Fuzzy Associative Memories: Identification and Control of Complex Systems," *IEEE Symposium on Intelligent Control*, Philadelphia, PA, pp. 910-915, 1990.

[87] Wasserman, P.D., *Adcanced Methods in Neural Computing*, Van Nostrand, New York, 1993.

[88] Werbos, P.J., "Overview of designs and capabilities," *Neural Networks for Control*, T. Miller et al. (Eds.) MIT Press, Cambridge, MA, 1990.

[89] Werbos, P.J., "A menu of designs for reinforcement Learning Over Time," *Neural Networks for Control*, T. Miller et al. (Eds.) MIT Press, Cambridge, MA, 1990.

[90] Werbos, P.J., "An overview of neural networks for control," *IEEE Control Systems Magazine*, Vol. 11, Iss. 1, pp. 40-41, 1991.

[91] Werbos, P.J., "Beyond regression: new tools for prediction and analysis in the behavior science," Ph.D. Thesis, Harward University, Cambridge, MA, 1974.

[92] Widrow, B. and Smith, F.W. (1964), "Pattern recognizing control systems," *Proc. of Computer and Information Sciences (COINS)*, Washington D.C., Spartan Books.

[93] Williams, R.J. and Zipser, D., "A learning algorithm for continually running fully recurrent neural networks," *Neural Computation*, Vol. 1, pp. 270-280, 1989.

[94] Williams, R.J., "Adaptive state representation and estimation using recurrent connectionist networks," *Neural Networks for Control*, T. Miller et al. (Eds.) MIT Press, Cambridge, MA, 1990.

[95] Wu, Q.M. and de Silva, C.W., "Model Identification for Fuzzy Dynamic Systems," *Proceedings of American Control Conference' 93*, San Francisco, CA, 1993.

[96] Wu, Q.M. and Stanley, K., "Modular Neural-Visual Servo using a Neural-Fuzzy Decision Network," *ICRA'97*, Albequerque, NM, 1997.

[97] Wu, Q.M. and Stanley, K., "Neural Network Based Robotic System with Image Compression Input," *WAC'98*, Anchorage, Alaska, 1998.

[98] Yamada, T. and Yabuta, T., "Dynamic system identification using neural networks," *IEEE Transactions on Systems, Man and Cybernetics*, Vol. 23, Iss. 1, pp. 204-11, 1993.

[99] Zomaya, A.Y., Suddaby, M.E., and Morris, A.S., "Direct neuro-adaptive control of robot manipulators," *Procs. of 1992 IEEE conf. on Robotics and Automation*, Nice, France, 1992.

[100] Bessi'ere, P., Chams, A., and Chol, P., "MENTAL : A virtual machine approach to artificial neural networks programming," *NERVES, ESPRIT B.R.A. project no 3049*, 1991.

[101] Bessiere, P., et al., "SMART – Sparse Matrix Adaptive and Recursive Transforms: From Hardware to Software: Designing a 'Neurostation'," *VLSI Design of Neural Networks*, pp. 311-335, 1990.

[102] Boser, B.E., et al., "Hardware Requirements for Neural Network Pattern Classifiers," *IEEE Micro*, Vol. 12, Iss. 1, pp. 32-40, 1992.

[103] Elias, J.G., Fisher, M.D., and Monemi, C.M., "A multiprocessor machine for large-scale neural network simulation," *IJCNN91-Seattle: Intl. Joint Conf. on Neural Networks*, Vol. 1, pp. 469-474, 1991.

[104] Hauser, R., "Architectural Considerations for NERV – a General Purpose Neural Network Simulation System," *Workshop on Parallel Processing: Logic, Organization and Technology – WOPPLOT 89*, pp. 183-195. Springer Verlag, 1989.

[105] Ienne, P. and Viredaz, M.A., "A Bit-Serial Processing Element for a Multi-Model Neural-Network Accelerator," *Procs. of the Intl. Conf. on Application Specific Array Processors*, Venezia, 1993.

[106] Jan N.H., et al., "Mindshape: a neurocomputer concept based on a fractal architecture," *Procs. of Intl. Conf. on Artificial Neural Networks*, Elsevier Science, 1992.

[107] Jan, N.H., et al., "The bsp400: A modular neurocomputer assembled from 400 low-cost microprocessors," *Procs. of Intl. Conf. on Artificial Neural Networks*, Elsevier Science, 1991.

[108] Jones, S., et al., "Toroidal Neural Network: Architecture and Processor Granularity Issues," *VLSI Design of Neural Networks*, pp. 229-254, 1990.

[109] Krikelis. A., "A novel massively associative processing architecture for the implementation artificial neural networks," *Procs. of 1991 Intl. Conf. on Acoustics, Speech and Signal Processing*, Vol. 2, pp. 1057-1060, 1991.

[110] Landsverk, O., et al., "RENNS - a Reconfigurable Computer System for Simulating Artificial Neural Network Algorithms," *Parallel and Distributed Computing Systems, Proceedings of the ISMM 5th International Conference*, pp. 251-256, 1992.

[111] McCartor, H., "Back Propagation Implementation on the Adaptive Solutions CNAPS Neurocomputer," *Advances in Neural Information Processing Systems*, Vol. 3, 1991.

[112] Michael, L., et al., "The mod 2 neurocomputer system design," *IEEE Transactions on Neural Networks*, Vol. 3, No.3, pp. 423-433, 1992.

[113] Morgan, N., et al. "The ring array processor: A multiprocessing peripheral for connectionist applications," *Journal of Parallel and Distributed Computing*, pp. 248-259, 1992.

[114] Mueller, P., et al., "Design and performance of a prototype analog neural computer," *Neurocomputing*, Vol. 4, No.6, pp. 311-323, 1992.

[115] Orrey, D.A., Myers, D.J., and Vincent, J.M., "A high performance digital processor for implementing large artificial neural networks," *Procs. of the IEEE 1991 Custom Integrated Circuits Conference*, pp. 16.3/1-4. 1991.

[116] Pazienti. F., "Neural networks simulation with array processors," *Advanced Computer Technology, Reliable Systems and Applications; Proceedings of the 5th Annual Computer Conference*, pp. 547-551, 1991.

[117] Ramacher, U., Beichter, J., and Bruls, N., "Architecture of a general-purpose neural signal processor," *IJCNN91-Seattle: Intl. Joint Conf. on Neural Networks*, Vol. 1, pp. 443-446, 1991.

[118] Viredaz, M.A., "MANTRA I: An SIMD Processor Array for Neural Computation," *Procs. of the Euro-ARCH'93 Conference*, Mü, 1993.

[119] Watanabe, T. et al., "Neural network simulation on a massively parallel cellular array processor: AAP-2.," *Procs. Intl. Joint Conference on Neural Networks*, 1989.

[120] Wojciechowski, E., "SNAP: A parallel processor for implementing real time neural networks," *Procs. of the IEEE 1991 National Aerospace and Electronics Conference; NAECON-91*, Vol. 2, pp. 736-742. 1991.

[121] Hashimoto, et al. "Self-Organizing Visual Servo System Based on Neural Networks," *American Control Conference, Boston MA, 1991.*

[122] Chan, C.C. and Lo, E.W.C., "Visual Servo Control with Artificial Neural Network," IEEE Publications CDROM.

[123] Hassoun, Mohamad H., *Fundamentals of Neural Networks*, MIT Press, Cambridge, 1995.

5 | FEATURE SPACE NEURAL FILTERS AND CONTROLLERS

Horia N. Teodorescu and **Cristian Bonciu**
Faculty of Electronics and Communications
Technical University "Gh. Asachi" Iasi,
11-22, Iasi, 6600, Romania

This chapter is devoted to the introduction of the "Features Space Processing" concept in artificial neural filtering and control. This general methodology deals with the adaptation in signal processing applications *in view* of pattern recognition. Merging the optimization into the data space and the optimization into the pattern space is the main technique employed by the features space processing schemes. Several neural hybrid systems configurations that implement this methodology are proposed, tested, and compared. These configurations are designed for filtering, identification, and control applications.

1   Introduction

Computational intelligence (CI) methods [1] are widely used today in complex signal processing, pattern recognition, identification, and control of dynamic systems. Hybrid CI methods combine neural networks, fuzzy systems, and evolutionary computing with classic data processing methods and symbolic artificial intelligence methods, thus being able to handle more complex problems. The challenge is to obtain flexible abstract concepts and intelligent systems behavior from rough and noisy data sets using simple and consistent basic rules.

The main features of CI methods emerge from their intrinsically adaptive nature, which accounts for the capacity to learn a desired or an optimal behavior in the environment context. Combining intelligent techniques to solve complex problems generally involves coupling of fundamentally different systems, which are driven by various adaptation laws. This yields increased complexity, which must be offset by better performance from the new system, namely, by the new features emerged from the system component's fusion.

Hybrid Neural Adaptive Systems

Functional and hierarchical decompositions are the main procedures employed to partition the tasks, reduce complexity, and increase control over the data processing systems. The former functionally separates different stages in order to design chains of cascaded processing systems. The latter organizes the sub-tasks as levels in a hierarchy, and qualitatively designs different processing stages, each of them ensuring the global synchronization and convergence for the inferior stage [2].

Whatever the structuring concept and the combination between the structuring procedures, the ability of such hybrid system to solve the problem at hand depends not only on the quality of each component, but also on the effectiveness of the links between the components. The hierarchical decomposition assures the intelligent link between processing stages using global information flows from inferior stages for the decision making, and drives the inferior stages by means of local information flows for synchronization and correction.

Without a supervisory integrating level, it is often difficult to obtain better overall results in a single chain of coupled data processors, compared with the processing quality of each component individually. One way to overcome this drawback is to design bi-directional local data paths between system components carrying mutual information. Tracking, understanding, and driving the inter-dependencies between the hybrid system chain components may play a crucial role in the overall system response quality.

The need for hierarchical task organization is well perceived in many neural hybrid system realizations [3]–[6], effectively increasing the quality of the system response. Unfortunately, the inter-dependencies between hybrid system chain components at the same processing level are not yet sufficiently investigated. A priori functional assignment principle is largely used in practice. Each data processor has a pre-determined function, and its processing capabilities are measured based only on its intrinsic characteristics at the system design stage. A decisive factor in increasing the quality of hybrid systems response is the design of efficient inter-links between system components, based on particular component mechanisms for acquiring data about its neighbors' functions and evolution.

The connectionist paradigm employed in neural systems [7] states that the information is stored by an adaptation procedure in the links between elementary processing cells. The inward and outward data flows in connectionist systems provide an efficient way to use this kind of information. However, this type of implicit information encoding in neural systems remains obscure or equivocal in many respects.

The need to control the artificial neural system behavior and the internal data coding schemes was a determining factor in the effort to limit the size of neural networks and to restrict the type of processing elements which contribute to the internal network representation of a given environment. Recently it is generally accepted that complex neural processing systems must be well structured on functional ensembles. Moreover, it is accepted that the learning environment for each ensemble should be carefully selected.

Quantitative accumulation of artificial neural masses performs poorly without clearly defined functionality, that is, without specific learning paradigms and representation restrictions. This is due to the lack of modularity and, more important, to the lack of direct and appropriate information sources for building useful internal representations during learning.

The chapter is focused on neural signal processing, specifically on merging classic neural filtering, identification, and control processes with the neural features extraction methods. Neural signal filtering, pattern recognition, and control are still viewed as subsequent, completely separate stages in the processing chains.

Modalities of merging these sub-fields and methods are investigated. The merging technique employed here refers mainly to the design of a knowledge-based intelligent link between the neural processor (filter, identification network, or controller) and the features extractor. The emphasis is made on the associated learning systems, which are able to form a useful semantic link, and to the degree of generality of the presented methods.

We employ here the general Pineda's approach to neural computation [8]. Pineda uses the concept of assembled primitive neural processors to design a general framework for neural networks coupled dynamics and learning, keeping the specificity of each neural system component. Our approach uses the same principles, but with a different aim: the superior integration of the neural adaptive filters, identification, and control networks with the features extraction systems.

The second part of the chapter introduces the features space filtering methodology. The technique is illustrated with two neural hybrid configurations and the associated learning systems are described in detail. An application example is presented for every architecture. In the third part, the features space identification and control methodologies are presented. Several general neural topologies are proposed and discussed. A neural hybrid control system with features space adaptation procedure is described. The capabilities of such a control scheme are illustrated in a classic neural control example.

2   Feature Space Filtering Using Neural Networks

Many neural approaches refer the integration of the features extractor and the data pre-processor to a single neural heterogeneous system. The integration effort is focused on attenuating the composite hybrid system characteristics by means of new global learning schemes. The starting components usually are modified to complete the system design constraints. This is the way to finally design new specific neural stand-alone processors [9], like principal component analysis networks [10]-[12], blind signal identification networks [13], [14], or wavelet networks [15].

Many researchers have reported the limits and drawbacks of the existing paradigms, and derived methodologies and systems for adaptive signal processing. Several ways to overcome the lack of connection between the imbedded semantics (features) of the signals and traditional signal processing methods were proposed. An important part of the recent signal processing literature focuses on finding different useful signal representations such as time-frequency representations employed in wavelet adaptive filters [16], general representation models like cone classes described in [17], or higher order cumulants as representation features in [18], [19].

Another trend looks at novel adaptation algorithms and optimization criteria, able to extract the desired semantics of the processed signal. A large class of adaptive signal processing methods may be considered as optimization problems. The optimization criterion is most often applied in the signal samples space. The only classic departure from this type of processing is that of the matched filter. This type of filter is widely used in signal detection applications such as radar and digital communication and in image processing. Matched filters generate an output corresponding to a measure of the spatial correlation between the input signal frame and a template function [20].

Nowadays, the adaptive filter is usually followed by an adaptive features extractor. The latter maps the filtered data space to a features space, using a features space optimization criterion. The extension of this general procedure of coupling basic filtering processes with features extraction processes is expanded by the concept of "features space filtering" (FSF) introduced in [21], [22]. According to this concept, filtering and pattern recognition are merged in order to design a single optimization criterion for both processes [23]-[25].

The first section outlines the differences between features space and data space filtering systems. The general structure of the proposed FSF systems is presented in the second section. Two specific FSF system neural realizations are presented in Sections 3 and 4. The last section contains a discussion of possible applications and concluding remarks.

2.1   Adaptive Signal Processing for Pattern Recognition

In this section, we present the differences between the adaptation criteria applied in classic signal processing and in features-oriented processing. The basic operation flow in classic adaptive signal processing (Figure 1) includes two optimization stages, one for filtering and for pattern extraction.

We propose another approach to adaptively optimized signal filtering *for pattern recognition purposes*. This approach consists of adaptation, at the same time, *as a whole process*, for both filtering and pattern recognition. Such an approach to adaptive complex processing could be closer to the way humans perform their tasks and allows a more meaningful approach to the optimality–as perceived by human experts–in filtering and control [26].

The main departure from the classic processing scheme is the exclusive use of the pattern space error in the filter adaptation process (Figure 2). Doing so, the sample space filtered signal may be freely used in subsequent processing and decision stages, without loss of specific information induced by an error criterion imposed in the sample space. Another advantage of this approach is the increased degree of freedom in choosing the pattern space, when both the processed signal and the extracted pattern are used in subsequent stages.

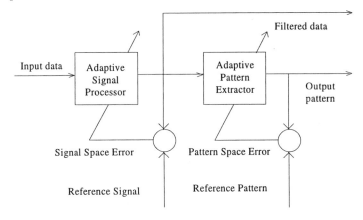

Figure 1: Classic coupled adaptive signal processor and pattern extractor.

The difficulty of this approach is that it leads to a nonlinear optimization problem, which is hard to solve by classic methods. The optimization in the samples space is generally performed based on an Euclidean distance-based criterion:

$$E_{DS} = \| X^* - X^d \|^2 \tag{1}$$

where X^* is the filter output signal and X^d is the target (desired) signal. The same applies to the features space. However, there is no guarantee for the convergence with respect to the error, Equation (1), when the samples space processor is adapted with a features space criterion. The relation between the signal space and the pattern spaces is nonlinear and difficult to obtain analytically in almost all applications. It is shown in this section that it is possible to obtain optimal adaptation with respect to the features space error criterion, without degrading the samples space representation, but obtaining a signal which contains the maximum of information with respect to its reference pattern space representation. Moreover, the samples space filtering driven by a pattern space criterion leads to a data space error, Equation (1), which is comparable with the cost obtained with a classic data space filtering technique.

There are several techniques addressing the problem of optimal filtering. These techniques are focused on finding various measures of optimality in the same input space (usually in the time-delay space) or in the frequency (integral transform) space. Our approach deals with a representation problem. The FSF techniques try to grasp those data characteristics that human experts perceive as meaningful in the pattern recognition task. Consequently, the FSF methods generate representations more meaningful for the users than those produced by methods based on the samples space or the frequency space [27]. The main idea behind the FSF methods is to select the features spaces according to their (supposed) relevance to the human expert and try to perform the filter adaptation using only the features space error criterion. This is possible only with a synthetic approach to signal processing, well represented by the neural network technique [28]. It is well known [29] that the neural network signal processors are able to effectively solve optimal filtering problems for specific tasks.

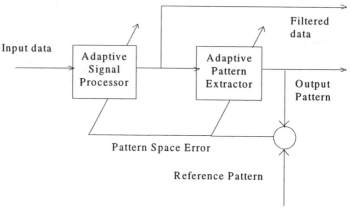

Figure 2: Proposed pattern space filtering scheme.

2.2   General Structure of the FSF System

The block diagram of the general FSF system discussed here includes an adaptive filter, followed by a feature extractor, and an error block (Figure 3). The error signal is fed back to the adaptation block, which adjusts the parameters of the adaptive filter to reduce error. Note that the feature extractor realizes the direct transformation from the input data space to the features space, and the input of the adaptation block is the error signal defined in the feature space, while its outputs are in the adaptive filter parameter space.

The FSF system is governed by the set of general equations:

$$X^*(n) = F(X(n), X(n-1), \ldots, X(n - N_I + 1)) \tag{2}$$

$$Y^*(n) = G(X^*(n), X^*(n-1), \ldots, X^*(n - N_X + 1)) \tag{3}$$

$$E(n) = E(Y^*(n), \ldots, Y^*(n - N_Y + 1), Y^{d*}(n), \ldots, Y^{d*}(n - N_Y + 1)) \tag{4}$$

$$\Delta W(n) = H(E(n), E(n-1), \ldots, E(n - N_E + 1)) \tag{5}$$

where X and X^* are the input data vector and the output data vector of the adaptive filter $F(.)$, respectively. Here, Y^* is the actual features vector at the output of the features extractor block $G(.)$. The error E is computed based on the Y^* vector and the desired feature vector Y^{d*}. The adaptation block $H(.)$ computes the variations ΔW of the adaptive system parameters, based on the error function E.

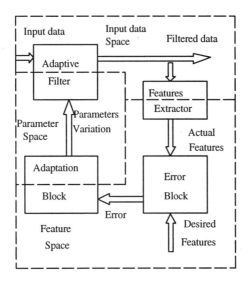

Figure 3: The general structure of the proposed features space filtering system.

2.3 A FSF Architecture Using Adaptive Linear Combiner Filter and Radial Basis Function Network Feature Extractor

A FSF system realization, based on an ALC filter and on an RBF neural network features extractor [30], [31], is described in this section. The associated learning system is described and simulation results on image contours enhancement are presented. Some other possible applications are discussed.

An adaptive linear combiner (**ALC**) is used as a transversal filter [32]. The features extractor realizes a local decomposition of the input signal with a radial-basis function (**RBF**) network. The features space is formed by the parameters of this local nonlinear function decomposition. The desired features are the reference data decomposition parameters. The RBF hidden processing elements (**PEs**) parameters and the ALC coefficients are adapted to minimize the distance in the features space between the reference features vector and the actual features vector obtained from a noisy data set. The features space filtering method can be formally regarded in this case as a generalization (in the nonlinear sense) of the linear parametric filtering, when the features vectors contain the adaptive parameters of the system. This approach is useful when the radial-basis function decomposition of the data set is motivated by physical evidence.

2.3.1   FSF Realization

The ALC has a simple structure of adaptive linear finite impulse response (**FIR**) filters. The discrete filter equation, according to Figure 4, is

$$X^*(n) = \sum_{i=0}^{N_I} w_i X(n - i) \tag{6}$$

where w_i are the N_I components of the filter coefficients vector, and X is the input data vector in the N_I-dimensional input space. Comparing Equations (2) and (6), one recognizes that the $F(.)$ operator in (2) is the scalar product between the input data vector and the filter coefficients, i.e., an affine function.

An RBF network (Figure 5) is used as a feature extractor. The local radial-basis function decomposition is well documented in the literature, from the points of view of the regularization theory [33], [34] and the nonlinear multivariate interpolation theory [35].

The RBF network performs a local nonlinear decomposition of the input signal, according to a linearly separable equation, as

$$Y(n) = \sum_{i=0}^{N} a_i f(X^*, b_i, c_i) \tag{7}$$

$$f(X^*, b_i, c_i) = e^{-\frac{\|X^* - c_i\|^2}{2 b_i^2}}$$

where c_i is the N_x-dimensional weights vector of the connections set from the network inputs to the i-th hidden PE, and b_i is the spread of the Gaussian function associated with the i-th hidden PE. The c_i vector is the center of the receptive field employed by the i-th hidden PE. The norm $\|.\|$ is used to measure the distance between this center and the input vector. The a_i scalar parameters are the weighting coefficient corresponding to the i-th PE contribution to the overall network response $Y(n)$. In Equation (7), we used the Gaussian local decomposition, but any other radial-basis function may be employed.

The output signal space is obtained using delayed versions of the radial-basis function decomposition signal $Y(n)$. The output vector Y lies in the N_y-dimensional space of the RBF network output

$$Y = \left[Y(n)\, Y(n-1)\, Y(n-2)...Y(n - N_y + 1) \right]^T \tag{8}$$

where the parameter N_y defines the maximum delay interval considered in the learning phase.

The relation between the input signal space and the output space is nonlinear and non-invertible, due to the RBF network decomposition. Using Equations (6) and (7), the direct propagation through the FSF system is

$$Y(n) = \sum_{i=0}^{N} a_i \exp\left(\frac{1}{2 b_i^2} \| X^* - c_i \|^2 \right) = \sum_{i=0}^{N} a_i \exp\left(-\frac{1}{2 b_i^2} \sum_{j=0}^{N_x - 1} (X^*(n-j) - c_{ij})^2 \right)$$

$$= \sum_{i=0}^{N} a_i \exp\left(-\frac{1}{2 b_i^2} \sum_{j=0}^{N_x - 1} \left(\sum_{k=0}^{N_i - 1} X(n - j - k) w_k - c_{ij} \right)^2 \right) \tag{9}$$

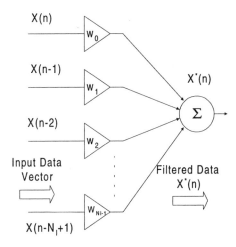

Figure 4: ALC as transversal filter.

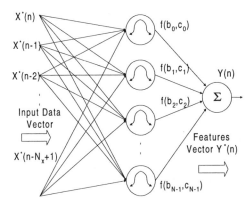

Figure 5: RBF network as features extractor.

where the dependencies between the output and the input of the FSF system are explicit.

In our approach, the features space is obtained using the Gaussian decomposition parameters, i.e., Y^* is a heterogeneous vector of the form

$$Y^* = [c_{00}, \ldots, c_{0T_x-1}, \ldots, c_{N0}, \ldots, c_{NT_x-1}, a_0, \ldots, a_N, b_0, \ldots, b_N]^T \quad (10)$$

2.3.2   ALC-RBF FSF System Learning

The relation between the feature space and the input signal delay space depends *on overall history* of the adaptation procedure of the RBF network parameters. A general explicit form of this dependence does not exist. The ALC coefficients depend on the RBF decomposition parameters. These parameters obey a nonlinear and discrete-time varying generic equation expressed by the functional operator H $(.)$ in Equation (5). H depends on the error function E, which, in turn, depends on the RBF network parameters $a(n)$, $b(n)$, $c(n)$, and on the reference features vector.

The filter parameters are adapted using both the gradient-based least-mean square (LMS) algorithm and the perturbation learning procedure. For the initial stage of the ALC learning algorithm, a quadratic error criterion in the signal delay space is used

$$E_{DS}(X^*, X^d) = \frac{1}{2} \sum_{n=0}^{P-1} [X^d(n) - X^*(n)]^2 \tag{11}$$

where the subscript DS identifies the input data space error, the P index defines the temporal error average (the batch training cycle dimension), and X^d is the target signal at the output of the ALC. For the perturbation learning procedure, used when the ALC is adapted following only a feature space constraint, the ALC training error may be considered as a binary function. This function is derived from the FSF system cost criterion, which informs that the perturbation is accepted / rejected at the system level.

The RBF decomposition parameters are adapted here using the gradient backpropagation technique for a quadratic error function defined in the output delay space

$$E_{RBF}(Y, Y^d) = \frac{1}{2} \sum_{n=0}^{Q-1} [Y^d(n) - Y(n)]^2 \tag{12}$$

The subscript RBF identifies the RBF network output error, and the index Q defines the temporal error average. Y^d is the target signal at the output of the RBF network.

The partial derivatives of the E_{RBF} quadratic error with respect to a generic RBF network parameter q are computed as

$$g_q(n) = \frac{\partial E_{RBF}}{\partial q} = \sum_{n=0}^{Q-1} [Y^d(n) - Y(n)] \frac{\partial Y(n)}{\partial q} \tag{13}$$

Replacing the output Y from Equation (7), the error gradient components for the weights a_i of the output layer are

$$g_{a_i}(n) = \frac{\partial E_{RBF}}{\partial a_i} = \sum_{n=0}^{Q-1} [Y^d(n) - Y(n)] f_i \tag{14}$$

The partial derivatives of the Gaussian function $f(\ .\ )$ for the i-th processing element with respect to the c_{ik} and b_i parameters are

$$\frac{\partial f(X^*(n), b_i, c_i)}{\partial c_{ik}} = \frac{X_k^*(n) - c_{ik}}{b_i^2} f_i \tag{15}$$

$$\frac{\partial f(X^*(n), b_i, c_i)}{\partial b_i} = \frac{1}{b_i^3} \| X^* (n) - c_i \| f_i \qquad (16)$$

The left-hand f_i notation identifies the output *value* for the i-th Gaussian PE, evaluated for the actual values of its parameters. The partial derivatives with respect to the other parameters c_{jk} and b_j, $(j \neq i)$ are zero. Replacing Equations (15) and (16) in Equation (13), the explicit forms of the gradient components are

$$g_{b_i}(n) = \sum_{n=0}^{R-1} \left[Y^d(n) - Y(n) \right] \frac{1}{b_i^3} \| X^*(n) - c_i \| f_i \qquad (17)$$

$$g_{c_{ik}}(n) = \sum_{t=0}^{R-1} \left[Y^d(n) - Y(n) \right] \frac{X_k^*(n) - c_{ik}}{b_i^2} f_i \qquad (18)$$

The above relations express the linear separability property of the RBF network architecture, which allows linear discrimination of the contribution of each PE in forming the network output signal. This property is reflected in Equations (14) to (18), i.e., the gradient associated with the PE depends only on the network response and the internal PE parameters.

The RBF gradient learning procedure is the same as for the ALC, following the scalar gradient descent update equation

$$\Delta q(n+1) = -\eta_q g_q(n) + \alpha_q \Delta q(n) \qquad (19)$$

where the argument n denotes the update time step and η_q and α_q are, respectively, the learning rate and the momentum coefficient associated with the generic adaptive parameter q.

The features space error criterion is defined as the weighted L_2 distance between the actual features vector Y^* and a reference features vector Y^{d*}.

$$E_{FS}(Y^*, Y^{d*}) = \gamma_a \sum_{i=0}^{N} (a_i - a_i^d)^2 + \gamma_b \sum_{i=0}^{N} (b_i - b_i^d)^2 + \gamma_c \sum_{i=0}^{N} \sum_{j=0}^{T_x-1} (c_{ij} - c_{ij}^d)^2 \qquad (20)$$

where the subscript *FS* identifies the features space error. The weighting factors γ_a, γ_b, γ_c are positive constants that control the contribution of each class of adaptive parameters to the feature space error.

However, the weighting factors γ in Equation (20) are difficult to estimate in a very crisp manner. Moreover, different users could indicate various values. Furthermore, the subjective effect of quality is not necessarily a quadratic function of the distance in the features space. Consequently, in such situations, fuzzy membership functions could be defined (such as in [27])

$$\mu[E_{FS}^-(Y^*, Y^{d*})] = \mu_a \left[\sum_{i=0}^{N} (a_i - a_i^d)^2 \right] \cup \mu_b \left[\sum_{i=0}^{N} (b_i - b_i^d)^2 \right] \cup \mu_c \left[\sum_{i=0}^{N} \sum_{j=0}^{T_x-1} (c_{ij} - c_{ij}^d)^2 \right] \qquad (21)$$

The features vectors Y^* are obtained at the end of each learning session and they completely describe the RBF network realization. The reference feature vector Y^{d*} is obtained using an ALC trained with a clean data set as both input and desired output. The same ALC is used as a starting filter for the iterating training procedure employed for the whole system. This time, a noisy data set is the input to the ALC. The ALC output signal is used to train an RBF network. After training, a first feature space error, Equation (20), is computed. A perturbation of the filter parameters generates a new ALC output signal.

An RBF network is trained and the feature error is computed. If this RBF network realization is superior to the first realization (smaller error), the RBF parameters and the perturbation in the ALC are accepted. If this realization is inferior to the previous one, in terms of the given error criterion, the RBF network and the perturbation in the ALC parameters are rejected. This procedure is continued until no better realizations are found for a pre-specified number of steps, or until the features error is acceptably small (smaller than a pre-defined value). The procedure may be considered as an iterative synthesis technique for the linear adaptive filter. The synthesis criterion is a features (Gaussian) decomposition as close as possible to a desired reference decomposition.

The following steps summarize the FSF system training procedure:

1. Obtain a reference filter ALC^* and a reference decomposition RBF^* using the clean (reference, original) signal (considered to be known).

2. Obtain an $ALC(0)$ and the associated Gaussian decomposition $RBF(0)$ using the noisy signal. $ALC(0)$ is obtained from ALC^* with a small parameter perturbation. The $RBF(0)$ is obtained using the $ALC(0)$ output signal and the clean desired signal, starting from a random initial state.

3. Repeat until the FSF system error is smaller than a given value, or the system is settled in a stable state (error changes are smaller than a given value, for a pre-specified number of steps).

3.1. Obtain an $ALC(i)$ from $ALC(i-1)$, with a small parameter perturbation.

3.2. Obtain a Gaussian decomposition $RBF(i)$, using the $ALC(i)$ output signal, the clean desired signal, and the $RBF(i-1)$ state as initial state.

3.3. Compute the FSF error $E_{FS}(i)$ with (20), using $RBF(i)$ and RBF^*.

3.4. If $E_{FS}(i) > E_{FS}(i-1)$

Replace $ALC(i)$ with $ALC(i-1)$, $RBF(i)$ with $RBF(i-1)$, and $E_{FS}(i)$ with $E_{FS}(i-1)$ (reject the whole i-th iteration).

If $E_{FS}(i) < E_{FS}(i-1)$

Go to 3.5.

3.5. i = i+1.

When the error E_{FS} decreases, one gets a better Gaussian decomposition for the linear filter output signal, and the perturbed ALC parameter is retained. When the E_{FS} increases, one restarts the algorithm with the previous state of the ALC and RBF. Note that the memory of the learning system refers, in this algorithm, only to the features space error. In other words, the retained initial states for the ALC and RBF system components depend only on the features space error.

2.3.3 Image Contour Enhancement and Recognition Optimization Using ALC-RBF FSF System

The task described here is related to the 2D contours extraction and recognition. We assume that the objects are recognized using the Hough transform. For an introduction to the Hough transform, as used here, see, for instance [36], with a fast neural network realization in [37]. The Hough transform allows the extraction of a contour characteristic function, that is independent of the scaling and orientation of the object. The contour is generally noisy, due to the image quality and the extraction algorithm. It is desirable to filter the resulted output of the Hough transform system to eliminate noise, while preserving the features space attributes of the contour.

The simulations use a synthetic contour formed from Gaussian signals, a FSF system realized with a 3:1 ALC network, and a 3:7:1 RBF network ($N_I = N_X = 3$). In the first step, we obtained the system reference components using a reference contour. An approximation performance measure, expressed in terms of absolute percent error lower than 1%, is obtained for both the ALC and RBF reference networks. For simulation, we used a sub-vector of the features vector

$$Y_c^* = \left[\; c_{00}, \; \cdots \; c_{0T_X-1}, \; \cdots, \; c_{N0}, \; \cdots, c_{NT_X-1} \;\right]^T \tag{22}$$

In all RBF networks realizations, we consider the reference decomposition parameters,

$$Y_{ab}^d = \left[\; a_0^d, \ldots, a_N^d, b_0^d, \ldots, b_N^d \;\right]^T \tag{23}$$

as constants. For system training, we used a corrupted contour (Figure 6). Each FSF learning system iteration i is built from the perturbation of a randomly chosen ALC parameter, followed by a complete training session of the RBF network, using the output signal of the ALC and the desired reference contour in Figure 7.

The perturbation value is $\eta_{ALC} = 0.001$ and the RBF network learning parameters are: $T_q = 200$ (the batch cycle period) and $\eta_c = 0.1$ (the centers learning rate). As usual, the learning rate is divided by the inset cardinality of each processing element. The number of training epochs for the RBF learning session was fixed, for simplicity, at 200 epochs.

A typical E_{FS} error evolution is plotted in Figure 8. The starting normalized error is $E_{FS}(1) = 0.11414$ and the ending error, after 100 iterations, is $E_{FS}(100) = 0.1016$. The final ALC output is plotted in Figure 9.

This simple experiment shows that the filtering capabilities in the samples space of the obtained ALC are comparable to those of a standard LMS realization, but the error in the features space is lower in our approach. Consequently, the "perception" of the result is enhanced, and the subsequent processing stages in features space will be more accurate.

Note that the FSF learning system adds important constraints to the ALC realizations. These constraints are imposed in a nonlinear features space, by means of Gaussian function decomposition parameters, which are targets for the FSF learning system.

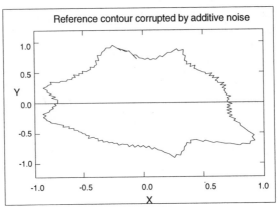

Figure 6: ALC noisy input data (additive uniform noise in the [0.0; 0.05] interval).

The noise filtering performance is higher for the well-shaped features (assimilated as the Gaussian peaks in Figure 7) and lower for the small radial amplitude samples. This property arises from the local nonlinear decomposition with radial basis function employed here and from the specific features constraints learning algorithm.

2.3.4   Discussion

The above FSF system works well only for small perturbations in the input delay space and, consequently, in the features space. Due to the nonlinear and time-dependent input-output equation, the system components must be pre-trained partially or totally before starting the FSF learning iterations. This FSF system is able to extract the imposed features from noisy data. The resulting data signal has a better features representation and a higher signal-to-noise ratio. This property makes the system appropriate for data pre-processing (pre-filtering), especially when the data space classical filtering techniques are susceptible to hiding some essential features of the original data set.

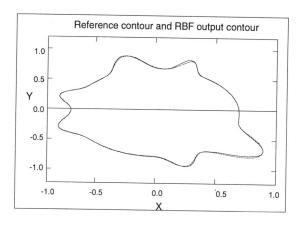

Figure 7: The reference contour (solid) and the reference RBF output contour (dashed).

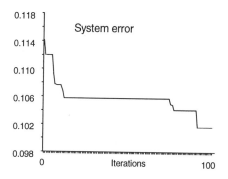

Figure 8: Feature space error for the ALC-RBF FSF system.

Although it was designed as a data conditioning system, by means of the imposed desired features, the proposed FSF system may also be used in pattern matching and classification, or in pattern detection applications. The same basic learning algorithm also works for features space optimal control systems, which have a structure similar to that of the filtering systems in Figure 2.

Poorer performances (lower noise reduction capabilities) over the regions of the signal that do not contain useful information (i.e., information related to a pattern, e.g., expressed as maximal radii of the embedded contour circle) are a drawback of the method. Classic methods behave in a uniform manner with respect to the existence of patterns in the signal. However, using a simple pre-processing technique, namely, by applying a piecewise linear transform of the input variable (time, distance, and angle) easily eliminates this drawback. The transform can be defined with only two slopes (Figure 10). In this way, the signal-to-noise ratio can be significantly increased.

Better results can be obtained with a refined FSF learning scheme, which tightly integrates both the delay signal space and the features space learning procedures. In actual realization, these learning components are coupled only by means of a binary decision function employed in step 3.4 of the FSF learning algorithm.

The features space constraints may be transformed to regularization constraints in a more elaborated ALC learning scheme. Another aspect that might be improved in the future is the temporal coupling during the FSF learning iterations. In the presented learning algorithm, the ALC and RBF training phases are de-coupled. It is possible to design an "on-line" training scheme that uses the above mentioned weak constraints. A similar decomposition by Gaussian functions, using adaptation in the features space, but using a conventional features extractor, produced encouraging preliminary results.

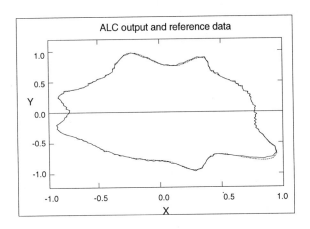

Figure 9: The ALC output after 100 iterations of the ALC-RBF System (solid line–ALC output, dashed line–ALC reference data).

2.4   FSF Architecture Using Multilayer Perceptron Filter and Principal Component Analysis Network Feature Extractor

In this section, a multilayer perceptron (MLP) neural network is used for the adaptive filter. The features extractor is based on information obtained from the second order statistics of the signal. The statistics are extracted with a principal component analysis (PCA) hierarchical network. The training procedure uses an error signal computed as the difference between the desired and actual largest eigenvalues. Advantages of the proposed method are illustrated by experiments on electrocardiographic (ECG) signal filtering (see also [38]).

2.4.1   MLP-PCA FSF System Realization

We propose an MLP neural network designed as an adaptive transversal filter. The network realizes a nonlinear moving average of the input signal, using the sigmoidal non-linearity of the form

$$x_j^L = f(a_j^L) = \frac{1}{1 + e^{-a_j^L}} \tag{24}$$

The activation a_j^L of the processing element (PE) j of the layer L follows the usual additive equation:

$$a_j^L = \sum_{i=1}^{n_{L-1}} x_i^{L-1} w_{ji}^L + b_j^L \tag{25}$$

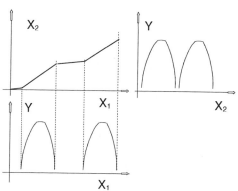

Figure 10: Piecewise linear x axis compression to eliminate zones where features are inactive (after [27]).

where $w_{j\,i}^{L}$ parameters form the weights inset to the processing element PE j of layer L, and b_j^L is the bias term. The index L denotes the layer of the network, and $L = 0$ is assigned to the inputs of the MLP.

In vector notation, the direct data flow through the MLP may be written, layer by layer, as

$$x^L = f(W^L x^{L-1}) \tag{26}$$

considering f as a punctual function, that operates identically on each component of its vector argument.

The MLP network has a structure appropriate for filtering operations, and many experiments have proved its capabilities. We used a refined form of the MLP, the asymmetric pyramidal ($n_L < n_{L-1}$) topology. Each processing element PE of the layer L is connected only with its right-handed counterparts of the layer above (Figure 11). This formally shows that the right quadratic weight sub-matrix $W^L_q \in R(n_L \times n_L)$ is lower triangular (when processing elements are indexed from left to right). This symmetry breaking topology allows a better performance of the transversal filter, measured in terms of learning convergence per connection [39].

The MLP network implicitly performs a weighting operation on each input from the signal delay space. In this particular case, this is driven not only by the symmetric backpropagation of the error function gradient, but also by the topological constraint. The triangular constraint allows a progressively reduced number of connections for the oldest delayed input signal samples. Consequently, this produces an attenuation of their contributions to the output signal.

As a feature extractor, we use a hierarchical PCA network [9]. The properties of such a system are attractive from several points of view. The Karhunen-Loéve transform (KLT), well known in multivariate analysis [40], is a powerful tool in the linear feature extraction. PCA is closely related to spectral analysis, and particularly to discrete cosine transform (DCT) used mainly in data compression applications. Its neural realization meets both biological and engineering requirements, if one considers only the Hebbian learning system employed in PCA and the global convergence properties well demonstrated in the literature (see, for example, a Lyapunov based analysis in [41]). In the above notations, the KLT realizes the linear optimal transform

$$Y = V X^*, \ Y \in R^P, \ X^* \in R^M \tag{27}$$

$P < M$ in the usual data compression case. Moreover, the weights matrix $V \in R^{P \times M}$ insures the optimality in the mean squared error sense of the KL transform, if its rows are the eigenvectors corresponding to the largest eigenvalues of the auto-correlation matrix R_{xx} of the signal X^*. The signal is supposed to be wide-sense stationary and with zero mean.

The PCA network is a simple linear layer of additive processing elements (Figure 12), with the weight matrix V, employing Equation (27). After training, its rows v_j approximate the P principal eigenvectors of the R_{xx} matrix, and, consequently, if the signal X^* has zero mean, the variance probes obtained at the outputs Y_j are estimates of the principal eigenvalues.

The variance estimator extracts the first P largest eigenvalues $Y_1^* > Y_2^* > ... > Y_P^* > 0$, with the usual approximation (in the ergodicity and zero mean hypothesis), by the temporal mean squared values of each output of the PCA layer. We consider an equal width R for each time average window associated with the PCA outputs.

2.4.2   Learning in MLP-PCA FSF System

The norms used to measure the distance between the desired vector Y^d and the actual output vector Y^* of the feature extractor, i.e., the error functions are

$$E_{FS1}(Y^d, Y^*) = \sum_{i=1}^{P} |Y^d_i - Y^*_i|$$

$$E_{FS2}(Y^d, Y^*) = \frac{1}{2} \sum_{i=1}^{P} (Y^d_i - Y^*_i)^2 \tag{28}$$

It may be pointed out here that the information extracted from the signal X^* at the output of the MLP is passed through the de-correlation PCA layer and is already averaged in the variance estimator. It is more likely that $Y^* > 0_p$ is a (very) slow varying signal, and the modulus of the difference between the desired and output vectors seems to be a sufficient error measure. On the other hand, the error function E_{FS2} has derivatives with respect to the system adaptive parameters. Both E_{FS1} and E_{FS2} are used in our simulations.

The relevance of the mapping $Y^* = Y^*(X)$ is dependent on both the MLP network and the PCA layer adaptation procedures. The latter uses an unsupervised Hebbian-like training algorithm, which proves its convergence to globally stable solutions in a relatively reduced number of iterations. In contrast, the MLP training procedure is more difficult and requires a considerable number of iterations.

Filtered signal X˙(t)

MLP network
transversal filter

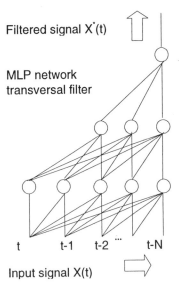

t   t-1   t-2 ···   t-N

Input signal X(t)

Figure 11: MLP as adaptive transversal filter.

Because of these different learning dynamics, we separate the training procedure of the system into two stages: an open-loop direct pre-training stage and a closed-loop, indirect adaptation stage. In the case of the MLP network, the open-loop pre-training is performed as a standard backpropagation procedure, based on a representative selection of input signals and desired output signals. We use the instantaneous mean squared error between the reference and actual output signal of the MLP, measured in the data space:

$$E_{DS} = \| X^* - X^d \|^2 \tag{29}$$

where the subscript DS differentiates between the pre-training error in signal space and the whole system adaptation errors of Equation (28). The weight update equation, in vector form is

$$\Delta W_j^{L^T}(t+1) = -\eta\, g_j^L(t)\, x^{L-1^T}(t) + \alpha\Delta W_j^{L^T}(t) \tag{30}$$

where g_j^L is the backpropagated gradient of the error function to the j-th processing element of the layer L, η is the learning rate parameter, and α the momentum coefficient. The transposed unfolding of the backpropagation paths is constrained in the asymmetric MLP topology as is shown in the scalar form of Equation (30)

$$\Delta w_{ji}^L(t+1) = -\eta\, g_j^L(t)\, x_i^{L-1}(t) + \alpha\Delta w_{ji}^L(t), \; i \leq j + n_{L-1} - n_L \tag{31}$$

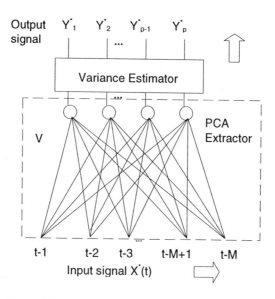

Figure 12: The PCA network as features extractor.

by the limited variation of the subscript index i. Certainly, elaborated efficient backpropagation methods may be exploited, the above basic technique being used only for illustrative purposes.

Once the pre-trained MLP network performs the reference input-output mapping in signal space with sufficient (given) accuracy, the MLP outputs are used to pre-train the PCA layer.

In the learning phase, the PCA layer becomes recurrent, due to the specificity of the Hebbian learning algorithm, which supposes that the strength of the connection between two processing elements depends on both processing elements' output values. We use the well-known Sanger's generalized Hebbian algorithm (GHA) [9], expressed by the following updating equation:

$$v_{ji}(t+1) = v_{ji}(t) + \eta_j(t) \; y_j(t) \left(X_i^*(t) - \sum_{k=1}^{j} v_{ki}(t) \; y_k(t) \right) \quad (32)$$

where $j = 1,...,P$ and $i = 1,...,M$. The y signal represents the direct output of the PCA layer and the input vector to the variance estimator. The first term of the right-hand side of Equation (32) (after multiplication) is the standard Hebbian correlation term and the second term is the active decay factor, expressing the competitive effects induced in the algorithm by the eigenvectors normalization constraint. The learning rate η is a positive small constant.

After the pre-training stage, when all the adaptive components of the FSF system are settled in a favorable state, the loop is closed and the system is fed with the real signal. The changes of detected eigenvalues of the signal autocorrelation matrix, relative to the reference vector Y^d, are used to drive the indirect adaptation of the MLP. In our experiments we use two techniques to train the FSF: a perturbation algorithm and a gradient backpropagation algorithm. The former technique uses the direct propagation paths of the system, with a small computational load. The latter technique exploits the inverse propagation path of the system as a whole.

The perturbation technique consists in changing a weight parameter of the MLP by a small amount. Each such perturbation is directly propagated through the system and modifies the cost function E_{FSI}. If the change produces an increase of the error, the change is discarded and another weight is perturbed. The weight selection procedure follows a uniform probability density.

The gradient backpropagation through the system follows the same steps as the standard BP algorithm. Starting at the output of the variance estimator, the gradient of the error function with respect to the weights of the MLP is computed. This computation involves the time unfolding of the MLP subsystem [42], [43], due to temporal averaging in the variance estimator and due to the serialization delay-line between the MLP and PCA layer. The instantaneous gradient for each weight parameter, at each time step, is cumulated and the weights update is made after a discrete time interval T.

2.4.3   Example of MLP-PCA Feature Space System in Signal Filtering

The electrocardiographic (ECG) signal characteristics are generally difficult to model, and elaborated pattern detection schemes must be used to recover this type of signal from noise [44]. The ECG filtering capabilities of the above FSF system was explored in [39]. The simulations use a FSF system realized with a 40:20:1 MLP network and a 40:10 PCA layer ($N_x = N_y = 40$ in Equations (2) and (3)). The variance estimator has a temporal window width $R = 100$. Following the procedure described in the previous section, we pre-train the system with an ECG reference signal (Figure 13, dashed line). The output signal of the MLP is used to train the PCA layer until convergence is achieved. The largest reference eigenvalue vector Y^d is stored and used in the second training stage. A signal corrupted with additive Gaussian noise is used (Figure 13, solid line). The noise has zero mean and its variance is $\sigma_n = 0.05$. The reference signal has the variance $\sigma_s = 0.38$.

The system is trained using both perturbation and gradient adaptation procedures. A comparison of the results reveals that the direct and the inverse system propagation paths are efficient. The perturbation technique is simpler, but the gradient backpropagation technique seems to be more robust and the filtering operation is more accurate [39] (Figure 14). The cleaned output signal noise component corresponds to the final absolute error in the feature space (see [39] for details).

The linear counterpart of the above FSF system replaces the MLP with an adaptive linear combiner (ALC). For performance comparison we use a 40:1 ALC pre-trained with the LMS procedure to approximate the same reference signal as the initial MLP.

The adaptive linear combiner behaves well for periodic signals (e.g., a single ECG period), but its performance degrades significantly for a quasi-periodic ECG signal (Figure 15). Instead, the performance of our hybrid system does not significantly degrade in such a case. This shows that the hybrid system definitely outperforms the classic ones in this type of application.

2.4.4   Discussion

The use of neural networks to extract signal from noise, based on parametric representations, was recently also addressed in [45]. The authors use PCA as a signal features extractor, and a neural network as a parameter estimator. This approach differs from ours in methodology and in the way of recovering the signal from noise.

Though the closed loop training uses perturbation and gradient technique only for moving the system in the desired features space point for a relatively small perturbation in the input signal space, the direct and inverse data flows in the FSF system prove to be efficient. The learning system consists of two stages: the open loop separate pre-training stage for each adaptive component, actually the MLP and PCA, and a closed-loop adaptation stage. Even in these conditions, when both the FSF and its associated learning system employ a leaky coupling between the adaptive components, the learning is convergent and the filter has a good performance.

Improved results could be obtained by refinements in the perturbation training procedure, e.g., by using the recent developments of the matrix perturbation theory applied in MLP learning [46]. Also, the orthogonal weight estimate techniques (OWE) (see, for example, [47]) may contribute to the improvement of learning in the FSF system.

In order to realize a more efficient integration of the PCA and MLP learning systems, the intimate relation that seems to emerge from the interaction between the asymmetric proposed MLP and the hierarchical PCA learning rule, in connection with the nonlinear generalizations of the PCA [11], can be used. Another promising approach might be the study of a fully symmetric FSF system using the standard MLP. The associated features extractor is based on the ortho-normalization system with equal output variance on each channel, as described in [10].

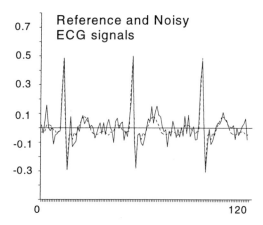

Figure 13: Reference (dashed line) and noisy (solid line) EGC signals.

2.5   Concluding Remarks

The "features space filtering" method, a novel approach to design complex adaptive systems for simultaneous and interlaced filtering and pattern extraction, was proposed and analyzed. We use hybrid neural configurations for the features space filtering (FSF) systems.

Two hybrid architectures, based on neural networks, were used to illustrate the method: the ALC-RBF system and the MLP-PCA system. The former uses patterns obtained with RBF neural network decomposition parameters to adapt the coefficients of a linear FIR filter. The latter uses the principal components of the auto-correlation matrix of the input data to adapt a multilayer asymmetric perceptron.

The ALC-RBF realization, which combines the advantages of linear filtering with the power of nonlinear radial basis functions local decomposition, shows that the features space filtering is a valid approach. The RBF features extractor may be replaced with another classic pattern detection system, for example, a Kohonen self-organizing network, as in [48], for the visuo-motor control in robotics. Moreover, the RBF extractor itself may be trained with an unsupervised learning procedure. Preliminary experiments in this direction, using an unsupervised RBF network, show good results in pattern auto-extraction and classification. The features space learning target in the ALC-RBF system experiments is chosen only for illustration purposes and for direct comparison between the samples and features spaces errors. Any application specific learning target is allowed, provided that the target signal differentiates the hidden input data features well. This is also an application specific constraint and a preliminary study of the input data properties is necessary. Without a good input data analysis, such a FSF system may fail to work.

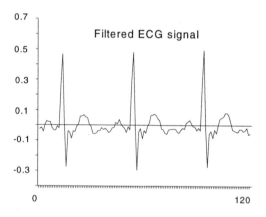

Figure 14: Filtered ECG signal (100 batch cycles with T = 25 and η = 100).

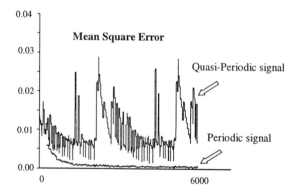

Figure 15: Typical MSE error plots for the pre-training stage of the ALC.

The MPL-PCA system uses the FSF principle in different ways. In this case, the filter is nonlinear and its training procedure is complex. The features extractor is fixed and an inverse gradient path for the whole system exists. The learning scheme is different, and needs a single pre-training stage for each component. The system adaptation with noisy data input fully exploits the direct and inverse data paths. The ECG signal conditioning is not an easy task, due to its specific frequency spectrum and classic linear adaptive filtering methods that fail to cope with non-periodic signals.

In contrast to the adaptive linear combiner, which behaves well only for periodic signals, the performance of the presented hybrid system does not significantly degrade in such a case. This shows the hybrid system outperforms the classic ones in a large class of applications.

The targeted applications area of the features space systems is a real time control system, where its increased capabilities of temporal processing, especially for discrete time event systems and sequences recognition, may be well employed. The feature space systems are especially useful in complex systems control, when the underlying dynamic model is hard to obtain analytically, as shown in the next section.

3   Feature Space Identification and Control Schemes Using Neural Networks

In recent years, there has been an increasing interest in the application of hybrid methods in the identification and control of dynamical systems. Neural and fuzzy systems are the most frequently implemented in intelligent control applications.

Various types of identification and control systems are based on composed neural structures. Model reference adaptive control schemes [49], [50], [51] use an identification neural network to extract the inverse plant dynamics, and to feed a neural controller during the learning phase. Recently, the optimal tracking problem was explored using neural networks [52], based on feedforward and feedback neural controllers and on neuro-identifiers.

The direct adaptive neuro-controllers, traditionally employed with Radial Basis Function network [53], are now extended to adaptive basis functions, like local multilinear splines [54], and to the dynamic structured RBF network [55]. Recurrent neural systems, with complex learning algorithms, are also currently investigated in identification control applications [56], [57].

The neuro-controllers learning algorithms are complex and the dynamic behavior of the various control system components is an essential research topic. Many efforts are focused on finding algorithms for stable neuro-controllers adaptation. Lewis and co-workers [58], [59] introduced correction terms and robustifying factors in the learning method, specifically in tracking robot arm neuro-control applications. Tanaka [60] introduces the linear differential inclusion models and parameter region representation, to investigate the stability properties of neural controllers. Stabilizing error terms were also recently proposed for single layer perceptron controllers [61].

In all these complex aspects investigated in the literature, the neuro-control problems may be thought of as representation problems. Then, the main target is to find representations for the underlying direct and inverse dynamics of an unknown plant. These representations are subsequently used to obtain the desired plant response.

The approach presented here extends the representation problem to the features space. The introduction of the features space restrictions in the neuro-control systems produce a qualitatively different plant response. Depending on the desired features space, various aspects of the control quality may be improved. The overall plant response may be improved with respect to a known reference features space response, without loss of information related to the data space response. This aspect will be covered in detail. On the other hand, the stability properties of the system may be addressed if the features extractor expresses accurately the dynamic characteristics of the system. For example, the stabilization error terms introduced in [61] may be directly extracted using a features extractor.

In the following sub-sections, several general neural topologies for identification and control, based on features space error measures, are described. Features space processing methods are added to classic neural identification and control methods, in order to obtain better characterization of the plants in specific features spaces. In complex systems, identification and control methods are integral parts in the global processing chain. The identification of plant features space characteristics and the extraction of pattern based control laws are determinant for the processing quality in the subsequent stages.

In the first section, data and features space identification topologies are presented and compared. In the second, features space control methods are discussed. A features space control example is presented in the third sub-section. Concluding remarks and discussion make up the last sub-section.

3.1   Feature Space Neural Identification Topologies

Classic identification schemes using neural networks are designed to find an input-output data space model as close as possible to the plant in terms of a simple Euclidean error criterion [49], expressed in the plant input and output data spaces. This identification approach is based on a known input-output data set. The series-parallel scheme in Figure 16 uses this known input-output data set during the network training. The plant output is coupled with the identification network input through a tapped delay line (TDL). The delay line length corresponds to the input-output propagation delay through the plant. The parallel scheme uses the on-line estimation (during adaptation) of the plant output as a function of its input. The plant output data set is used only to compute the identification error (Figure 17). This latter scheme is generally unstable in the initial stage of the adaptation procedure, and it is not used without a pre-training series-parallel stage. On the other hand, the generalization capabilities of the parallel configuration are better for certain plant classes.

The features space identification scheme assumes that another type of input-output relation should be approximated, namely, a relation between a data space set and a specific features space set. This type of problem arises in many practical circumstances, when a data-space error criterion is difficult to express in a semantic form useful for the general modeling context.

The representations in Figures 18 and 19 add to the classical neural identification configurations the features extractor blocks. In this case, the error criterion is related to that features space representation and hybrid approaches are more feasible, if the neural model inputs and desired output data sets contain all the information available in the modeling context. The input and output data space vectors and the additional features space error constraint are simultaneously used in the adaptation procedure, to ensure the maximum accuracy for the identification process.

In the case of the hybrid data and features space approach, the error criterion is a combination of two cost functionals (but it may be based only on the features space measure). A functional measures the output space error and the other functional measures the features space error. A parametric regulation technique may be employed to improve the accuracy of the model in both spaces, with a temporal scheduler for weighting the cost functionals during learning.

With the additional nonlinearities induced by the features extractor, the adaptation process is more difficult. The temporal scheduler must accentuate the data space error in the initial stage of the adaptation process. When the data space identification loop is stabilized, the features space outer loop gain must be increased.

The above general features of space identification configurations must be particularized for the problem at hand. Depending on the features extractor in use, these configurations may employ specific identification tasks. The possible applications range from dynamic output data space conditioning to automatic selection and validation of complex model hypothesis.

3.2   Feature Space Neural Control Topologies

The approach presented in this section is based on the traditional reference model control schemes. This type of scheme assumes that a sufficiently precise model of the plant response for a given input reference signal is available. The neural control strategy consists in finding an estimation of this model, using the plant or an identification model of the plant. The first section compares the data and the features space neural control topologies. A synthetic control example is solved in the features space in the second section. The last section contains concluding remarks.

3.2.1   Data Versus Feature Space Neural Control Topologies

The classic approaches to direct and indirect neural control schemes, according to [51] are presented in Figures 20 and 21. The direct schemes assume that the plant control parameters are accessible in the control process context. The indirect control strategies are based on the identification model of the plant, and assume that the control parameters are good estimates for the real plant parameters. The indirect control schemes are more favored. However, direct adaptive neural control schemes are also used for specific classes of unknown plants. Recently, a direct adaptive control scheme using a RBF neural network controller for a large class of plants was proposed [53].

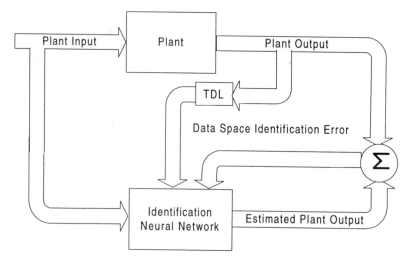

Figure 16: Series parallel identification scheme in data space.

In the case of the direct adaptive topology, the neural controller synthesizes the input of the plant, following the reference model input. The plant response is compared with the reference model output, and the error is formed in the plant output space and not in the neural controller output space. This is the main difficulty in training the controller with classic gradient backpropagation methods, because of the unknown plant input-state-output equations.

The indirect adaptive topology (Figure 21) faces the same difficulty. Gradient methods use, in this case, the backpropagation through the identification model, whenever possible. The identification model must be smooth and the derivatives of its output, with respect to the input, must exist.

Using the arguments from the previous section, it is possible to design the features space direct or indirect control topologies, as shown in Figure 22. The features extractor compares the plant response with the reference model response. The features space error contributes to the neural controller adaptation. Again, it is possible to use various combinations of cost functionals to realize effective control strategies.

The indirect control schemes are more elaborate and the control effectiveness is directly related to the quality of the available identification model. They are, however, the only choice in many practical control problems. An identification model with accurate features space behavior may significantly improve the control effectiveness, even for the classical indirect approach.

In many situations, the features extractor block (Figure 22) may be thought of as a known component of the controlled system. Depending upon its characteristics, the significance of the real plant output may be sensibly different from the point of view of the global feedback control loop, allowing different approaches to the design of the neural controller. When the reference control model is available in the data space, the features space component of the overall control system is considered to act as a correction module. This component brings supplemental constraints in the desired response of the controlled system. These feature constraints may improve significantly the global control performance. The outer error loop in Figure 22 accelerates the neural controller learning process, due to the expert quality of the information extracted in the features space. An accurate reference control model is not always available in the data space for complex plants. In this case, the desired plant behavior may be better expressed in the features spaces. The neural controller learning procedure is driven exclusively by the features space error, as in Figure 23.

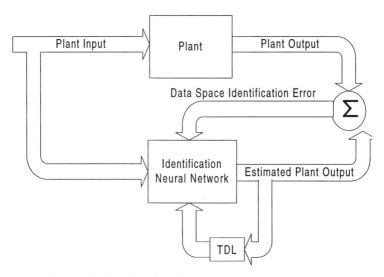

Figure 17: Parallel identification scheme in data space.

The features extractor in Figure 23 is directly inserted into the feedback control loop. The reference control model becomes an expert model, depending on the desired feature space behavior. The overall control system is directly dependent now on the quality of the reference features model and on the accuracy of the features extractor.

3.2.2   Feature Space Control Example

To demonstrate the capabilities of the general features space control architecture, we use a control example described in [49]. The unknown plant is given by the equation

$$y_p(k+1) = \frac{y_p(k)}{1+y_p(k)^2} + u^3(k) \qquad (33)$$

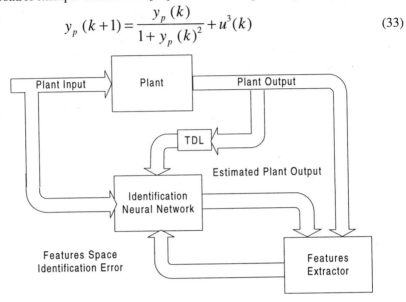

Figure 18: Series-parallel identification scheme in features space.

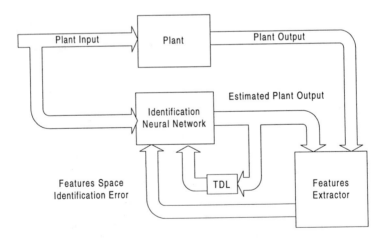

Figure 19: Parallel identification scheme in features space.

The reference model is described by the data space input-output relation

$$y_m(k+1) = 0.6y_m(k) + r(k) \qquad (34)$$

and the control input (the permanent excitation) is a composed sinusoid signal of the form

$$r(k) = \sin\frac{2\pi k}{25} + \sin\frac{2\pi k}{10} \tag{35}$$

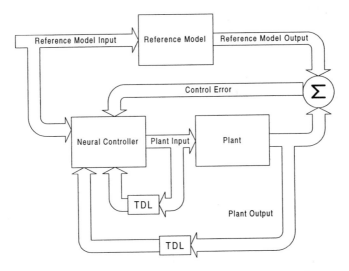

Figure 20: Reference model direct adaptive control.

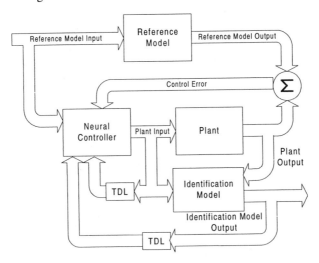

Figure 21: Reference model indirect control.

The plant, Equation (33), belongs to the model III discrete-time nonlinear system [49], described by the difference equation of the form

$$y_p(k+1) = H(y_p(k),..., y_p(k-n+1), u(k),..., u(k-m+1)) \tag{36}$$

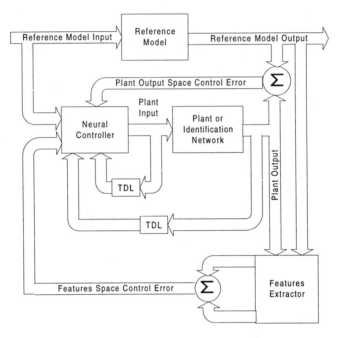

Figure 22: Data space direct and indirect control topologies using data and features space control errors.

where F and G are the separable output and state nonlinear operators. In this experiment we consider the model IV general discrete-time difference equation of the form

$$y_p(k+1) = F(y_p(k),..., y_p(k-n+1)) + G(u(k),...,u(k-m+1)) \qquad (37)$$

The model IV approach is motivated by the lack of information related to the state of the plant in many practical contexts. Our control strategy does not use the state space information. The identification and control schemes may be applied without restriction to any unknown plant, described by a representative input-output data set.

On the other hand, the model III approach allows a more accurate design of the control system, using distinct identification modules for each nonlinear operator in Equation (36), when state space information is available.

3.2.3   Identification Neural Network

A data space series-parallel model identification scheme is used to obtain a neural model for the plant. The overall neural network input-output relation is expressed as

$$\hat{y}_p(k+1) = N(y(k),u(k)) \qquad (38)$$

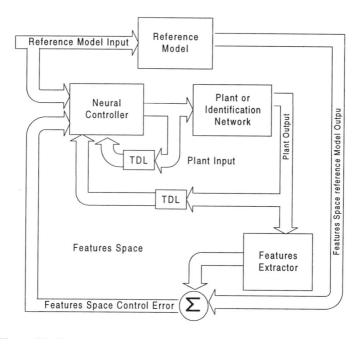

Figure 23: Features space reference model direct and indirect control.

In the neural identification scheme, we use a multilayer perceptron (MLP) network. The processing elements (PEs) in the network layers have linear finite impulse response (FIR) filters as weights. The PE is described by a set of equations.

$$o_j(k+1) = f\left(\sum_{i=1}^{p} a_{ji}(k) - \theta_j(k)\right) \tag{39}$$

where o_j is the output signal of the i-th PE, a_{ji} is the activation of the link between the i-th PE of the previous layer and the j-th PE of the current layer, and θ_j is the activation threshold of the current PE.

The output nonlinear PE function f is generally monotonic and its derivative, with respect to the activation, vanishes when the modulus of activation grows infinitely. While, for the static PE, the link activation signal is the weighted output of the i-th PE in the previous layer, for the FIR PE, the link activation signal is formed as the convolution between the output of the i-th PE and the weights vector of the link FIR filter.

$$a_{ji}(k) = \sum_{n=1}^{N} w_{ji}(n)o_i(k-n) = W_{ji}^T O_i(k) \tag{40}$$

where N is the filter order and $w_{ji}(n)$ are the filter coefficients.

For the FIR MLP network, the specific learning algorithm is the temporal backpropagation. The method is based on the particular expansion of the partial derivatives of the mean error cost functional E with respect to the FIR weights $w_{ji}(n)$ expressing the changes in the error produced by the changes in the activation of the link between PEs.

$$\frac{\partial E}{\partial w_{ji}(k)} = \sum_k \frac{\partial E}{\partial a_j(k)} \frac{\partial a_j(k)}{\partial w_{ji}(k)} \qquad (41)$$

The general instantaneous weight adaptation equation is

$$W_{ji}(k+1) = W_{ji}(k) - \eta \frac{\partial E}{\partial a_j(k)} \frac{\partial a_j(k)}{\partial W_{ji}(k)} = W_{ji}(k) + \eta \delta_j(k) O_i(k) \qquad (42)$$

where η is the learning rate parameter and δ_j denotes the instantaneous anti-gradient for the j-th PE. The specific expansion in Equation (31) leads to the following local gradient expressions [40]:

$$\delta_j(k) = e_j(k) f'(a_j(k)) \qquad (43)$$

when the j-th PE is in the output layer of the network and

$$\delta_j(k) = f'(a_j(k)) \sum_{m \in I} \sum_{l=0}^{N} \delta_m(k+l) w_{mj}(l) = f'(a_j(k)) \sum_{m \in I} \Delta_m(k) W_{mj} \qquad (44)$$

when the j-th PE is in a hidden layer of the network.

The $e_j(k)$ in Equation (43) denotes the derivative of the error function with respect to the output of the j-th PE in the output layer and I in Equation (44) denotes the set of PEs which are driven by the output of the j-th PE.

The implementation of the above learning algorithm must be slightly modified due to the causality constraints. The derivation of the causal algorithm is similar to the derivation of the delayed LMS algorithm, and supposes a temporal shift of the input vector $O_i(k)$ in Equation (42). The delay is computed taking into account the maximum number of unity delays in the inverse propagation paths between the output of the network and the current PE.

The neural model N used in our experiment is a 2:20:1 FIR neural network, with two unity delay taps for each weight filter ($N = 2$). The input signal that fed the network during the learning phase is obtained from uniform noise in the $[-1;+1]$ interval. The noise samples are linearly interpolated over ten time steps. Hence, the frequency characteristics of the input signal match better with the 'low pass' characteristic of the neural network FIR synapses. This type of signal is more appropriate to the FIR learning algorithm. The convergence speed in the learning process with interpolated signal is greater than the convergence speed of the learning sessions that use only additive noise as input to the neural model and to the plant.

In the learning phase, the identification FIR MLP network is considered a one-step-ahead predictor with exogenous input. This means that the plant output is delayed with one time step when feeding the FIR MLP network. The learning algorithm control procedure uses cross-validation between the training error formed in the series-parallel learning stage and the testing error formed in the auto-feedback parallel testing stage. A first initial learning stage is assumed. Then, the learning process stabilizes and the learning error oscillations decrease. An increase in the auto-feedback testing error indicates that the learning process must be stopped. This kind of cross-validation assures optimal performance when the network totally replaces the plant in the control loop. This is the case of many complex plants, for which it is not possible to make on-line adjustments of the control parameters.

3.2.3.1   Feature Extractor

A 5:3 principal component analysis (PCA) network is used as a features extractor. The network is trained with the output of the identification network fed with uniform noise signal with the same characteristics as the input used to train the identification network. The time evolution of the three principal eigenvectors norms is plotted in Figure 24. The unsupervised adaptation procedure of the PCA network is stopped when all the principal eigenvector norms are close to unity.

3.2.3.2 Control Network

A classic 1:5:1 multilayer perceptron is used for synthesizing the control input associated with the reference model input signal. The network is trained using a perturbation procedure. In order to obtain an acceptable high closed-loop gain in the control system initial adaptation stage, the network was pre-trained to approximate the reference model response to the reference input signal. The initial control system output after the MLP pre-training is plotted in Figure 25.

Two training experiments are presented. The first experiment uses the FIR identification network model of the plant and the MLP control network to obtain the data space reference model output as the desired response. The second experiment uses the features space error between the reference model output and the identification model output. The error is obtained using the PCA features extractor, fed with the above signals.

The features space error is plotted in Figure 26 and the corresponding data space errors are plotted in Figure 27. The data space error is better for the plain experiment, but the final system response is approximately the same.

While the data space responses are comparable (see Figure 28), the features space behavior is clearly better expressed when the features extractor is used during learning. This allows improved features space processing results in the subsequent stages, which are based on the features space representation of the plant output.

This experiment shows that it is possible to obtain the desired behavior of the plant in both data and features spaces, following the general features space control strategy. The FIR synapses MLP network models accurately the data space plant behavior. The MLP identification network synthesizes well the plant input, following the data space in the first experiment and the features space in the second experiment. All system components need pre-training stages to assure the initial closed-loop gains.

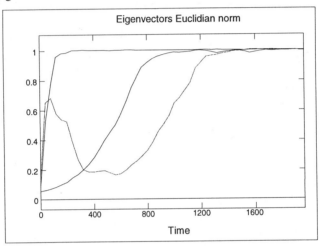

Figure 24: Principal eigenvector norms time evolution during PCA learning.

3.3   Discussion and Concluding Remarks

Depending on the type of features extractor, the proposed control schemes may be used in various applications. When the outer feature space error loop is employed, the features space plant response carries specific quantitative and qualitative information that may be directly used in the subsequent processing stages. Taking into account the features space behavior in the design of the neuro-controller may bring major benefits in real-time control systems. When the built-in features extractor response contributes to the design of the neural controller, the specific features may express various running context attributes, like complex dynamic output bounds for the data space response or alert prescribed temporal patterns. Another promising application area is the design of self-stabilized controllers, by means of stabilization features extractors.

4   Summary

The "features space processing" methodology is a novel approach to design complex adaptive systems for filtering, identification, and control using features space information. Hybrid neural configurations for illustrating this approach were proposed and analyzed.

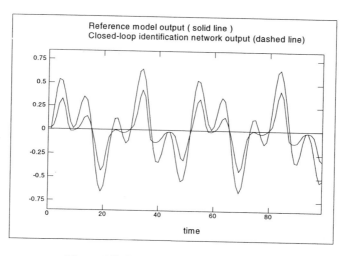

Figure 25: Control system initial output.

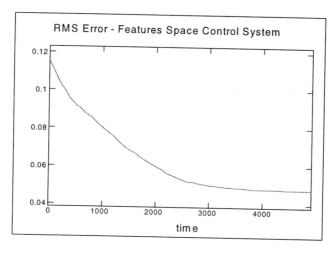

Figure 26: Features space training error.

For the features space filtering schemes, performance in the samples space is comparable to that obtained with classic methods, but the "perception" of the result is enhanced, and the subsequent processing stages in features space are more accurate. Consequently, in applications where the filtered signals are to be further recognized by humans or by machines using a specific pattern-extraction procedure, the recognition results are enhanced. Moreover, the features space filtering systems clearly outperform the classic two-stages adaptive systems in complex, dedicated tasks, when a filtering criterion in the samples space can hide the features of interest, i.e., the signal specificity.

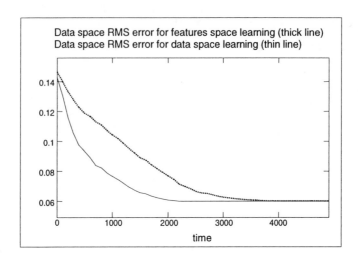

Figure 27: Data space RMS error when training is performed.

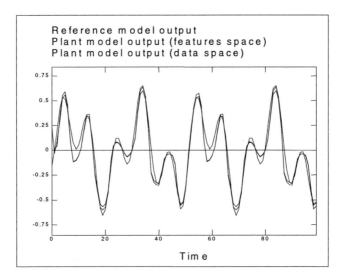

Figure 28: Reference model output (solid line), the plant model output (dashed line) for the features space driven learning, and the plant model output (dotted line) for the data space driven learning.

For the features space filtering schemes, performance in the samples space is comparable to that obtained with classic methods, but the "perception" of the result is enhanced, and the subsequent processing stages in features space are more accurate. Consequently, in applications where the filtered signals are to be further recognized by humans or by machines using a specific pattern-extraction procedure, the recognition results are enhanced. Moreover, the features space filtering systems clearly outperform the classic two-stages adaptive systems in complex, dedicated tasks, when a filtering criterion in the samples space can hide the features of interest, i.e., the signal specificity.

The features space processing principle is applied to the identification and control of unknown plants, when the underlying dynamic model is difficult to obtain analytically. Features space neural identification schemes are discussed. Depending on the features extractor in use, these configurations may employ specific identification tasks. The possible applications range from dynamic output data space conditioning to complex model hypothesis automatic selection and validation. The proposed neural control topologies add an outer control loop on the model reference control schemes. The reference control model becomes an expert model, depending on the desired feature space behavior. Consequently, the overall control system strategy is directly dependent on the desired   reference features model. This allows a broad range of applications. The major aspect treated here refers the accurate features representation of the plant response.

The proposed features space processing architectures are general and may produce many effective implementations. Our neural hybrid approach has the advantage of exploiting the intrinsic nonlinear features of the input data sets and allowing a large class of application specific pattern extraction procedures. Consequently, following the proposed features space processing methodology, many other data-based learning applications are allowed.

References

[1] J.C. Bezdek (1994), "What is computational intelligence?" in *Computational Intelligence: Imitating Life*, J.M. Zurada, R.J. Marks, C.J. Robinson, Eds., Piscataway, NJ, IEEE Press, pp. 1-11.

[2] T. Fukuda and T. Shibata (1992), "Theory and applications for neural networks for industrial control systems," *IEEE Transactions on Industrial Electronics*, vol. 39, no. 6, pp. 472-489.

[3] R.A. Jacobs, M.I. Jordan, S.J. Nowlan, G.E. Hinton (1991), "Adaptive mixtures of local experts," *Neural Computation*, vol. 3, pp. 78-97.

[4] S.-B. Cho, J.H. Kim (1995), "Multiple network fusion using fuzzy logic," *IEEE Transactions on Neural Networks*, vol. 6, no. 2, pp. 497-501.

[5] C. Rodriguez, S. Rementeira, J.I. Martin, A. Lafuente, J. Muguerza, J. Perez (1996), "A modular neural network approach to fault diagnosis," *IEEE Transactions on Neural Networks*, vol. 7, no. 2, pp. 326-340.

[6] Y. Matsuyama (1996), "Harmonic competition: a self-organizing multiple criteria optimization," *IEEE Transactions on Neural Networks*, vol. 7, no. 3, pp. 652-669.

[7] F. Rosenblatt (1958), "The perceptron: a probabilistic model for information storage and organization in the brain," *Psychological Review*, Vol. 65, no. 6., pp. 386-408.

[8]   A. Cichocki, R. Unbehauen (1993), "Neural networks for optimization and signal processing," New York, John Wiley & Sons.

[9]   T.D. Sanger (1989), "Optimal unsupervised learning in a single-layer feedforward neural network," *Neural Networks*, vol. 2., pp. 459-473.

[10] F.M. Silva, and L.B. Almeida (1991), "A distributed decorrelation algorithm," *Neural Networks: Advances and Applications*, E. Gelenbe (Editor), Elsevier Science Publishers B.V., pp. 145-163.

[11] J. Karhunen and J. Joutsensalo (1995), "Generalizations of principal component analysis, optimization problems and neural networks," *Neural Networks*, vol. 8, no. 4, pp. 549-562.

[12] C. Jutten, J. Herault (1991), "Blind separation of sources, part I: an adaptive algorithm based on neuromimetic architecture," *Signal Processing*, vol. 24, no. 1, pp. 1-10.

[13] J. Karhunen, E. Oja, L. Wang, R. Vigario, J. Joutsensalo (1997), "A class of neural networks for independent component analysis," *IEEE Transactions on Neural Networks*, vol. 8, no. 3, pp. 486-504.

[14] Q. Zang, A. Benveniste (1992), "Wavelet networks," *IEEE Transactions on Neural Networks*, vol. 3, no. 6, pp. 889-898.

[15] F.J. Pineda (1988), "Dynamics and architecture for neural computation," *Journal of Complexity*, vol. 4, pp. 216-275.

[16] N. Erdol, F. Basbug (1996), "Wavelet transform based adaptive filters: analysis and new results," *IEEE Transactions on Signal Processing*, vol. 44, no. 9, pp. 2163-2171.

[17] S. Ramprashad, T.W. Parks, R. Shenoy (1996), "Signal modeling and detection using cone classes," *IEEE Transactions on Signal Processing*, vol. 44, no. 2, pp. 329-338.

[18] J.J. Rajan, P.J.W. Rayner (1996), "Generalized features extraction for time-varying autoregressive models," *IEEE Transactions on Signal Processing*, vol. 44, no. 10, pp. 2498-2508.

[19] B.M. Sadler (1996), "Detection in cumulated impulsive noise using fourth- and second-order cumulants," *IEEE Transactions on Signal Processing*, vol. 44, no. 11, pp. 2793-2800.

[20] G.L. Turin (1960), "An introduction to matched filters," *IRE Transactions on Information Theory*, vol. 6, no. 3, pp. 311-329.

[21] H.N. Teodorescu (1986), "Pattern-oriented robust filtering, pattern electro-cardiography: a new method," *13th International Congress on Electrocardiology*, George Washington University, Medical Center, (summary in the Book of Abstracts).

[22] H.N. Teodorescu (1987),"Robustness in terms of fuzziness," *Proc. 2nd IFSA Congress*, Tokyo, Japan, vol. 2, pp. 733-736.

[23] H.N. Teodorescu (1988) "Pattern oriented robust fuzzy filtering," *Proc. of IIZUKA Conference on Fuzzy Systems*, Kyushu, Japan, pp. 57-58.

[24] H.N. Teodorescu (1988), "Pattern-oriented robust fuzzy filtering - digital methods," *Proc. of IIZUKA Conference on Fuzzy Systems*, Kyushu, Japan, pp. 59-60.

[25] H.N. Teodorescu (1988), "Pattern-oriented filtering of images," *Archives of the 10th ISPRS Congress*, Kyoto, Japan.

[26] H.N. Teodorescu (1988), "Robust digital signal conditioning by pattern-oriented filtering," *MESSCOMP'88*, Network GmbH, Wiesbaden (Germany), pp. 2B.71-2B.75.

[27] H.N. Teodorescu, T. Yamakawa, T. Uchino, T. Miki, N. Inagawa, E. Sofron, S. Suceveanu (1994), "Pattern oriented neuro-fuzzy filtering," *Proc. of the 3rd Int. Conf. on Fuzzy Logic, Neural Nets and Soft Computing*, Iizuka, Japan, pp. 657-660.

[28] T. Masters (1993), "Signal and image processing with neural networks," John Wiley & Sons, New York.

[29] J. T.-H. Lo (1994), "Synthetic approach to optimal filtering," *IEEE Transactions on Neural Networks,* vol. 5, no. 5, pp. 803-811.

[30] H.N. Teodorescu, C. Bonciu (1997) "Learning algorithm for RBF networks as features extractors," *First Int. Conf. on Conventional & Knowledge-Based Intelligent Electronic Systems KES'97*, Adelaide, Australia.

[31] C. Bonciu, H.N. Teodorescu (1997), "Features space filtering with RBF networks," *EUFIT'97*, Aachen, Germany.

[32] B. Widrow, A. Stearns (1985), "Adaptive signal processing," Prentice Hall.

[33] T. Poggio, F. Girossi (1990), "Networks for approximation and learning," *Proceedings of the IEEE,* vol. 78, no. 9, pp. 1481-1497.

[34] T. Chen, H. Chen (1995), "Approximation capabilities for the function of several variables, nonlinear functionals, and operators by radial basis function neural network," *IEEE Transactions on Neural Networks*, vol. 6, no. 4, pp. 904-910.

[35] R.M. Sanner, J.J.E. Slotine (1992), "Gaussian networks for direct adaptive control," *IEEE Transactions on Neural Networks,* vol. 3, no. 6, pp. 837-862.

[36] W.K. Pratt (1991), "Digital image processing," (second edition). Wiley Interscience, John Wiley & Sons, New York.

[37] G.L. Dempsey and E.S. McVey (1992), "A Hough transform system based on neural networks," *IEEE Transactions on Industrial Electronics*, vol. 39, no. 6, pp. 522-528.

[38] H.N. Teodorescu, C. Bonciu (1996), "Feedforward neural filter with learning in features space. Preliminary results," *Conference Report of the First Int. Symposium on Neuro-Fuzzy Systems AT'96*, Lausanne, Switzerland.

[39] H.N. Teodorescu, C. Bonciu (1993), "A neural ECG noise filter," *Proc. Int. Symposium on Signals, Circuits & Systems SCS'93*, Iasi, Romania, pp. 345-348.

[40] S. Haykin (1994), "Neural networks: a comprehensive foundation," *IEEE Press.*

[41] M.D Plumbey (1995) "Lyapunov functions for convergence of principal component algorithms," *Neural Networks*, vol. 8., no. 1, pp. 11-23.

[42] K.J. Lang, G.E. Hinton (1988), "The development of the TDNN architecture for speech recognition," *Technical Report CMU-CS-88-152*, Carnegie-Mellon University.

[43] A. Waibel, T. Hanazawa, G.E. Hinton, K. Shikano, K.J. Lang (1987), "Phoneme recognition using time delay neural networks," *Technical Report TR–1-0006*, Japan, Advanced Telecommunications Research Institute.

[44] Y. Suzuki (1995), "Self-organizing QRS-wave recognition in ECG using neural networks," *IEEE Transactions on Neural Networks*, vol. 6, no. 6, pp. 1469-1477.

[45] A. Materka, S. Mizuchina (1996), "Parametric signal restoration using artificial neural networks," *IEEE Transactions on Biomedical Engineering*, vol. 43, no. 4, pp. 357-372.

[46] S.D. Hunt, J.R. Deller (1995), "Selective training of feedforward artificial neural networks using matrix perturbation theory," *Neural Networks*, vol. 8, no. 6, pp. 931-944.

[47] N. Pican, J.C. Fort, F. Alexandre (1994) "An on-line learning algorithm for the orthogonal weight estimation of MLP," *Neural Processing Letters*, vol. 1, no. 1, pp. 21-24.

[48] J.A. Walter, K.J. Schulten (1993), "Implementation of self-organizing neural networks for visuo-motor control of an industrial robot," *IEEE Transactions on Neural Networks*, Vol. 4, no. 1.

[49] K.S. Narendra, K. Parthasarathy, (1990), "Identification and control of dynamical systems using neural networks," *IEEE Transactions on Neural Networks*, Vol. 1, pp. 4-27.

[50] A.U. Levin, K.S. Narendra (1996), "Control of nonlinear dynamical system using neural networks-part II: observability, identification, and control," *IEEE Transactions on Neural Networks*, vol. 7, no. 1, pp. 30-42.

[51] K.S. Narendra and S. Mukhopadhyay (1997), "Adaptive control using neural networks and approximate models," *IEEE Transactions on Neural Networks*, vol. 8, no. 3, pp. 475-485.

[52] Y.-M. Park, M-.S. Choi, K.Y. Lee (1996), "An optimal tracking neuro-controller for nonlinear dynamic system," *IEEE Transactions on Neural Networks*, vol. 7, no. 5, pp. 1099-1110.

[53] R.M. Sanner, J.J.E. Slotine (1992), Gaussian networks for direct adaptive control, *IEEE Transactions on Neural Networks*, vol. 3, pp. 837-862.

[54] N. Sadegh, (1995), "A nodal link perceptron network with applications to control of nonholonomic system," *IEEE Transactions on Neural Networks*, Vol. 6, no. 6, pp.1516-1523.

[55] S.Fabri, V. Kadirkamanathan (1996), "Dynamic structure neural networks for stable adaptive control of nonlinear systems," *IEEE Transactions on Neural Networks*, vol. 7, no. 5, pp. 1151-1168.

[56] E.B. Kosmatopoulos, M.M. Polycarpou, M.A. Christodoulo, P.A. Ioannou, (1995), "High-order neural networks for identification of dynamical systems," *IEEE Transactions on Neural Networks*, vol. 6, no. 2, pp. 422-432.

[57] C.-C. Ku, K.Y. Lee (1995), "Diagonal recurrent neural networks for dynamic system control," *IEEE Transactions on Neural Networks*, vol. 6, no. 1, pp. 144-156.

[58] F.L. Lewis, K. Liu, A. Yesildirek (1995), "Neural net robot controller with guaranteed tracking performance," *IEEE Transactions on Neural Networks*, vol. 6, no. 3, pp. 703-715.

[59] F.L. Lewis, A. Yesildirek, K. Liu, (1996), "Multilayer neural-net robot controller with guaranteed tracking performance," *IEEE Transactions on Neural Networks*, vol. 7, no. 2, pp. 388-399.

[60] K. Tanaka (1996), "An approach to stability criteria of neural-network control systems," *IEEE Transactions on Neural Networks*, vol. 7, no. 3, pp. 629-642.

[61] S. Kuntanapreeda, R.R. Fullmer (1996), "A training rule which guarantees finite-region stability for a class of closed-loop neural-network control systems," *IEEE Transactions on Neural Networks*, vol. 7, no. 3, pp. 745-751.

6 DISCRETE-TIME NEURAL NETWORK CONTROL OF NONLINEAR SYSTEMS

S. Jagannathan
Automated Analysis Corporation
423 S. W. Washington St.
Peoria, IL 61602

F. L. Lewis
Automation and Robotics Research Institute
7300 Jack Newell Blvd. South
Fort Worth, TX 76118

Discrete-time implementation of controllers is of importance since virtually all implementations are processed on a digital computer. This chapter provides a comprehensive treatment of digital controller design for closed-loop control applications that use neural networks (NN). For a class of multi-input–multi-output (MIMO), continuous and discrete-time nonlinear systems, a neural network-based controller in discrete-time which linearizes the system through feedback is presented; in other words, digital control of a class of well-defined nonlinear systems that use NN is presented.

Control action is defined to achieve tracking performance for a state-feedback linearizable, but unknown, nonlinear system. The neural network controller exhibits alearning-while-functioning feature, instead of learning-then-control. The structure of the NN controller is derived using filtered error notions and passivity approach. A uniform ultimate boundedness of the closed-loop system is given in the sense of Lyapunov. The notions of discrete-time passive and dissipative NN are defined and used to show the boundedness of NN weight estimates.

1 Introduction

Adaptive control is an important area of research that is widely pursued. Progress of the adaptive control theory and the availability of microprocessors have led to a series of successful applications in the last two decades in the areas including robotics, aircraft control, process control, and estimation. However, most of the adaptive controllers are designed for systems that are expressed as linear in the unknown parameters. Uncertainty in the regression matrix, disturbances, and nonlinear behavior in a system will cause the performance of an adaptive controller to deteriorate considerably [1,7].

Real-time control of nonlinear plants with unknown dynamics remains a very challenging area of research. Traditionally, the plant dynamics were first modeled and verified through experiments, and then controllers were designed. These controllers are limited by the complexity of the identified model and cannot accommodate variations in parameters. However, these controllers guarantee good tracking performance and robustness.

On the other hand, robust adaptive controllers attempt to learn the plant characteristics while simultaneously achieving the control objectives. Even though the conventional methods guarantee stability for a large class of systems, the system still needs to satisfy assumptions such as linearity in the parameters. In addition, a regression matrix must be computed for each system by often tedious on-line techniques.

In recent years, learning-based control such as neural network and fuzzy logic based controllers has emerged as an alternative to adaptive control. In NN controllers, the learning capability is achieved by training the network with input-output pairs. Therefore, research in NN for control applications is being pursued by several researchers. Notably Narendra et al. [12] emphasizes the use of dynamic backpropagation for tuning neural networks. Sadegh [15] employs an approximate gradient to perform stability analysis, while Polycarpou and Ioannou [13], and Chen and Khalil [2] offer rigorous proofs of performance in terms of tracking error stability and bounded NN weights. All these works use feedforward neural networks, whereas Christodolou [14] uses a recurrent NN structure to establish stability results for the closed-loop system. While some work in the design of NN controllers with system identification has been performed, until recently there have been no results for closed-loop control of nonlinear systems using multilayer NNs, either in the continuous or the discrete-time domain.

Note that even if the system is stable, it must be shown that inputs, outputs, and states remain bounded for a certain class of systems that are feedback linearizable. For example, if the controller is given in the form of $u = \dfrac{N(x)}{D(x)}$, then $D(x)$ must be nonzero for all time, in which case it is termed as a well-defined controller. For feedback linearization, this type of controller is usually needed. Moreover, if any adaptation scheme is implemented to estimate D with $\hat{D}$, then extra precaution is required to guarantee that $\hat{D} \neq 0$ for all time.

In [5-6], it has been shown that NN can effectively control, in discrete-time, a complex nonlinear system without the necessity of a regression matrix. There, the nonlinear systems under consideration are of the form $x(k+1) = f(x(k)) + u(k)$, with input matrix being an identity matrix. Even though in [2] a multilayer NN controller, designed in discrete-time, is employed to control a nonlinear system of the form $x(k+1) = f(x(k)) + g((k))u(k)$, off-line learning phase is initially needed. In addition, if $g(x(k))$ is reconstructed by an adaptive scheme as $\hat{g}(x(k))$, then a local solution can be given by assuming that initial estimates are close to the actual values and they do not leave a feasible invariant set in which $\hat{g}(x(k)) \neq 0$. Even with a good knowledge of the system, it is not easy to choose the initial weights so the NN properly approximates it.   Therefore, off-line learning phase is used to identify the system in order to choose the initial weight values.   In addition, the higher order terms of Taylor series expansion in [2] are considered to be bounded, by neglecting very important terms.   In [2], a multilayer NN is designed in continuous-time to perform feedback linearization, by taking into account all the problems without tight assumptions.

The motivation of this chapter is to extend the results presented in [8,9], where an adaptive controller was designed in discrete-time with the linearity-in-the-parameters (LIP) assumption and tedious computation of a regression matrix.   In this chapter, assumptions used in [2,8,9] are relaxed by using an NN that is linear in the tunable weights and has no off-line learning phase. However, this assumption is far milder than the adaptive control requirement of LIP, since the universal approximation property of NN means any smooth nonlinear function can be reconstructed. Further, this chapter also avoids the problem of $\hat{g}(x(k)) \neq 0$ by appropriately selecting the weight updates as well as the control input.

The controller is composed of a neural net incorporated into a dynamical system, where the structure comes from filtered error notions standard in robotics control literature. It is shown through Lyapunov stability analysis that the tracking error is guaranteed to be bounded. However, the NN weights are guaranteed to be bounded only under a persistency of excitation and the weight update laws being passive. It is extremely difficult to demonstrate that the error in weight updates are passive. Modified update laws are proposed so that the PE condition is relaxed.   This concludes that tracking error and the weight estimates are bounded, implying that all the signals in the closed-loop system are guaranteed to be bounded. Finally, a continuous-time nonlinear system control is considered by using a digital controller designed using a NN and it is shown that the desired tracking performance is guaranteed.

2 Background

Let R denote the real numbers, R^n denote the real n-vectors, and R^{mxn} the real mxn matrices.   Let S be a compact simply connected set of R^n.   With maps $f: S \rightarrow R^k$, define $C^k(S)$ as the space such that f is continuous. We denote by $\| . \|$

any suitable vector norm. Given a matrix $A = [a_{ij}]$, $A \in R^{nxm}$ the Frobenius norm is defined by

$$\| A \|_F^2 = tr(A^T A) = \sum_{i,j} a_{ij}^2, \text{ with } tr() \text{ denoting the trace operation. The associated}$$

inner product is $< A, B >_F = tr(A^T B)$. The Frobenius norm $\| A \|_F$, which is denoted by $\| . \|$ throughout this chapter unless specified explicitly, is the vector 2-norm over the space defined by stacking the matrix columns into a vector, so that it is compatible with the vector 2-norm, that is $\| Ax \| \leq \| A \| . \| x \|$.

2.1 Neural Networks

Given $x_k \in R$, a three-layer neural network as shown in Figure 1 has a net output

$$y_i = \sum_{j=1}^{N_2} [w_{ij} \sigma [\sum_{k=1}^{N_1} v_{jk} x_k + \theta_{vj}] + \theta_{wi}]; \quad i = 1, ..., N_3 \qquad (2.1)$$

with $\sigma(.)$ the activation functions, v_{jk} the first-to-second layer interconnection weights, and w_{ij} the second-to-third layer interconnection weights. The θ_{vm}, θ_{wm}, $m = 1, 2, ..$, are threshold offsets and the number of neurons in layer 1 is N_1, with N_2 the number of hidden-layer neurons. The weights of the neural net are adapted on-line in order to achieve a desired performance.

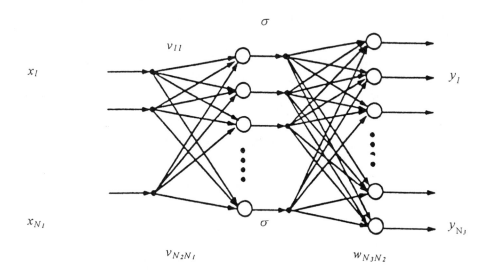

Figure 1: A multilayer neural network.

The neural net equation may be conveniently expressed in the matrix form by defining $x = [x_0, x_1, ..., x_{N_1}]^T$, $y = [y_0, y_1, ..., y_{N_3}]^T$, and weight matrices $W^T = [w_{ij}]^T$, $V^T = [v_{jk}]^T$. Including $x_0 \equiv 1$ in x allows one to include the threshold vector $\theta = [\theta_{v1}, \theta_{v2}, ..., \theta_{vN_1}]^T$ as the first column of V^T, so that V^T contains both weights and thresholds of the first-to-second layer connections. Then,

$$y = W^T \sigma(V^T x),\tag{2.2}$$

with the vector of activation functions $\sigma(z) = [\sigma(z_1), ..., \sigma(z_a)]^T$ defined for a vector $z = [z_1, ..., z_a]^T$. Including a 1 as the first element in $\sigma(z)$, allows one to incorporate the thresholds θ_{wj} as the first column of W^T. Any tuning of W and V then includes tuning of the thresholds as well.

A general function $f(x) \in C^k(S)$ can be written as [3]

$$f(x) = W^T \sigma(V^T x) + \varepsilon(k),\tag{2.3}$$

with $\varepsilon(k)$ a neural net functional reconstruction error vector. If there exists N_2 and constant ideal weights W and V such that $\varepsilon = 0$ for all $x \in S$, then $f(x)$ is in the functional range of the neural net. In general, given a constant real number $\varepsilon \geq 0$, $f(x)$ is within ε_N of the neural net range, if there exists N_2 and constant weights such that for all $x \in R^n$, Equation (2.3) holds with $\|\varepsilon\| \leq \varepsilon_N$. Typical activation functions $\sigma(.)$ are bounded, measurable, and non-decreasing from real numbers onto [0,1], which include, for example, the sigmoid function. Note that the choice of N_2 for a specified $S \subset R^n$, and the neural net reconstruction error bound ε_N are current topics of research.

Define $\phi(x(k)) = \sigma(V^T(k)x(k))$ so that the NN output is

$$\hat{y}(k) = \hat{W}^T(k)\phi(x(k))\tag{2.4}$$

In addition, by choosing the first-to-second layer interconnection weights to be identity (V=I), a two-layer neural network can be obtained. Then, for suitable two-layer NN approximation properties, $\phi(x(k))$ must be a basis [15]. It is easy to choose a basis function $\phi(x(k))$; for instance, it is well-known that radial basis functions (RBF), for example, sigmoids form a basis. If the activation functions $\phi(x(k))$ are selected as a basis [15], then one has the following approximation property for the two-layer NN: let S be a compact simply-connected set of $\Re^n$ and $f(.): S \to \Re^m$. Define $C^m(S)$ as the space of continuous functions $f(.)$. Then, for all $f(.) \in C^m(S)$, there exist weights W such that Equation(2.3) holds. A basis set is defined as follows.

Definition 2.1 (Basis Set of Functions) [15]: Let S be a compact simply connected set of $\Re^n$ and $\phi(x(k)): S \to \Re^L$ be integrable and bounded. Then $\phi(x(k))$ is said to provide a basis for $C^m(S)$, if

a)   constant function can be expressed as Equation (2.3) for finite L.

b)   the functional range of Equation (2.3) is dense in $\Re^m$ for countable L.

In the remainder of this chapter, we use the two-layer net in Equation (2.4) for controls purposes.

2.2 Advantage of NN Over Adaptive Controllers

The contrast between the property, Equation (2.3), and the LIP assumption in adaptive control should be clearly understood. Both are linear in the tunable parameters, but the former is linear in the tunable NN weights, while the latter is linear in the unknown system parameters. The former holds for all functions $f(x) \in C^m(S)$, while the latter holds only for a specific function $f(x)$. In the NN property, the same basis set $\phi(x(k))$ suffices for all $f(x) \in C^m(S)$, while, in the LIP assumption, the regression matrix depends on $f(x)$ and must be computed for each $f(x)$. Therefore, for each dynamical system, for example, a robot arm, one must compute the regression matrix. Therefore, the two-layer NN controller is significantly more powerful than an adaptive controller.

2.3 Stability of Dynamical Systems

In order to formulate the discrete-time controller, the following stability notions are needed [19]. Consider the nonlinear system given by

$$x(k+1)=f(x(k), u(k)), \; y(k)=h(x(k)), \tag{2.5}$$

where $x(k)$ is a state vector, $u(k)$ is the input vector, and $y(k)$ is the output vector. The solution is uniformly ultimately bounded (UUB) if for all $x(k_0) = x_0$, there exists $\varepsilon > 0$ and a number $N(\varepsilon, x_0)$ such that $\|x(k)\| \le \varepsilon$ for all $k \ge k_0 + N$.

Consider now the linear discrete-time varying system given by

$$x(k+1)=A(k)x(k)+B(k)u(k), \; y(k) = C(k)x(k), \tag{2.6}$$

with matrices A(k), B(k), C(k) appropriately dimensioned.

Lemma 2.2: Define $\psi(k_1, k_0)$ as the state-transition matrix corresponding to $A(k)$ for the system, Equation (2.6), i.e., $\psi(k_1, k_0) = \prod_{k=k_0}^{k_1-1} A(k)$. Then, if $\|\psi(k_1, k_0)\| \le 1$, $\forall k_1, k_0 \ge 0$, the system, Equation (2.6) is exponentially stable.

Proof: See [15].

2.4 MIMO Dynamical Systems

Consider an mnth-order multi-input and multi-output state-feedback linearizable, discrete-time, nonlinear system in the multi-variable form

$$x_1(k+1) = x_2(k)$$

$$\cdot$$
$$\cdot$$
$$\cdot$$

$$x_{n-1}(k+1) = x_n(k)$$
$$x_n(k+1) = f(x(k)) + g(x(k))u(k) + d(k),$$ (2.7)
$$y(k) \;=\; x_1(k),$$

where $x(k) = [\, x_1(k),.., x_n(k) \,]^T$ with $x_i(k) \in R^m$; $i=1,...,n,$ $u(k) \in R^m$, and $d(k) \in R^m$ denoting a disturbance vector that acts on the system at the instant k, with $\|d(k)\| \le d_M$ a known constant and $f(.): \Re^{mxn} \to \Re^m$ and $g(.): \Re^{mxn} \to \Re^{mxm}$ unknown smooth functions satisfying the mild assumption

$$\|g(x(k))\| \ge \underline{g} > 0, \forall x,$$ (2.8)

with the assumption that $g(.)$ is invertible and $\underline{g}$ a known lower bound.   The assumption given above on the smooth function $g(x(k))$ implies that $g(x(k))$ is strictly either positive or negative for all x.   From now on, without loss of generality, we will assume that g is strictly positive. Note that at this point there is no general approach to analyze this class of unknown nonlinear systems. Adaptive control, for instance, needs an additional LIP assumption [1].

Note, most physical systems, including robotic systems, both in continuous and discrete-time, can be expressed in the above form except that we would like to design a controller in discrete-time (digital controller), usually for a continuous-time nonlinear system. In fact, it is extremely difficult to discretize a nonlinear system or to model a system in discrete time.

2.5 Tracking Problem

Feedback linearization will be used to perform output tracking, whose objective can be described as: given a desired trajectory in terms of output, $x_{nd}(k)$, and its delayed values, find a control input $u(k)$, so that the system tracks the desired trajectory with an acceptable bounded error in the presence of disturbances while all the states and controls remain bounded. In order to continue, the following mild assumptions are employed.

Assumptions:

1. The sign of $g(x(k))$ is known.

2. The desired trajectory vector with its delayed values is assumed to be available for measurement and bounded by an upper bound.

Given the desired trajectory $x_{nd}(k)$ and its delayed values, define the tracking error as

$$e_n(k) = x_n(k) - x_{nd}(k). \tag{2.9}$$

It is typical in robotics to define a so-called filtered tracking error, as $r(k) \in R^m$ and given by

$$r(k) = e_n(k) + \lambda_1 e_{n-1}(k) + ... + \lambda_{n-1} e_1(k), \tag{2.10}$$

where $e_{n-1}(k), ..., e_1(k)$ are the delayed values of the error $e_n(k)$, and $\lambda_1, ..., \lambda_{n-1}$ are constant matrices selected so that $|Z^{n-1} + \lambda_1 Z^{n-2} + ... + \lambda_{n-1}|$ is stable. Equation (2.10) can be further expressed as

$$r(k+1) = e_n(k+1) + \lambda_1 e_{n-1}(k+1) + ... + \lambda_{n-1} e_1(k+1). \tag{2.11}$$

Using (2.7) in (2.11), the dynamics of the mnth-order MIMO system can be expressed in terms of the tracking error as

$$r(k+1) = f(x(k)) - x_{nd}(k+1) + \lambda_1 e_n(k) + ... + \lambda_{n-1} e_2(k) + g(x(k))u(k) + d(k) \tag{2.12}$$

Equation (2.12) can be rewritten as

$$r(k+1) = f(x(k)) + g(x(k))u(k) + d(k) + Y_d, \tag{2.13}$$

where

$$Y_d \equiv -x_{nd}(k+1) + \sum_{i=0}^{n-2} \lambda_{i+1} e_{n-i}. \tag{2.14}$$

If we exactly knew the nonlinear functions $f(x)$ and $g(x)$, and when no disturbances are present, the control input u(k) can be selected as

$$u(k) = \frac{1}{g(x)}(-f(x) + v(k)), \tag{2.15}$$

where $v(k)$ is another auxiliary input which is given by

$$v(k) = k_v r(k) - Y_d. \tag{2.16}$$

Then the filtered tracking error $r(k)$ goes to zero exponentially by properly selecting the gain matrix k_v. Since, in our problem, these functions are not known *a priori*, the control input, u(k) can be given by

$$u(k) = \frac{1}{\hat{g}(x)}(-\hat{f}(x) + v(k)), \tag{2.17}$$

with $\hat{f}(x)$ and $\hat{g}(x)$ being the estimates of $f(x(k))$ *and* $g(x(k))$, respectively. It is well-known that, even in adaptive control of linear systems, guaranteeing the boundedness of $\hat{g}(x)$ away from zero is an important issue and it is discussed in the next section.

Equation (2.13) can be rewritten using
$$r(k+1) = v(k) - v(k) + f(x(k)) + g(x(k))u(k) + d(k) + Y_d. \tag{2.18}$$
Substituting (2.16),(2.17) for v(k), (2.18) can be rewritten as
$$r(k+1) = k_v r(k) + \tilde{f}(x(k)) + \tilde{g}(x(k))u(k) + d(k), \tag{2.19}$$
where the functional estimation errors are given by
$$\tilde{f}(x(k)) = f(x(k)) - \hat{f}(x(k)), \tag{2.20}$$
and
$$\tilde{g}(x(k)) = g(x(k)) - \hat{g}(x(k)). \tag{2.21}$$
This is an error system wherein the filtered tracking error is driven by the functional estimation error.

In the remainder of this chapter, (2.19) is used to focus on selecting NN tuning algorithms that guarantee stability of the filtered tracking error $r(k)$. Then, since (2.10), with the input considered as $r(k)$ and the output $e_n(k)$ and its delayed values, describes a stable system and it is clear that standard techniques [8-9] guarantee that $e_n(k)$ and their delayed values exhibit stable behavior.

3 Neural Network Controller Design

In the remainder of this chapter, a two-layer NN, which is linear in the tunable weights, is considered as a first step to bridging the gap between adaptive and NN control. However, this is a far milder assumption than the adaptive control requirement of linearity in the parameters, since the universal approximation property of the NN means that any smooth nonlinear function can be reconstructed. This case is treated in [2] using discrete-time update laws for NN weights along with certainty equivalence. In addition, the bound on the higher order terms of the Taylor series in [2] is neglected. Furthermore, the bound on the tracking error and NN weights depend upon the size of the dead zone, and this choice of the size of the dead zone is arbitrary. In general, the assumptions made in this chapter are far milder and also the Lyapunov stability analysis is presented similar to that in continuous-time.

In this section, stability analysis by Lyapunov's direct method is performed for weight tuning algorithms developed based on the delta rule. It is possible to show only the boundedness of filtered tracking error using the stability analysis. In order to show a bound on the NN weights, one needs to select appropriate weight update laws and the persistency of excitation condition is required. This is extremely difficult. However, it is shown in the next section that the NN weight update laws are passive by appropriately selecting them. Then using the persistency of excitation condition and invoking the passivity theorem, the boundedness of NN weights is guaranteed.

Assume that there exists constant ideal weights W_f and W_g so that the nonlinear functions in (2.7) can be written as

$$f(x(k)) = W_f^T \phi_f(k) + \varepsilon_f(k), \tag{3.1}$$

and

$$g(x(k)) = W_g^T \bar{\phi}_g(k) + \varepsilon_g(k), \tag{3.2}$$

where $\phi_f(k)$ and $\bar{\phi}_g(k)$ provides a suitable basis and $\|\varepsilon_f(k)\| \le \varepsilon_{Nf}$, $\|\varepsilon_g(k)\| \le \varepsilon_{Ng}$, with the bounding constants ϕ_{Nf}, ϕ_{Ng} known. This allows one to select a simple NN structure based on a reduced basis and thereafter compensating for the increased magnitude of ε_{Nf}, and ε_{Ng} by using the gain matrix k_v.

In general, the required size of the NN for given ε_{Nf}, and ε_{Ng} is currently unknown, but an estimated size (the number of hidden-layer neurons) is commonly used. Therefore, we define the problem as: given the size of the NN, there exists constant ideal weights (may not be unique [17-18]) such that (3.1) and (3.2) hold. Note that ε_{Nf}, and ε_{Ng} become small as the number of neurons in the hidden-layers for both NNs, one for estimating $f(x(k))$ and the other for $g(x(k))$, increases.

Lemma 3.1: The extremum of $y e^{-\gamma y}$ for $\forall y > 0$ can be found as $y = \dfrac{1}{\gamma}$.

Proof: The extremum of $y e^{-\gamma y}$ for $\forall y > 0$ can be found by solving following equation:

$$\frac{\partial(y e^{-\gamma y})}{\partial y} = (1 - \gamma y) e^{-\gamma y} = 0, \tag{3.3}$$

Solving (3.3) results in $y = \dfrac{1}{\gamma}$, and it is a maximum value.

Lemma 3.2 (Bound on NN input $x(k)$): For each time k, $x(k)$ is bounded by

$$\|x(k)\| \le d_{01} + d_{11}\|r(k)\| \le q_B + l_1\|r(0)\| + d_{11}\|r(k)\|, \tag{3.4}$$

for computable positive constants l_1, d_{01}, d_{12}.

Proof: The solution of the LTI system (2.10) for an initial value vector is

$$e(k) = \Phi(k)e(0) + \sum_{i=0}^{k-1} \Phi(k-i-1)r(i), \quad k > k_0, \tag{3.5}$$

where $e(0) = x^1(0) - x^d(0)$. Thus,

$$\|e(k)\| \leq \|e_0(k)\| + \frac{\|r(k)\|}{\sigma_{min}(L)} , \tag{3.6}$$

where $\sigma_{min}(L)$ is the minimum singular value of Λ. Using the NN inputs $x(k)$, the bound can be obtained as

$$\|x(k)\| \leq (1 + \sigma_{max}(L))\|e(k)\| + q_B + \|r(k)\| \tag{3.7}$$

which results in (3.4) where

$$d_{01} = [1 + \sigma_{max}(L)]\|e_0\| + q_B , \tag{3.8}$$

and

$$d_{11} = 1 + \frac{1}{\sigma_{min}(L)} + \frac{\sigma_{max}(L)}{\sigma_{min}(L)} \tag{3.9}$$

with

$$l_1 = \frac{1 + \sigma_{max}(L)}{\sigma_{min}(L)} . \tag{3.10}$$

Let $x \in U$ be a compact subset of R^n. Assume that $h(x(k)) \in C^\infty [U]$, i.e., a smooth function $U \to R$, so that the Taylor series expansion of h exists. One can derive, using (2.4), that $\|x(k)\| \leq d_{01} + d_{11}\|r(k)\|$. Then using the bound on $x(k)$ and expressing $h(x(k))$ as (3.2), yields an upper bound for $h(x)$ as

$$\|h(x(k))\| = \|W_h^T \phi(k) + \varepsilon_h(k)\| \leq C_{01} + C_{11}\|r\|, \tag{3.11}$$

with C_{01} and C_{11} computable constants. In addition, the hidden layer activation functions, such as radial basis functions, sigmoids, are bounded by an known upper bound

$$\|\phi(k)\| \leq \phi_{max} . \tag{3.12}$$

3.1 NN Controller Structure and Error System

Define the functional estimates given by

$$\hat{f}(x(k)) = \hat{W}_f^T(k)\phi_f(k) , \tag{3.13}$$

and

$$\hat{g}(x(k)) = \hat{W}_g^T(k)\overline{\phi}_g(k) , \tag{3.14}$$

with $\hat{W}_f(k)$ and $\hat{W}_g(k)$ representing the actual weights. Then the closed-loop filtered dynamics (2.19) become

$$r(k+1) = k_v r(k) + \tilde{W}_f^T(k)\phi_f(k) + \tilde{W}_g^T(k)\overline{\phi}_g(k)u(k) + (\varepsilon_f(k) + \varepsilon_g(k)u(k)) + d(k) \tag{3.15}$$

where

$$\tilde{W}_f(k) = W_f - \hat{W}_f(k),$$ (3.16)

and

$$\tilde{W}_g(k) = W_g - \hat{W}_g(k).$$ (3.17)

The proposed NN controller structure is shown in Figure 2. The output of the plant is processed through a series of delays in order to obtain the past values of the output, and fed as inputs to both NNs for constructing $f(x)$ and $g(x)$ so that the nonlinear functions in (2.7) can be suitably approximated. Thus, the NN controller derived in a straightforward manner using filtered error notion naturally provides a dynamical structure. The next step is to determine the weight updates for both NN so that the tracking performance of the closed-loop filtered error dynamics is guaranteed.

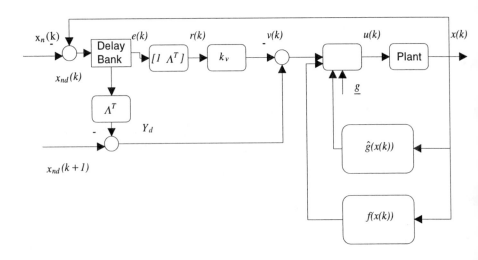

Figure 2: Neural network controller structure.

3.2 Well-Defined Control Problem

In general, boundedness of $x(k)$, $\hat{W}_f(k)$ and $\hat{W}_g(k)$ does not imply stability of the closed-loop system, because control law (2.17) is not well defined when $\hat{g}(\hat{W}_g, x) = 0$. Therefore, some attention must be taken to guarantee the boundedness of the controller as well. To overcome this problem, several techniques exist in the literature that assure local or global stability with additional knowledge. First, if the bounds on the function $g(x)$ are known, then $\hat{g}(\hat{W}_g, x)$ may be set to a constant, and a robust-adaptive controller will avoid this problem. This is not an accurate method because the bounds on the function are not known *a priori*.

If $g(x)$ is reconstructed by an adaptive scheme, then a local solution can be generated by assuming that the initial estimates are close to the actual values. These values do not leave a feasible invariant set in which the $\hat{g}(\hat{W}_g, x)$ is not equal to zero [11], or lie inside a region of attraction of a stable equilibrium point which forms a feasible set [10]. Unfortunately, even with good knowledge of the system, it is not easy to pick initial weight values such that NN would approximate it.

The most popular way to avoid the problem is to project $\hat{W}_g(k)$ inside an estimated feasible region by properly selecting the weight values [13]. A shortcoming of this approach is that the actual $\hat{W}_g(k)$ does not necessarily belong to this set, which then renders a suboptimal solution.

3.3 Proposed Controller

In order to guarantee the boundedness of $\hat{g}(x)$ away from zero, for all well-defined values of $x(k)$, $\hat{W}_f(k)$ and $\hat{W}_g(k)$, the control input in (2.17) is selected in terms of another control input, $u_c(k)$, and a robustifying term, $u_r(k)$ as

$$u(k) = u_c(k) + \frac{u_r(k) - u_c(k)}{2} e^{\gamma(\|u_c(k)\| - s)} \quad if \quad I = 1,$$

$$= u_r(k) - \frac{u_r(k) - u_c(k)}{2} e^{-\gamma(\|u_c(k)\| - s)} \quad if \quad I = 0, \quad (3.18)$$

where

$$u_c(k) = \frac{1}{\hat{g}(x)}(-\hat{f}(x) + v(k)), \quad (3.19)$$

$$u_r(k) = -\mu \frac{\|u_c(k)\|}{\underline{g}} sgn(r(k)), \quad (3.20)$$

the indicator I in (3.18) is

$$I \quad = \quad 1, If \; \|\hat{g}(x(k))\| \geq \underline{g} \text{ and } \|u_c(k)\| \leq s$$
$$= \quad 0, Other\, wise \quad (3.21)$$

with $\gamma < ln\dfrac{2}{s}$, $\mu > 0$, and $s > 0$ are design parameters. These modifications in the control input are necessary in order to ensure that the functional estimate $\hat{g}(x)$ is bounded away from zero.

The intuition behind this controller is as follows: when $\|\hat{g}(x(k))\| \geq \underline{g}$ and $\|u_c(k)\| \leq s,$ then the total control input is set to $u_c(k)$; otherwise, the control is smoothly switched to the auxiliary input u_r due to the additional term in (3.18).

This results in a well-defined control everywhere, and the uniformly ultimately boundedness of the closed-loop system can be shown by appropriately selecting the NN weight update algorithms. First, an assumption is presented here which specifies the region of convergence of the NN controller that is designed in this section.

Assumption 3.3 (Initial Condition Requirement): Suppose that the desired trajectory is bounded by q_B. Define constants l_1 and d_{11} by Lemma 3.2. Let the NN approximation property (2.2) hold for the functions $f(x)$ and $g(x)$ with an accuracy of ε_{Nf}, and ε_{Ng}, respectively, for all x inside the ball of radius $b_x > q_B$.

Let the initial tracking error satisfy $\|r(0)\| < \dfrac{(b_x - q_B)}{(l_1 + d_{11})}$.

The set S_r specifies the set of allowed initial tracking errors $r(0)$. Note that the approximation accuracy of the NN determines the allowed magnitude of the initial tracking error $r(0)$. By suitably selecting the number of membership functions, the NN accuracies ε_{Nf} and ε_{Ng} will be smaller for a large b_x. Note the dependence of S_r on the proportional and derivative (PD) gain ratio Λ —both l_1 and d_{11}.

3.4 Weight Updates for Guaranteed Performance

It is necessary to demonstrate that the tracking error r(k) is suitably small and that the NN weights $\hat{W}_f(k)$ and $\hat{W}_g(k)$ remain bounded. In order to proceed, the following are needed.

Lemma 3.4:   If $A(k) = I - \alpha\phi(k)\phi^T(k)$ in (2.6), where $0 < \alpha < 2$ and $\phi(k)$ is a vector of basis functions, then $\|\psi(k_1, k_0)\| < 1$ is guaranteed if there is an $L > 0$ such that $\sum\limits_{k=k_0}^{k_1+L-1} \phi(k)\phi^T(k) > 0$ for all k. Then Lemma 2.2 guarantees stability of the system (2.6).

Proof: See[15].

Definition 3.5: An input sequence $x(k)$ is said to be persistently exciting if there are $\lambda > 0$ and an integer $k_1 \geq 1$ such that

$$\lambda_{min}[\sum_{k=k_0}^{k_1+L-1} \phi(k)\phi^T(k)] > \lambda, \forall k_0 \geq 0, \tag{3.22}$$

where $\lambda_{min}(P)$ represents the smallest eigen value of P.

Note: PE is exactly the stability condition needed in Lemma 3.1.

In the following, it is taken that the NN reconstruction error bounds ε_{Nf} and ε_{Ng} and the disturbance bound d_M are nonzero. Let the NN weight updates for $f(x(k))$ be given by

$$\hat{W}_f(k+1) = \hat{W}_f(k) + \alpha \phi_f(k) r^T(k+1),\tag{3.23}$$

and NN weight updates for $g(x(k))$ are provided by

$$\hat{W}_g(k+1) = \hat{W}_g(k) + \beta \bar{\phi}_g(k) u_c(k) r^T(k+1), \qquad \text{If } I = 1$$

$$= \hat{W}_g(k), \qquad\qquad\qquad\qquad Otherwise \tag{3.24}$$

where $\alpha > 0$ and $\beta > 0$ are adaptation gains or learning rate parameters.

Theorem 3.6:   Assume that the state feedback linearizable system has a representation in the controllability canonical form as in (2.7) and the control input is given by (3.18). Let the mild assumptions given in this chapter hold. Assume that the error in weight updates (3.23) and (3.24) yield passive NN, and the vectors $\phi_f(k), \bar{\phi}_g(k) u_c(k)$ of the NN for $f(x(k))$ and $g(x(k))$, respectively, are persistently exciting. Let the gain satisfy

$$K_{vmin} > \frac{(\varepsilon_N + d_M)(l_1 + d_{11})}{(b_x - q_B)}.$$

Then, the filtered tracking error $r(k)$ and the weight estimates $\hat{W}_f(k)$ and $\hat{W}_g(k)$, are UUB provided that the following conditions hold:

(a) $\beta \left\| \bar{\beta}_g(k) u_c(k) \right\|^2 = \beta \left\| \phi_g(k) \right\|^2 < 1$ $\qquad\qquad\qquad\qquad\qquad$ (3.25)

(b) $\eta < 1,$ $\qquad\qquad\qquad\qquad\qquad\qquad\qquad\qquad\qquad\qquad\qquad$ (3.26)

where

$$\eta = \alpha \left\| \phi_f(k) \right\|^2 + \beta \left\| \bar{\phi}_g(k) u_c(k) \right\|^2, \qquad\qquad \text{for } I = 1,$$

$$= \alpha \left\| \phi_f(k) \right\|^2, \qquad\qquad\qquad\qquad\qquad\qquad \text{for } I = 0.$$

and

(c) $Max((a_4, b_0) < 1,$ $\qquad\qquad\qquad\qquad\qquad\qquad\qquad\qquad\qquad$ (3.27)

with a_4, b_0 design parameters chosen using the gain matrix k_{vmax}.

Remarks:

1. For practical purposes, (A.53) can be considered as bounds for $r(k)$, $\tilde{W}_f(k)$ and $\tilde{W}_g(k)$.

2. The NN reconstruction errors and the bounded disturbances are all embodied in the constants given by $\delta_r, \delta_f, and\ \delta_g$. Note that the bound on the tracking error will be kept small if the closed-loop poles are placed closer to the origin.

3. If the switching parameter s is chosen small, it will limit the control input and will result in a large tracking error, which will result in undesirable closed-loop performance. Larger value of s results in saturation of the control input $u(k)$.
4. Global stability of the closed-loop system is shown without making assumptions on the initial NN weight values. The NNs can be easily initialized as $\hat{W}_f(k) = 0$ and $\hat{W}_g(k) > g^{-1}(g)$. In addition, the NNs presented here do no need an initial off-line learning phase. No assumptions such as the existence of a invariant set, region of attraction, or a feasible region is needed.

In the following theorem, the weight update laws presented in (3.24) and (3.25) are modified to avoid the PE condition. Note that in adaptive control literature [1,4,16], several modifications to the parameter tuning methods have been suggested which include dead zone, σ, and ε. In [24], the ε-modification approach is employed for NN in continuous-time. In this chapter, an approach similar to ε-modification employed in discrete-time [9], is used for the NN control. Let the new NN weight updates for $f(x(k))$ be given by

$$\hat{W}_f(k+1) = \hat{W}_f(k) + \alpha\,\phi_f(k)r^T(k+1) - \delta\left\|I - \alpha\,\phi_f(k)\phi_f(k)^T\right\|\hat{W}_f(k), \qquad (3.28)$$

and NN weight updates for g(x(k)) be provided by

$$\hat{W}_g(k+1) = \hat{W}_g(k) + \beta\,\phi_g(k)r^T(k+1) - \rho\left\|I - \beta\,\phi_g(k)\phi_g^T(k)\right\|\hat{W}_g(k), \;\; If\; I = 1$$

$$= \hat{W}_g(k), \qquad\qquad\qquad\qquad\qquad Otherwise \quad (3.29)$$

where, $á > 0, â > 0, ä > 0$ and $ñ > 0$, are adaptation gains or learning rate parameters.

Theorem 3.7:   Assume that the state feedback linearizable system has a representation of the form (2.7) and the control input is given by (3.18). Let the assumptions given in this chapter hold, and consider the weight tuning laws (3.28) and (3.29).   Then the filtered tracking error $r(k)$ and the weight estimates, $\hat{W}_f(k)$ and $\hat{W}_g(k)$, are UUB provided that the following conditions hold:

(a) $\beta\left\|\phi_g(k)\right\|^2 < 1,$ (3.30)

(b) $\eta + max(\,P_1, P_3, P_4) < 1,$ (3.31)

(c) $0 < \delta < 1,$ (3.32)

(d) $0 < \rho < 1,$ (3.33)

(e) $Max(\,a_4, b_0) < 1,$ (3.34)

where

$$\eta = \alpha\left\|\phi_f(k)\right\|^2 + \beta\left\|\phi_g(k)\right\|^2, \qquad\qquad for\; I = 1,$$

$$= \alpha\left\|\phi_f(k)\right\|^2, \qquad\qquad\qquad\qquad for\; I = 0, \qquad (3.35)$$

with

$$P_1 = 2(\delta\left\|I - \alpha\,\phi_f(k)\phi_f^T(k)\right\| + \rho\left\|I - \beta\phi_g(k)\phi_g^T(k)\right\|). \qquad (3.36)$$

$$P_3 = (\eta + \delta \| I - \alpha \phi_f(k) \phi_f^T(k) \|)^2,$$

(3.37)

$$P_4 = (\eta + \rho \| I - \alpha \phi_g(k) \phi_g^T(k) \|)^2,$$

(3.38)

and a_4, b_0 are design parameters selected using the gain matrix k_{vmax}, and the relationship is given during the proof.

Remark: No PE condition is needed to show the boundedness of parameter estimates.

4 Passivity of Dynamical Systems

Passivity is essential in a closed-loop system as it guarantees the boundedness of signals, and, hence, suitable performance even in the presence of additional unforeseen bounded disturbances (i.e., NN robustness). In addition, the assumption made in the previous theorem can be justified by showing that the NN error in weight updates for $f(x)$ and $g(x)$ is passive and subsequently showing that the NN outputs are bounded. Therefore, in this section, passivity properties of the NN are explored using the weight updates (3.23), (3.24), (3.28) and (3.29). The closed-loop error system (2.19) is shown in Figure 3. Note that now the NN is in the standard feedback configuration as opposed to the NN controller of Figure 2.

4.1 Passive Systems

Some aspects of passivity [4] are important. The set of time instants that are of interest is $Z_+ = \{0,1,2...\}$. Consider the Hilbert space $l_2^n(Z_+)$ of sequences $y: Z_+ \rightarrow \Re^n$ with the inner product $<.,.>$ defined by

$$< y, u > = \sum_{k=0}^{\infty} y(k)^T u(k).$$

(4.1)

A norm on $l_2^n(Z_+)$ is defined by $\|u\| = \sqrt{<u,u>}$. Let P_T denote the operator that truncates the signal u at time T. We have

$$P_T u(k) = u(k), k < T \qquad\qquad = 0, k \geq T.$$

(4.2)

The basic space $l_{2e}^n(Z_+)$ is given by an extension of $l_2^n(Z_+)$ according to

$$l_{2e}^n(Z_+) = \{u: Z_+ \rightarrow \Re^n | \forall T \in Z_+, P_T u \in l_2^n(Z_+)\}.$$

(4.3)

It is convenient to use the notation $u_T = P_T$ and $< y, u >_T = < y_T, u_T >$.

Define the energy supply function $E: l_{2e}^m(Z_+) \times l_{2e}^p(Z_+) \times Z_+ \rightarrow \Re$. A useful energy function E is in the quadratic form defined in [4] as

$$E(u, y, T) = < y, Su >_T + < u, Ru >_T,$$

(4.4)

with S and R appropriately dimensioned matrices. In general, a system with input $u(k)$ and output $y(k)$ is said to be passive if it satisfies an equality defined in [5-6] as the power form

$$\Delta V = y^T(k)Su(k) + u^T(k)Ru(k) - h(k), \tag{4.5}$$

with $V(k)$ lower bounded and $h(k) \geq 0$. In other words,

$$E(u, y, T) \geq \sum_{k=0}^{T} h(k) - \gamma_0^2, \quad \forall T \geq 0. \tag{4.6}$$

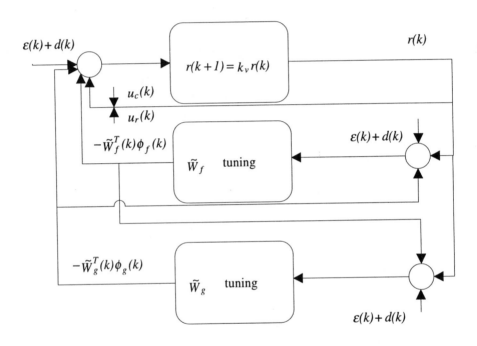

Figure 3: Neural network closed-loop system.

A system is defined to be dissipative if it is passive and, in addition,

$$E(u, y, T) \neq 0 \Rightarrow \sum_{k=0}^{T} h(k) > 0, \quad \forall T > 0. \tag{4.7}$$

4.2 Passivity of the Closed-loop System and NN

In general, the closed-loop tracking error system (2.19) can also expressed as

$$r(k+1) = k_v r(k) + \varsigma_0(k), \tag{4.8}$$

where

$$\varsigma_0(k) = \tilde{f}(x(k)) + \tilde{g}(x(k))u(k) + d(k). \tag{4.9}$$

The next dissipativity result holds for this system.

Theorem 4.1: The closed-loop tracking error system (4.8) is dissipative from $\varsigma_0(k)$ to $k_v r(k)$ provided that

$$\lambda_{max}(k_v^T k_v) < 1. \tag{4.10}$$

Proof: Select a Lyapunov function candidate

$$V = r^T(k)r(k). \tag{4.11}$$

The first difference is given by

$$\Delta V = r^T(k+1)r(k+1) - r^T(k)r(k). \tag{4.12}$$

Substituting (4.8) in (4.12) yields

$$\Delta V = -r^T(k)[I - k_v^T k_v]r(k) + 2r^T(k)k_v \varsigma_0(k) + \varsigma_0^T \varsigma_0. \tag{4.13}$$

Note that (4.13) is in the power form defined in (4.5), with $h(k) > 0$ as long as the condition (4.10) holds. Hence, (4.8) is a dissipative system.

Even though the closed-loop filtered tracking error system (4.8) is dissipative, the overall system is not passive unless the weight update laws guarantee the passivity of the lower block in Figure 3. It is usually difficult to demonstrate that the error-in-weight updates are passive. However, in the upcoming theorem, it is, in fact, shown that the delta-rule-based tuning algorithms (3.23) and (3.24) yield a passive NN.

Theorem 4.2: The weight tuning algorithms (3.23) and (3.24) make dissipative map from, $k_v r(k) + \bar{e}_g(k) + g u_d(k) + \varepsilon(k) + d(k)$, for the case of (3.23) and $k_v r(k) + \bar{e}_f(k) + g u_d(k) + \varepsilon(k) + d(k)$

for the case of (3.24) to $-\tilde{W}_f^T(k)\phi_f(k) and -\tilde{W}_g^T(k)\phi_g(k)$, respectively.

Proof: Define a Lyapunov function candidate

$$V = \frac{1}{\alpha} tr[\tilde{W}_f^T(k)\tilde{W}_f(k)], \tag{4.14}$$

whose first difference

$$\Delta V = \frac{1}{\alpha} tr[\tilde{W}_f^T(k+1)\tilde{W}_f(k+1) - \tilde{W}_f^T(k)\tilde{W}_f(k)]. \tag{4.15}$$

Using the error in update law in (3.23), and on simplification one obtains

$$\Delta V = -(2 - \alpha\phi_f^T(k)\phi_f(k))(-\tilde{W}_f^T(k)\phi_f(k))^T((-\tilde{W}_f^T(k)\phi_f(k)) + \alpha\phi_f^T(k)\phi_f(k)$$

$$(k_v r(k) + \bar{e}_g(k) + g u_d(k) + \varepsilon(k) + d(k))^T(k_v r(k) + \bar{e}_g(k) + g u_d(k) + \varepsilon(k) + d(k))$$

$$+2(1 - \alpha\phi_f^T(k)\phi_f(k))(-\tilde{W}_f^T(k)\phi_f(k))(k_v r(k) + \bar{e}_g(k) + g u_d(k) + \varepsilon(k) + d(k)), \tag{4.16}$$

which is in the power form given by (4.5) as long as the conditions (3.25) through (3.27) hold. This condition is exactly the PE condition given in (3.22).

Similarly, we can prove that the error in weight update law in (3.24) is dissipative as long as the PE condition given in (3.22) is satisfied. In fact, one chooses the first difference as

$$\Delta V = \frac{1}{\beta} tr[\widetilde{W}_g^T(k+1)\widetilde{W}_g(k+1) - \widetilde{W}_g^T(k)\widetilde{W}_g(k)] .$$

Using the error in update law (3.24), and on simplification one obtains

$$\Delta V = -(2 - \beta \phi_g^T(k)\phi_g(k))(-\widetilde{W}_g^T(k)\phi_g(k))^T((-\widetilde{W}_g^T(k)\phi_g(k)) + \alpha \phi_g^T(k)\phi_g(k)$$

$$(k_v r(k) + \overline{e}_f(k) + g\, u_d(k) + \varepsilon(k) + d(k))^T (k_v r(k) + \overline{e}_f(k) + g\, u_d(k) + \varepsilon(k) + d(k))$$
$$+2(1 - \alpha \phi_g^T(k)\phi_g(k))(-\widetilde{W}_g^T(k)\phi_g(k))$$
$$(k_v r(k) + \overline{e}_f(k) + g\, u_d(k) + \varepsilon(k) + d(k)),$$ (4.17)

which is in power form defined in (4.5).

Thus, the weight error block is dissipative and the closed-loop filtered tracking error system (4.8) in Figure 3 is dissipative; this guarantees the dissipativity of the closed-loop system [16]. By employing the passivity theorem [16], one can conclude that the input/output signals of each block are bounded as long as the external inputs are bounded. ince the external inputs, ε_{Nf} and ε_{Ng}, are bounded, the outputs of each block will be bounded.

An NN is defined to be passive, if, in the error formulation, it guarantees the passivity of the lower blocks in Figure 3. To show that the NN are dissipative, an extra PE condition is needed to guarantee boundedness of the weight estimates. This passivity property of the NN is very important in order to show the boundedness of weight estimates when extra bounded disturbances are present.

Theorem 4.3: The weight tuning algorithms (3.28) and (3.29) make SSP map from
$$k_v r(k) + \overline{e}_g(k) + g\, u_d(k) + \varepsilon(k) + d(k),$$
for the case of (3.28), and
$$k_v r(k) + \overline{e}_f(k) + g\, u_d(k) + \varepsilon(k) + d(k)$$
for the case of (3.29), to
$$-\widetilde{W}_f^T(k)\phi_f(k) \, and \, -\widetilde{W}_g^T(k)\phi_g(k),$$
respectively.

Proof: Define a Lyapunov function candidate

$$V = \frac{1}{\alpha} tr[\, \tilde{W}_f^T(k) \tilde{W}_f(k)], \tag{4.18}$$

whose first difference is

$$\Delta V = \frac{1}{\alpha} tr[\, \tilde{W}_f^T(k+1) \tilde{W}_f(k+1) - \tilde{W}_f^T(k) \tilde{W}_f(k)]. \tag{4.19}$$

Using the error in update law in (3.28) and on simplification, one obtains

$$\Delta V = -(2 - \alpha \phi_f^T(k) \phi_f(k))(-\tilde{W}_f^T(k) \phi_f(k))^T((-\tilde{W}_f^T(k) \phi_f(k)) + 2[(1 - \alpha \phi_f^T(k) \phi_f(k))$$

$$- \delta \| I - \alpha \phi_f(k) \phi_f^T(k) \|](-\tilde{W}_f^T(k) \phi_f(k)) \, (k_v r(k) + \bar{e}_g(k) + g \, u_d(k) + \varepsilon(k) + d(k))$$

$$+ (k_v r(k) + \bar{e}_g(k) + g \, u_d(k) + \varepsilon(k) + d(k))^T \, (\alpha \phi_f^T(k) \phi_f(k)$$

$$(k_v r(k) + \bar{e}_g(k) + g \, u_d(k) + \varepsilon(k) + d(k)) + 2\delta \| I - \alpha \phi_f(k) \phi_f^T(k) \| W_f^T \phi_f(k))$$

$$- \vartheta - \gamma_0^2, \tag{4.20}$$

where

$$\vartheta = \frac{1}{\alpha} \| I - \alpha \phi_f(k) \phi_f^T(k) \|^2 \delta(2 - \delta)[\| \tilde{W}_f(k) \|^2 - 2 \frac{(1 - \delta)}{(2 - \delta)} \| \tilde{W}_f(k) \| W_{fmax}], \tag{4.21}$$

and

$$\gamma_0^2 = \frac{\delta^2}{\alpha} \| I - \alpha \phi_f(k) \phi_f^T(k) \|^2 W_{fmax}^2, \tag{4.22}$$

which is in the power form given by (4.5).

Similarly, we can prove that the error in weight updates (3.29) yield SSP NN. In fact, suppose that if one chooses the first difference as

$$\Delta V = \frac{1}{\beta} tr[\, \tilde{W}_g^T(k+1) \tilde{W}_g(k+1) - \tilde{W}_g^T(k) \tilde{W}_g(k)]. \tag{4.23}$$

Using the error in update law in (3.29), and on simplification, one obtains

$$\Delta V = -(2 - \beta \phi_g^T(k) \phi_g(k))(-\tilde{W}_g^T(k) \phi_g(k))^T((-\tilde{W}_g^T(k) \phi_g(k)) + 2[(1 - \beta \phi_g^T(k) \phi_g(k))$$

$$- \rho \| I - \beta \phi_g(k) \phi_g^T(k) \|](-\tilde{W}_g^T(k) \phi_g(k)) \, (k_v r(k) + \bar{e}_f(k) + g \, u_d(k) + \varepsilon(k) + d(k))$$

$$+ (k_v r(k) + \bar{e}_f(k) + g \, u_d(k) + \varepsilon(k) + d(k))^T \, (\beta \phi_g^T(k) \phi_g(k)$$

$$(k_v r(k) + \bar{e}_f(k) + g \, u_d(k) + \varepsilon(k) + d(k)) + 2\rho \| I - \beta \phi_g(k) \phi_g^T(k) \| W_g^T \phi_g(k))$$

$$- \vartheta - \gamma_0^2, \tag{4.24}$$

where

$$\vartheta = \frac{1}{\beta} \| I - \beta \phi_g(k) \phi_g^T(k) \|^2 \rho(2 - \rho)[\| \tilde{W}_g(k) \|^2 - 2 \frac{(1 - \rho)}{(2 - \rho)} \| \tilde{W}_g(k) \| W_{gmax}], \tag{4.25}$$

and

$$\gamma_0^2 = \frac{\rho^2}{\beta} \left\| I - \beta \phi_g(k)\phi_g^T(k) \right\|^2 W_{gmax}^2 ,$$ (4.26)

which is in the power form defined in (4.5).

Thus, the weight error block is SSP and the closed-loop filtered tracking error system (4.8) in Figure 3 is SSP. This guarantees the state-strict passivity of the closed-loop system [20]. By employing the passivity theorem [20], one can conclude that the input/output signals of each block are bounded as long as the external inputs are bounded. In addition, note that the internal states are also bounded. If one chooses to use the update laws (3.23) and (3.24), then internal states are guaranteed to be bounded with only the PE condition. On the other hand, PE condition is relaxed with the new weight tuning paradigms (3.28) and (3.29).

An NN is defined to be passive if, in the error formulation, it guarantees the passivity of the lower blocks in Figure 3. To show that the NN are dissipative, an extra PE condition is needed. Furthermore, to obtain SSP of the closed-loop system, the weight tuning laws based on delta-rule have to be modified. This SSP property of the NN is very important in order to show the input-output boundedness of the closed-loop system when extra bounded disturbances are present or when systems are connected in parallel and driving one another.

5 Simulation Results

As an example, consider the MIMO continuous-time system given by

$$\dot{x}_1 = x_2 + (1 + x_2^2)u_1$$
$$\dot{x}_2 = -x_1 - (0.1 - e^{-(x_1^2 + x_2^2)})x_2 + (1 + x_1^2)u_2 ,$$ (5.1)

with $\|g(x(k))\| \geq 1, \forall x(k)$. The objective is to employ two neural networks, one for the function $f(x)$ and the other for $g(x)$, which satisfy

$$\dot{x} = f(x) + g(x)u ,$$
$$y = x,$$ (5.2)

where

$$f(x) = \begin{bmatrix} x_2 \\ -x_1 + 2 e^{-(x_1^2 + x_2^2)} x_2 - 0.1 x_2 \end{bmatrix} ,$$ (5.3)

$$g(x) = \begin{bmatrix} 1 + x_2^2 & 0 \\ 0 & 1 + x_1^2 \end{bmatrix} ,$$ (5.4)

$$x = \begin{bmatrix} x_1 \\ x_2 \end{bmatrix} ,$$ (5.5)

and

$$u = \begin{bmatrix} u_1 \\ u_2 \end{bmatrix} ,$$ (5.6)

with y being the output. The system (5.1) is a first order multi-input–multi-output system. When discretized using Euler's approximation, (5.1) can be rewritten as

$$x(k+1) = F(x(k)) + G(x(k))u(k), \tag{5.7}$$

where

$$x(k) = \begin{bmatrix} x_1(k) \\ x_2(k) \end{bmatrix}, \tag{5.8}$$

$$F(x(k)) = x(k) + Tf(x(k)), \tag{5.9}$$

and

$$G(x(k)) = Tg(x(k)). \tag{5.10}$$

Note that $f(x(k))$ and $g(x(k))$ are respectively obtained from $f(x)$ and $g(x)$ by substituting $x(t)$ with $x(k)$. The magnitudes of the desired trajectories are

$$y_{d1}(t) = \begin{cases} 0.5 & 0 < t \le 0.5 \\ -0.5 & 0.5 < t \le 1 \end{cases},$$

and

$$y_{d2}(t) = 0.5cos(2\pi t).$$

This nonlinear system was discretized with a sampling interval of 10 milliseconds. Fifteen sigmoidal activation functions were employed for all the nodes in each of the NN. The initial conditions for states are taken as [-0.5,0] and the weight estimates for f(x) are initialized at zero whereas the weight estimates for g(x) are initialized at 0.5. Here the lower bound for g(x) denoted by $\underline{g}$ is chosen to be 1.0.

Design parameters are set to $s=10$, $\gamma = 0.05, k_v = \begin{bmatrix} 0.1 & 0 \\ 0 & 0.1 \end{bmatrix}$, $\mu=4$, $\alpha = \beta = \delta = \rho = 0.1$.

Figure 4 shows the desired and actual outputs for the MIMO system using the controller derived in this chapter using the new weight tuning updates. From the results, it can be inferred that the performance of the NN controller is impressive even though the NN controller knows none of the dynamics *a priori*.

Example II:

Consider a planar two link robot arm whose state equations in continuous-time are written as

$$\dot{X}_1 = X_2,$$
$$\dot{X}_2 = F(X_1, X_2) + G(X_1, X_2)U, \tag{5.11}$$

where $X_1 = [q_1, q_2]^T$ are the joint angles, U is the input vector, and the nonlinear functions are described by

$$F(X_1, X_2) = -[M(X_1)]^{-1}[V(X_1, X_2)X_2 - G_1(X_1)], \qquad (5.12)$$

$$G(X_1, X_2) = [M(X_1)]^{-1}, \qquad (5.13)$$

with

$M(X_1)$ being the inertia matrix, $V(X_1, X_2)$ being the Coriolis and Centripetal vector, and $G_1(X_1)$ being the gravitational vector. The actual elements of the inertia matrix and Coriolis, Centripetal, and gravitational vectors are given in [16]. The system, when discretized using Euler's approximation, can be written as

$$Z_1(k+1) = Z_2(k),$$

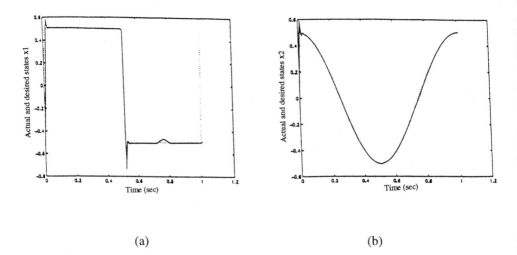

(a)   (b)

Figure 4: Response of the neural network controller. (a) State x_1 (b) State x_2.

$$Z_2(k+1) = F_2(Z_1(k), Z_2(k)) + G_2(Z_1(k), Z_2(k))U(k), \qquad (5.14)$$

where

$$F_2(Z_1(k), Z_2(k)) = -Z_1(k) + 2Z_2(k) + T^2 F(Z_1(k), Z_2(k)), \qquad (5.15)$$

and

$$G_2(Z_1(k), Z_2(k)) = T^2 G(Z_1(k), Z_2(k)) \qquad (5.16)$$

with the transformation matrix given by

$$\begin{bmatrix} Z_1(k) \\ Z_2(k) \end{bmatrix} = \begin{bmatrix} 1 & 0 \\ 1 & T \end{bmatrix} \begin{bmatrix} X_1(k) \\ X_2(k) \end{bmatrix}. \qquad (5.17)$$

Note that the Equations (5.11) are in the form of (2.7) in view of the transformation matrix (5.17). By appropriately choosing the desired trajectories for the joint angles and velocities, one should be able to track the desired joint angles.

A sampling interval of 10 ms was chosen and the magnitudes of the desired trajectories were selected as $y_{d1}(t) = 1.0sin(\frac{2\pi t}{25})$ and $y_{d2}(t) = 1.0cos(\frac{2\pi t}{25})$. Sigmoidal activation functions are employed for all the nodes. Fifteen nodes are selected for each NN.

The initial conditions for the states are taken as $[0.5, 0.1]^T$ and the parameter estimates for $f(x)$ are initialized at zero whereas the parameter estimates for $g(x)$ are initialized at 0.5. Here $\bar{g}$ is chosen to be 1.0. Design parameters are set to $s = 10$,

$$\gamma = 0.05, k_v = \begin{bmatrix} 0.1 & 0 \\ 0 & 0.1 \end{bmatrix}, \mu = 4, \alpha = \beta = \delta = \rho = 0.1.$$

Figure 5 shows the desired and actual outputs for the MIMO system and the representative NN outputs using the controller derived in this chapter, and given by (3.23) and (3.24). Figure 6 shows the performance of the PD controller alone without the NN system. From the results, it can be inferred that the performance of the NN controller is impressive even though the dynamics are not known *a priori* by the NN controller.

6 Summary

This chapter provided a comprehensive treatment of designing digital controllers for a well-defined class of nonlinear systems so that feedback linearization in discrete-time could be achieved. This should result in better understanding of the design of digital controllers for nonlinear systems. Since most implementations of controllers have been done on a digital computer, the discrete-time implementation of a controller is important. This chapter provided a comprehensive treatment of digital controller design for closed-loop control applications that use neural networks.

For a class of multi-input–multi-output (MIMO), continuous and discrete-time nonlinear systems, a neural network-based controller in discrete-time which linearizes the system in feedback was presented. In other words, digital control, using neural networks, of a class of well-defined nonlinear systems was presented. Control action was defined so as to achieve tracking performance for a state-feedback linearizable, but unknown, nonlinear system. The neural network controller was forward to exhibit learning-while-functioningfeature instead of learning-then-control. The structure of the NN controller was derived using a filtered error notions and the passivity approach. A persistency of excitation condition was defined for the inputs that were fed into these neural networks. A uniform ultimate boundedness of the closed-loop system was given in the sense of Lyapunov. The concepts of discrete-time passive NN and a dissipative NN were defined and used with persistency of excitation condition, to show the boundedness of NN weight estimates.

The PE condition was found to be extremely difficult to verify or guarantee even for linear systems. Therefore, modified update laws were presented so that PE condition would not be needed. The notion of state-strict passivity was defined with modified update laws.

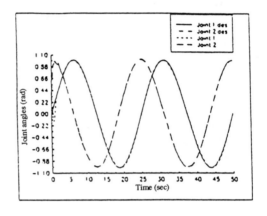

(a)

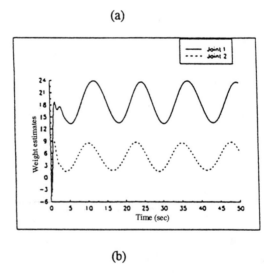

(b)

Figure 5: Response of the neural network controller; (a) desired and actual joint angles; (b) neural network outputs.

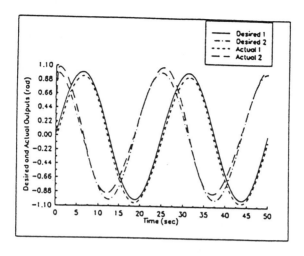

Figure 6: Response of the PD controller without neural network.

References

[1] Astrom, K. J. and Wittenmark, B., (1989), *Adaptive control*, Addison-Wesley Company, Reading, Massachusetts.

[2] Chen, F. and Khalil, K.H. (1992), Adaptive control of a Class of nonlinear discrete-time systems using neural networks, *IEEE Trans. on Automatic Control*, vol. 40, no. 5, 791-801.

[3] Cybenko, G. (1989), Approximation by superpositions of a sigmoidal functions, *Mathematics of Control, Signals, and Systems*, vol. 2, no. 4, 303-314.

[4] Goodwin, G.C. and Sin, K. S. (1984), *Adaptive Filtering, Prediction, and Control*, Prentice-Hall Inc., Englewood Cliffs, NJ.

[5] Jagannathan, S. and Lewis, F. L. (1996), Multilayer discrete-time NN Controller with guaranteed performance, *IEEE Trans. on Neural Networks*, vol. 7, no.1, pp. 107-130, January.

[6] Jagannathan, S. and Lewis, F. L. (1996), Discrete-time model reference adaptive control of nonlinear dynamical systems using neural networks, *International Journal of Control*, vol. 64, no. 2, pp. 217-239, 1996, March.

[7] Jagannathan, S. and Lewis, F. L. (1996), Discrete-time neural network controller for a class of nonlinear dynamical systems, *IEEE Trans. on Automatic Control*, Vol.41, No.11, pp. 1693-1699, November.

[8] Jagannathan, S. and Lewis, F. L. (1996), Robust implicit STR/MRAC of a class of nonlinear systems, *Automatica*, vol. 32, no. 12, pp.1629-1644, December.

[9]   Jagannathan, S. (1996), Adaptive Control of Unknown Feedback Linearizable Nonlinear Systems, *Proc. of the IEEE Conference on Decision and Control*, pp. 4747-4752, December.

[10] Kanellakopoulos, I., Kokotovic, P. V., and Morse, A. S. (1991), Systematic design of adaptive controllers for feedback linearizable systems, *IEEE Trans. on Automatic Control*, vol. 36, pp. 1241-1253.

[11] Liu, C. C. and Chen, F. C. (1991), Adaptive control of nonlinear continuous systems using neural networks-general relative degree and MIMO cases, *International Journal of Control*, vol. 58, pp. 317-335.

[12] Narendra, K. S. and Parthasarathy, M. (1990), Identification and control of dynamical systems using neural networks, *IEEE Trans. on Neural Networks*, vol. 1, pp. 4-27.

[13] Polycarpou, M. and Ioannou, P. A. (1991), Identification and control using neural network models: design and stability analysis, Tech. Report 91-09-01, Dept. Elect. Eng. Sys., Univ. S. California.

[14] Ravithakis, G. A. and Christodoulou, T. (1994), Adaptive control of unknown plants using dynamical neural networks, *IEEE Trans. on Systems, Man, and Cybernetics*, vol. 24, no. 7, pp. 971-981.

[15] Sadegh, N. (1993), A perceptron network for functional identification and control of nonlinear systems, *IEEE Trans. on Neural Networks*, vol. 4, no. 6, pp. 982-988.

[16] Slotine, J, J. and Li, W. (1992), *Applied Nonlinear Control*, Prentice-Hall Inc., Englewood Cliffs, NJ.

[17] Sontag, E., D. (1992), Feedback stabilization using two-hidden layer nets, *IEEE Trans. on Neural Networks*, vol. 3, no. 6, pp. 981-990.

[18] Sussmann, H., J. (1992), Uniqueness of the weights for minimal feedforward nets with a given input-output map, *Neural Networks*, vol. 5, pp. 589-593.

[19] Vidyasagar, M. (1989), *Nonlinear Systems Analysis*, Prentice-Hall Inc., Englewood Cliffs, NJ.

7 | ROBUST ADAPTIVE CONTROL OF ROBOTS BASED ON STATIC NEURAL NETWORKS

Shuzhi S. Ge
Department of Electrical Engineering
National University of Singapore
Singapore 119260
E-mail: elegesz@nus.edu.sg

In this chapter, a two-stage robust adaptive controller for robots is presented based on static neural network models [1,2]. Since the inertia matrix $D(q)$ and the gravitational force vector $G(q)$ are functions of joint variable q only, static neural networks are sufficient to approximate them. The neural networks for the Coriolis and centrifugal forces can be systematically constructed based on the static network for $D(q)$. The dynamics of robots, thus, can be described by pre-determined static neural networks for $D(q)$ and $G(q)$, and the dynamic variables q, $\dot{q}$, and $\ddot{q}$. The size of the resulting neural network (NN) models of robots is much smaller than the usual dynamic ones. The main advantages of the resulting controller based on the static neural network model include (1) systematic construction of controllers owing to its well structured nature and (2) low computational power required in actual implementation. It can be shown that all the closed-loop signals are bounded and the tracking error goes to zero asymptotically. While similar controllers have been successfully implemented in controlling a free gyro stabilized mirror system, intensive simulation results are also provided to show the effectiveness of the controller for robots.

1 Introduction

For the design of neural network controllers, in general, a sufficiently accurate approximation of the dynamics of the plant has to be obtained first (usually carried out off-line). Then an appropriate control strategy that utilizes this approximation can be constructed. This approach has been successfully applied to many robotic systems [3-6], however, it does not have any built-in capability to handle system

0-8493-9805-3/99/$0.00+$.50
©1999 by CRC Press LLC

changes. By directly parameterizing the control laws using suitable neural networks (NNs), some recent works have successfully incorporated adaptive techniques to show good overall closed-loop stability properties [7-13].

It is known that NNs have only the static mapping capability [14]. However, when they are used as controllers, they must be able to realize the dynamics. There are basically three methods to realize dynamic NNs [6]. The first method feeds back the dynamic information, such as velocity and acceleration as inputs. The second method includes past information, such as $x(t-\tau)$ and $u(t-\tau)$ in the inputs. The third method uses an inner recurrent type of NNs. Despite the fact that NNs are very powerful in learning complicated dynamics, the sizes of the networks are known to be very large, especially the dynamic ones, which will subsequently lead to the need for powerful computational facilities.

It is true that neural networks have learning capability and can be used to approximate any dynamical function to any desired accuracy, as long as the size of the networks is large enough. However, for engineering applications, we should not abuse their learning capability and let them learn all the characteristics of the systems. It has been shown that, by fully exploiting the structural properties of a particular system, the full dynamics of the system can be approximated by static neural networks of pre-determined structures [1,2]. As a result, the size of an NN model of a robot would be smaller than the usual dynamical model. This is very desirable for actual implementation as the required computational power is low and, accordingly, the cost will be low as well. The well known structural properties of robots, such as the linear-in-the-parameters dynamics, are retained in the NN models that are developed. In this chapter, robust adaptive neural network controller design is investigated for the case where the nonlinear functions of the system are not in the functional range of the corresponding neural network emulators. A two-stage adaptive controller is given, and it is shown that the closed-loop tracking error is asymptotically stable, and all signals in the closed-loop are bounded. The novelties of the chapter include: (i) unlike conventional neural network controllers given in the literature, only static neural networks are used for dynamic modeling and controller design, (ii) the system is analyzed under the more realistic assumption that the nonlinear functions being approximated are not in the functional range of the approximating neural networks, and (iii) system analysis and presentation are carried out using the new GL matrix product for easy analysis. While similar controllers have been successfully implemented in controlling a free gyro stabilized mirror system [23,24], intensive simulation results are presented here to show the effectiveness of the controller for robots as well.

This chapter is organized as follows: some of the notations used in the chapter are introduced in Section 2. In Section 3, problems of neural network approximations are discussed. Lagrange-Euler formulation of robots is presented in Section 4. Neural network modeling of robots is presented in Section 5. Static neural network based controller design is presented in Section 6, followed by a case study in Section 7, and conclusion in Section 8.

2 Notation

For clarity of presentation of the dynamic models based on static NN, the permutation operator is defined first. Then, the Ge-Lee (GL) matrix and its operator which have been introduced in [7-9] are presented for completeness and the ease of system stability analysis.

2.1 Permutation Operator "$\otimes$"

The permutation operator, denoted by "$\otimes$," is defined as

$$A \otimes B = BA, \forall A, B \in \mathfrak{R}^{n \times n} \tag{2.1}$$

Note that the permutation product should be computed first in a mixed matrix product. For example, in $A \otimes BC$, $A \otimes B$ should be computed first to get a conventional matrix $[A \otimes B] = BA$ and then compute $[A \otimes B]C$. For an expression of the form

$$z = a + \otimes Bb, \forall a, b \in \mathfrak{R}^n, B \in \mathfrak{R}^{n \times n}$$

the "dummy" vector z is not well defined (or is meaningless by itself) because of the presence of the permutation operator $\otimes$. However, for $A \in \mathfrak{R}^{n \times n}$, Az is well defined by

$$Az = Aa + A \otimes Bb = Aa + BAb \tag{2.2}$$

The permutation operator is used to describe **Property 5.4**: the linear-in-the-parameter dynamics, in Section 5. Note that it is not a simple priority operator as can be seen from Equation (2.2).

2.2 GL Product Operator "$\bullet$"

Let I_0 be the set of integers, and $\theta_{ij}, \xi_{ij} \in \mathfrak{R}^{n_{ij}}$, $n_{ij} \in I_0$, $i = 1,...,n, j = 1,...,k$. For function approximation, θ_{ij} can be taken as the weight vector while ξ_{ij} is the network basis function vector. Define ω_i and ω_i^T as

$$\omega_i = \begin{bmatrix} \theta_{i1} \\ \theta_{i2} \\ ... \\ \theta_{ik} \end{bmatrix}, \qquad \omega_i^T = \begin{bmatrix} \theta_{i1}^T & \theta_{i2}^T & \cdots & \theta_{ik}^T \end{bmatrix}$$

in the conventional way. Let us briefly discuss the definition of the GL vectors, denoted by {*} and the corresponding GL product operator, "$\bullet$", introduced in [7-

9]. Should any confusion arise in the text, [*] is used to denote an ordinary matrix, and {*} for a GL matrix explicitly.

Define

$$\{\omega_i\} = \{\theta_{i1} \quad \theta_{i2} \quad \cdots \quad \theta_{in}\} \tag{2.3}$$

$$\{\omega_i\}^T = \{\theta_{i1}^T \quad \theta_{i2}^T \quad \cdots \quad \theta_{in}^T\} \tag{2.4}$$

$$\{\varphi_i\} = \{\xi_{i1} \quad \xi_{i2} \quad \cdots \quad \xi_{in}\} \tag{2.5}$$

The GL product is defined as

$$\left[\{\omega_i\}^T \bullet \{\varphi_i\}\right] = \left[\theta_{i1}^T \xi_{i1} \quad \theta_{i2}^T \xi_{i2} \quad \cdots \quad \theta_{in}^T \xi_{in}\right] \tag{2.6}$$

By stacking $[\{\omega_i\}^T \bullet \{\varphi_i\}]$ in a column, we will have the corresponding GL matrix product:

$$\left[\{\Omega\}^T \bullet \{\Xi\}\right] = \begin{bmatrix} \{\omega_1\}^T \bullet \{\varphi_1\} \\ \{\omega_2\}^T \bullet \{\varphi_2\} \\ \cdots \\ \{\omega_n\}^T \bullet \{\varphi_n\} \end{bmatrix} \tag{2.7}$$

where

$$\{\Omega\} = \begin{Bmatrix} \theta_{11} & \theta_{12} & \cdots & \theta_{1n} \\ \theta_{21} & \theta_{22} & \cdots & \theta_{2n} \\ \cdots & \cdots & \cdots & \cdots \\ \theta_{n1} & \theta_{n2} & \cdots & \theta_{nn} \end{Bmatrix} = \begin{Bmatrix} \{\omega_1\} \\ \{\omega_2\} \\ \cdots \\ \{\omega_n\} \end{Bmatrix} \tag{2.8}$$

$$\{\Xi\} = \begin{Bmatrix} \xi_{11} & \xi_{12} & \cdots & \xi_{1n} \\ \xi_{21} & \xi_{22} & \cdots & \xi_{2n} \\ \cdots & \cdots & \cdots & \cdots \\ \xi_{n1} & \xi_{n2} & \cdots & \xi_{nn} \end{Bmatrix} = \begin{Bmatrix} \{\varphi_1\} \\ \{\varphi_2\} \\ \cdots \\ \{\varphi_n\} \end{Bmatrix} \tag{2.9}$$

The transpose of the GL matrix $\{\Omega\}$ is defined as

$$\{\Omega\}^T = \begin{Bmatrix} \theta_{11}^T & \theta_{12}^T & \cdots & \theta_{1n}^T \\ \theta_{21}^T & \theta_{22}^T & \cdots & \theta_{2n}^T \\ \cdots & \cdots & \cdots & \cdots \\ \theta_{n1}^T & \theta_{2n}^T & \cdots & \theta_{nn}^T \end{Bmatrix} \tag{2.10}$$

which only transposes its elements locally. The GL product can be regarded as a generalization of the Hadamard matrix product [15,16].

The GL product of a square matrix and a GL row vector are defined as follows: let a GL row vector S_i defined as $S_i = \{S_{i1}\ S_{i2}\ ...\ S_{in}\}$, $\forall\ S_{ij} \in \Re^{n_{ij}}$, $j = 1,2,...,n$, and a matrix $\{\Gamma\}_I = \{\Gamma\}_i^T = [\{\Gamma\}_{i1}\ \{\Gamma\}_{i2}\ ...\ \{\Gamma\}_{in}]$, $\forall\ \{\Gamma\}_{ij} \in \Re^{m_i \times n_{ij}}$, $m_i = \sum_{j=1}^{n} n_{ij}$, $j = 1,2,...,n$, then, we have

$$\Gamma_i \bullet \{S_i\} = \{\Gamma_i\} \bullet \{S_i\} = [\Gamma_{i1} S_{i1}\quad \Gamma_{i2} S_{i2}\quad ...\quad \Gamma_{in} S_{in}] \in \Re^{m_i \times n} \tag{2.11}$$

Note that the GL product should be computed first in a mixed matrix product. For example, in $\{A\}\bullet\{B\}C$, $\{A\}\bullet\{B\}$ should be computed first to get a conventional matrix $[\{A\}\bullet\{B\}]$ and then compute $[\{A\}\bullet\{B\}]C$.

3 Neural Network Approximation

It is well known that a multilayer neural network, $\hat{f}(x)$, can approximate a continuous nonlinear function, $f(x)$, to an arbitrary accuracy on a given compact set A, i.e.,

$$\left|f(x) - \hat{f}(x)\right| \le \varepsilon, \forall x \in A, \varepsilon > 0 \tag{3.1}$$

In general, it can be expressed as

$$\left|f(x) - \hat{f}(x)\right| \le \varepsilon + \alpha(x) \tag{3.2}$$

with $\alpha(x) = 0$ if $x \in A$.

It has been demonstrated that a linear superposition of Gaussian radial basis functions (RBFs) results in an optimal mean square approximation to an unknown function which is infinitely differentiable and whose values are specified at a finite set of points in $\Re^n$ [17]. Further, it has been proven that any continuous functions, not necessarily infinitely smooth, can be uniformly approximated by a linear combination of Gaussian [18,19].

Each Gaussian RBF network consists of three layers: the input layer, the hidden layer that contains the Gaussian radial basis function, and the output layer. At the input layer, the input space is divided into grids with a Gaussian function at each node defining a receptive field in $\Re^n$ with μ_i as its center and σ as its "width." The Gaussian RBF neural network can be expressed as [17]

$$\hat{f}(\theta, x) = \sum_{i=1}^{k} \theta_i \xi_i(x, \mu) = \theta^T \xi(x, \mu) \tag{3.3}$$

$$\xi_i(x,\mu) = \exp\left(-\frac{\|x-\mu\|^2}{\sigma^2}\right) = \exp\left(-\frac{(x-\mu)^T(x-\mu)}{\sigma^2}\right) \tag{3.4}$$

where k is the number of Gaussian radial basis functions, $x = (x_1,...,x_n)^T \in \Re^n$ is the input vector, $\sigma^2 \in \Re$ is the variance, and $\mu = (\mu_1,...\mu_n)^T \in \Re^n$ is the center vector. Given any function $y = f(x)$ that is sufficiently regular, it is known that an approximation can be provided by the Gaussian RBF neural network $y = \hat{f}(\theta, x)$ with a sufficient number of nodes [17].

In this chapter, we shall take the approximation function $\hat{f}(\theta, x)$ for $f(x)$ to be of the form given by Equation (3.3) and the basis functions to be radial basis functions (3.4). Networks based on other kinds of bounded basis functions can be similarly carried out. The GL matrix and its operator provide a convenient tool for system analysis, and will be used later to express the matrix and vector emulators.

4 Lagrange-Euler Formulation of Robots

The material given in this section is well known. It is provided here for the ease of neural network dynamic modeling of robots, which will be developed in the next section. Consider an n degrees-of-freedom robot, and let $q \in \Re^n$ and $\tau \in \Re^n$ be the joint variables and the applied torque/forces, respectively. The closed-form dynamic Equations of robots can be obtained by applying the Lagrange-Euler formulation. It is shown in the literature that robotic manipulators satisfy the following two conditions [20]:

(1)   The kinetic energy $E_k = E_k(q,\dot{q})$ is a quadratic function of the vector $\dot{q}$ of the form

$$E_k = \frac{1}{2}\dot{q}^T D(q)\dot{q} = \frac{1}{2}\sum_{i=1}^{n}\sum_{j=1}^{n} d_{ij}(q)\dot{q}_i\dot{q}_j$$

where $D(q)$ is the inertia matrix which is symmetric, positive definite, and $d_{ij}(q)$ is the ij-th element of $D(q)$, and $\dot{q}_i$ is the i-th element of $\dot{q}$.

(2)   The gravitational potential energy $E_p = E_p(q)$ is independent of $\dot{q}$.

The Lagrangian, L, of the system is given by

$$L = E_k - E_p = \frac{1}{2}\dot{q}^T D(q)\dot{q} - E_p = \frac{1}{2}\sum_{i=1}^{n}\sum_{j=1}^{n} d_{ij}(q)\dot{q}_i\dot{q}_j - E_p \tag{4.1}$$

which leads to

$$\frac{\partial L}{\partial \dot{q}_k} = \sum_{j=1}^{n} d_{kj}(q)\dot{q}_j \tag{4.2}$$

$$\frac{d}{dt}\frac{\partial L}{\partial \dot{q}_k} = \sum_{j=1}^{n} d_{kj}(q)\ddot{q}_j + \sum_{j=1}^{n}\frac{d}{dt}d_{kj}(q)\dot{q}_j = \sum_{j=1}^{n} d_{kj}(q)\ddot{q}_j + \sum_{j=1}^{n}\sum_{i=1}^{n}\frac{\partial d_{kj}(q)}{\partial q_i}\dot{q}_i\dot{q}_j \tag{4.3}$$

$$\frac{\partial L}{\partial q_k} = \frac{1}{2}\sum_{i=1}^{n}\sum_{j=1}^{n}\frac{\partial d_{ij}}{\partial q_k}\dot{q}_i\dot{q}_j - \frac{\partial E_p}{\partial q_k} \tag{4.4}$$

where $k = 1, 2, ..., n$. Thus, the Lagrange-Euler Equations can be written as

$$\sum_{j=1}^{n} d_{kj}(q)\ddot{q}_j + \sum_{i=1}^{n}\sum_{j=1}^{n}\left(\frac{\partial d_{kj}(q)}{\partial \dot{q}_i} - \frac{1}{2}\frac{\partial d_{ij}(q)}{\partial q_k}\right)\dot{q}_i\dot{q}_j + \frac{\partial E_p}{\partial q_k} = \tau_k \tag{4.5}$$

By interchanging the order of the summation in the second term above and taking advantage of the symmetry of the inertia matrix, it can be seen that

$$\sum_{j=1}^{n}\sum_{i=1}^{n}\frac{\partial d_{kj}(q)}{\partial q_i}\dot{q}_i\dot{q}_j = \frac{1}{2}\sum_{i=1}^{n}\sum_{j=1}^{n}\left[\frac{\partial d_{kj}(q)}{\partial q_i} + \frac{\partial d_{kj}(q)}{\partial q_i}\right]\dot{q}_i\dot{q}_j \tag{4.6}$$

$$= \frac{1}{2}\sum_{i=1}^{n}\sum_{j=1}^{n}\left[\frac{\partial d_{kj}(q)}{\partial q_i} + \frac{\partial d_{ki}}{\partial q_i}\right]\dot{q}_i\dot{q}_j \tag{4.7}$$

Therefore, the Lagrange-Euler Equations can be expressed as

$$\sum_{j=1}^{n} d_{kj}(q)\ddot{q}_j + \sum_{i=1}^{n}\sum_{j=1}^{n} c_{ijk}(q)\dot{q}_i\dot{q}_j + g_k(q) = \tau_k \tag{4.8}$$

where

$$c_{ijk}(q) = \frac{1}{2}\left[\frac{\partial d_{kj}(q)}{\partial q_i} + \frac{\partial d_{ki}(q)}{\partial q_j} - \frac{\partial d_{ij}(q)}{\partial q_k}\right] \tag{4.9}$$

$$g_k(q) = \frac{\partial E_p}{\partial q_k} \tag{4.10}$$

The coefficients $c_{ijk}(q)$ are known as **Christoffel Symbols** [20]. It is common to write the Equation above in a matrix form as

$$D(q)\ddot{q} + C(q,\dot{q})\dot{q} + G(q) = \tau \tag{4.11}$$

with the kj-th element of $C(q,\dot{q})$ defined by

$$c_{kj}(q,\dot{q}) = \sum_{i=1}^{n} c_{ijk}\dot{q}_i = \sum_{i=1}^{n}\frac{1}{2}\left[\frac{\partial d_{kj}}{\partial q_i} + \frac{\partial d_{ki}}{\partial q_j} - \frac{\partial d_{ij}}{\partial q_k}\right]\dot{q}_i \tag{4.12}$$

Advanced control laws have been obtained from the literature by thoroughly exploiting the structural properties of the system [20,21]. The most commonly used properties are as follows:

Property 4.1 $D(q)$ is symmetric positive definite.

Property 4.2 $N \equiv \dot{D}(q) - C(q, \dot{q})$ is skew-symmetric, i.e., $x^T N x = 0, \forall x \in \Re^n$.

Property 4.3 $D(q)a + C(q, \dot{q})v + G(q) = Y(q, \dot{q}, v, a)P$, where Y is the so-called regressor of known functions, and P is the robot parameter vector of interest, i.e., the unknown coefficients of variables $a, v, \dot{q}$, and q, which are directly related to the parameters of the system such as masses, moments of inertia, and the dimensions of the constituent bodies.

The model (4.11) is very simple in its representation, but it is very difficult, time consuming, and error prone to obtain, except for simple cases. The modeling process, in general, starts with assignment of the initial reference coordinate system and the relative coordinate systems, formulation of transformation matrices, through the Lagrangian function and Lagrange-Euler formulation, and ends at the problem of parameter identification and controller design. Neural networks, however, can be used directly as a generic modeling tool for system identification and controller design without getting into too many details of the system.

5   Dynamic Modeling of Robots Using Neural Networks

In this section, we shall investigate the problem of dynamic modeling of robots based on static neural networks. Because the inertia matrix $D(q)$ and $G(q)$ are functions of q only, static neural networks, where the inputs to the neural networks are q only, are sufficient to approximate them. Since $C(q, \dot{q})$ is derived from the matrix $D(q)$, it will be shown later that its corresponding neural network emulator can be constructed by the neural networks for $D(q)$. The size of the resulting neural network models of robots is much smaller compared with the dynamical ones, which is very desirable for implementation because less computational power is required. The resulting dynamic equations based on NNs retain several attractive structural properties, such as linear-in-the-parameters dynamics of the original dynamic equations, which can be used to facilitate controller design.

Throughout this chapter, we shall (1) let $|X|>|Y|$ denote $|X_i|>|Y_i|$ for vectors $X, Y \in R^n$, and (2) let $(*)_N$ be the corresponding neural network emulator of function $(*)$. It is well known that $d_{kj}(q)$ and $g_k(q)$ are infinitely differentiable. Therefore, for $q \in A$, $d_{kj}(q)$ and $g_k(q)$ can be expressed as

$$d_{kj}(q) = d_{Nkj}(q) + \eta_{kj}(q) \tag{5.1}$$

$$g_k(q) = g_{Nk}(q) + \eta_{k0}(q) \tag{5.2}$$

where

$$d_{Nkj}(q) = \theta_{kj}^T \xi_{kj}(q) \tag{5.3}$$

$$g_{Nk}(q) = \theta_{k0}^T \xi_{k0} \tag{5.4}$$

with

$$\xi_{kjl}(q,\mu) = \exp\left(-\frac{\|q-\mu\|^2}{\sigma^2}\right) = \exp\left(-\frac{(q-\mu)^T(q-\mu)}{\sigma^2}\right) \tag{5.5}$$

$$\left|\eta_{kj}(q)\right| \le \varepsilon_{kj}, j = 0, 1, ..., \text{n} \tag{5.6}$$

The corresponding networks for $d_{Nkj}(q)$ and $g_{Nk}(q)$ are shown in Figure 1. Define $\rho_{ijk} \equiv \partial\eta_{ij}(q)/\partial q_k$, and assume that $|\rho_{ijk}|$ are upper bounded by ε_{ijk}. Then, we have

$$\frac{\partial d_{ij}}{\partial q_k} = -\frac{2}{\sigma^2}\theta_{ij}^T \xi_{ij}(q_k - \mu_k) + \rho_{ijk}$$

$$= -\frac{2}{\sigma^2}d_{Nij}(q_k - \mu_k) + \rho_{ijk} \tag{5.7}$$

and from Equation (4.9), we obtain

$$c_{ijk}(q) = \frac{1}{\sigma^2}c_{Nijk}(q) + \rho_{cijk} \tag{5.8}$$

where the network emulator $c_{Nijk}(q)$ of function $c_{ijk}(q)$ is given by

$$c_{Nijk} := \left[-\theta_{kj}^T \xi_{kj}(q_i - \mu_i) - \theta_{ki}^T \xi_{ki}(q_j - \mu_j) + \theta_{ij}^T \xi_{ij}(q_k - \mu_k)\right]$$

$$= \left[-d_{Nkj}(q_i - \mu_i) - d_{Nki}(q_j - \mu_j) + d_{Nij}(q_k - \mu_k)\right] \tag{5.9}$$

and the corresponding approximation error is

$$\rho_{cijk} := \rho_{kji} + \rho_{kij} - \rho_{ijk} \tag{5.10}$$

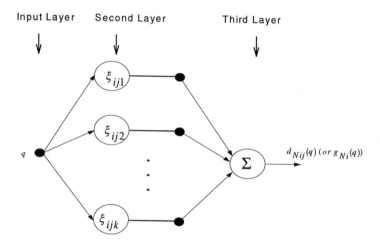

Figure 1: Neural network implementation for $d_{Nij}(q)$ and $g_{Ni}(q)$.

From Equation (4.12), we have

$$c_{kj}(q,\dot{q}) = \sum_{i=1}^{n} c_{ijk}(q)\dot{q}_i = c_{Nkj}(q,\dot{q}) + \rho_{kj} \tag{5.11}$$

where

$$c_{Nkj}(q,\dot{q}) = \frac{1}{\sigma^2} \sum_{i=1}^{n} c_{Nijk}\dot{q}_i$$

$$= \frac{1}{\sigma^2} \sum_{i=1}^{n} \left[-d_{Nkj}(q_i - \mu_i) - d_{Nki}(q_j - \mu_j) + d_{Nij}(q_k - \mu_k) \right]\dot{q}_i \tag{5.12}$$

$$\rho_{kj} = \sum_{i=1}^{n} \rho_{cijk}\dot{q}_i \tag{5.13}$$

Therefore, the corresponding dynamics of robots based on static neural networks can be described as

$$D_N(q)\ddot{q} + C_N(q,\dot{q})\dot{q} + G_N(q) + E_0 = \tau \tag{5.14}$$

where

$$D_N(q) = [d_{Nkj}(q)] \in R^{n\times n}$$

$$C_N(q,\dot{q}) = [c_{Nkj}(q,\dot{q})] \in R^{n\times n}$$

$$G_N(q) = [g_{Nk}(q)] \in R^n$$

$$E_0 = E_D \ddot{q} + E_C \dot{q} + E_G \in R^n \qquad (5.15)$$

and E_D, E_C, and E_G are the corresponding approximation error matrices, defined as

$$E_D = \begin{bmatrix} \eta_{11} & \eta_{12} & \cdots & \eta_{1n} \\ \eta_{21} & \eta_{22} & \cdots & \eta_{2n} \\ \cdots & \cdots & \cdots & \cdots \\ \eta_{n1} & \eta_{n2} & \cdots & \eta_{nn} \end{bmatrix} \qquad (5.16)$$

$$E_C = \begin{bmatrix} \rho_{11} & \rho_{12} & \cdots & \rho_{1n} \\ \rho_{21} & \rho_{22} & \cdots & \rho_{2n} \\ \cdots & \cdots & \cdots & \cdots \\ \rho_{n1} & \rho_{n2} & \cdots & \rho_{nn} \end{bmatrix} \qquad (5.17)$$

$$E_G = \begin{bmatrix} \eta_{10} \\ \eta_{20} \\ \cdots \\ \eta_{n0} \end{bmatrix} \qquad (5.18)$$

Thus, it can be easily seen that the full dynamics of robots can be approximated by using the neural networks $D_N(q)$ and $G_N(q)$ for $D(q)$ and $G(q)$, because $C(q, \dot{q})$ can be approximated by $C_N(q, \dot{q})$ based on the parameters for $D_N(q)$, as shown by Equation (5.12) and property 2 below. There is no need to use separate NNs to approximate $C(q, \dot{q})$. If the nonlinear functions are in the functional range of the neural networks, i.e., E_D, E_C, and $E_G = 0$, then we have the results first presented in [1]. Controllers designed based on this model can be constructed using static neural networks as will be shown later.

In order to facilitate static neural network controller design, the structural properties of the new model (5.14) are studied. It can be seen that most of the important properties of the robotic dynamic model (4.11) are retained in the corresponding NN model as listed below.

Property 5.1:

Since $D(q)$ is symmetric positive definite, i.e., $d_{kj}(q) = d_{jk}(q)$, we may have

$$d_{Nkj} = d_{Njk}, \ \eta_{kj} = \eta_{jk} \qquad (5.19)$$

if the corresponding off-diagonal elements are approximated using the same network. Based on the definitions for GL matrices and GL product in Section 2, we have

$$D(q) = D_N(q) + E_D \qquad (5.20)$$

$$D_N(q) = \left[\{\Omega\}^T \bullet \{\Xi\} \right]$$

$$= \begin{bmatrix} \theta_{11}^T \xi_{11} & \theta_{12}^T \xi_{12} & \cdots & \theta_{1n}^T \xi_{1n} \\ \theta_{21}^T \xi_{21} & \theta_{22}^T \xi_{22} & \cdots & \theta_{2n}^T \xi_{2n} \\ \cdots & \cdots & \cdots & \cdots \\ \theta_{n1}^T \xi_{n1} & \theta_{n2}^T \xi_{n2} & \cdots & \theta_{nn}^T \xi_{nn} \end{bmatrix} = \begin{bmatrix} \{\omega_1\}^T \bullet \{\varphi_1\} \\ \{\omega_2\}^T \bullet \{\varphi_2\} \\ \cdots \\ \{\omega_n\}^T \bullet \{\varphi_n\} \end{bmatrix} \qquad (5.21)$$

where $\{\omega_i\}$, $\{\varphi_i\}$, $\{\Omega\}$, and $\{\Xi\}$ are as defined by Equations (2.3), (2.5), (2.8), and (2.9), respectively. Therefore, we can have $D^T{}_N(q) = D_N(q)$ and $E^T{}_D = E_D$.

Property 5.2:

From Equation (5.12), we have

$$c_{Nkj}(q,\dot{q}) = -\frac{1}{\sigma^2} \sum_{i=1}^{n} \left[d_{Nkj}[q_i - \mu_i] + d_{Nki}[q_j - \mu_j] - d_{ji}[q_k - \mu_k] \right] \dot{q}_i$$

$$= -\frac{1}{\sigma^2} d_{Nkj} \sum_{i=1}^{n} [q_i - \mu_i] \dot{q}_i - \frac{1}{\sigma^2} \sum_{i=1}^{n} d_{Nki}[q_j - \mu_j] \dot{q}_i$$

$$+ \frac{1}{\sigma^2} \sum_{i=1}^{n} d_{Nji}[q_k - \mu_k] \dot{q}_i$$

$$= -\frac{1}{\sigma^2} d_{Nkj}[q - \mu]^T \dot{q} - \frac{1}{\sigma^2} [d_{Nk1} \quad d_{Nk2} \quad \cdots \quad d_{Nkn}] \dot{q}[q_j - \mu_j]$$

$$+ \frac{1}{\sigma^2} [d_{N1j} \quad d_{N2j} \quad \cdots \quad d_{Nnj}] \dot{q}[q_k - \mu_k]$$

$$= -\frac{1}{\sigma^2} d_{Nkj}[q - \mu]^T \dot{q} - \frac{1}{\sigma^2} [d_{Nk1} \quad d_{Nk2} \quad \cdots \quad d_{Nkn}] \dot{q}[q_j - \mu_j]$$

$$+ \frac{1}{\sigma^2} [q_k - \mu_k] \dot{q}^T [d_{N1j} \quad d_{N2j} \quad \cdots \quad d_{Nnj}]^T \qquad (5.22)$$

To write in matrix form, we have the corresponding neural network emulator $C_N(q,\dot{q})$ for $C(q,\dot{q})$ as follows:

$$C_N(q,\dot{q}) = -\frac{1}{\sigma^2} \left[D_N(q)[q - \mu]^T \dot{q} + D_N(q)\dot{q}[q - \mu]^T - [q - \mu]\dot{q}D_N(q) \right] \qquad (5.23)$$

Therefore, $C(q, \dot{q}) = C_N(q, \dot{q}) + E_C$ with E_C defined as in (5.17). It can be seen that $C_N(q,\dot{q})$ can be directly constructed by $D_N(q)$.

Property 5.3:

It can be shown that $\dot{D}_N(q) - 2C_N(q,\dot{q})$ is skew-symmetric. Differentiating $D_N(q) = [\{\Omega\} \bullet \{\Xi\}]$, we have

$$\dot{D}_N(q) = [\{\Omega\} \bullet \{\dot{\Xi}\}] = -\frac{2}{\sigma^2}[\{\Omega\} \bullet \{\Xi\}](q - \mu)^T \dot{q} = -\frac{2}{\sigma^2} D_N(q - \mu)^T \dot{q} \quad (5.24)$$

Therefore, we have

$$\dot{D}_N - 2C_N(q,\dot{q}) = \frac{2}{\sigma^2}\left[D_N(q)\dot{q}[q - \mu]^T - [q - \mu]\dot{q}^T D_N(q)\right] \quad (5.25)$$

which is obviously skew-symmetric.

Property 5.4:

Neural network-based robot dynamics (5.14) is linear-in-the-parameters as given by Equation (5.31).

Proof: From Equation (5.23), we have, for $v \in \mathfrak{R}^n$,

$$C(q,\dot{q})v = C_N(q,\dot{q})v + E_C v \quad (5.26)$$

where

$$C_N(q,\dot{q})v = -\frac{1}{\sigma^2}\left(D_N(q)[q - \mu]^T \dot{q}v + D_N(q)\dot{q}[q - \mu]^T v - [q - \mu]^T \dot{q}D_N(q)v\right)$$

$$= D_N(q)x \quad (5.27)$$

Here, based on the definition of the permutation operator in Section 2, the "dummy" vector x is defined by

$$x = \frac{1}{\sigma^2}\left[-[q - \mu]^T \dot{q}v - \dot{q}[q - \mu]^T v + \otimes[q - \mu]\dot{q}^T v\right] \quad (5.28)$$

Therefore, $C_N(q,\dot{q})v$ can be conveniently written as

$$C_N(q,\dot{q})v = D_N(q)x + E_C v \quad (5.29)$$

Note that $G_N(q)$ can be conveniently expressed, using the GL operator, as

$$G_N(q) = [\{\theta_0\}^T \bullet \{\xi_0(q)\}] = \begin{bmatrix} \theta_{10} \\ \theta_{20} \\ \dots \\ \theta_{n0} \end{bmatrix}^T \bullet \begin{bmatrix} \xi_{10}(q) \\ \xi_{20}(q) \\ \dots \\ \xi_{n0}(q) \end{bmatrix} \quad (5.30)$$

It follows that, for v and $a \in \Re^n$, the static neural network model can be expressed in the following *linear-in-the-parameters form*:

$$D(q)a + C(q,\dot{q})v + G(q) = D_N(q)a + C_N(q,\dot{q})v + G_N(q) + E_1$$

$$= D_N(q)(a+x) + G_N(q) + E_1$$

$$= [\{\Omega\}^T \bullet \{\Xi\}](a+x) + [\{\theta_0\}^T \bullet \{\xi_0\}] + E_1 \qquad (5.31)$$

where $\{\Omega\}$ and $\{\theta_0\}$ are the unknown constant neural network parameters. Also, $\{\Xi\}(\ddot{q}_d + z)$ and $\{\xi_0\}$ are the known regressors, and $E_1 = E_D a + E_C v + E_G$ with E_D, E_C, and E_G being the NN approximation errors. Attractive adaptive controllers will be designed based on this property.

As the individual network given in (5.3) and (5.4) can only approximate its corresponding nonlinear function to a specific accuracy on a compact set A, i.e., $q \in A$, Equation (5.31) is valid only on a given compact set A. Let A be a unit ball with respect to an appropriate weighted norm function, and region A_d completely contain A as in [12]. Thus A and A_d can be defined as

$$A = \left\{ q: \ \|q - q_0\|_{p,\omega} \le 1 \right\} \qquad (5.32)$$

$$A_d = \left\{ q: \ \|q - q_0\|_{p,\omega} \le 1 + \beta \right\} \qquad (5.33)$$

where q_0 fixes the absolute location of the set in the workspace of the robot, β is a positive constant which is large enough to include all possible configurations, and $\|x\|_{p,\omega}$ is a weighted *p-norm* of the form

$$\|x\|_{p,\omega} = \left\{ \sum_{i=1}^{n} \left(\frac{x_i}{\omega_i} \right)^p \right\}^{1/p} \qquad (5.34)$$

or, in the limiting case $p \to \infty$,

$$\|x\|_{\infty,\omega} = \max\left(\frac{x_1}{\omega_1}, \frac{x_2}{\omega_2}, \cdots, \frac{x_n}{\omega_n} \right) \qquad (5.35)$$

for a set of strictly positive weights $\omega_i, i = 1,...,n$. When $q \notin A$, Equation (5.31) does not hold, because neural network approximation is not guaranteed as explained.

6 Controller Design

Let $q_d, \dot{q}_d, \ddot{q}_d \in \Re^n$ be the position, velocity, and acceleration vectors of the desired trajectory. Let $\Lambda \in R^{n \times n}$ be a positive definite matrix. Define

$$e = q_d - q \tag{6.1}$$

$$\dot{q}_r = \dot{q}_d + \Lambda e \tag{6.2}$$

$$r = \dot{q}_r - \dot{q} = \dot{e} + \Lambda e \tag{6.3}$$

Therefore, r is the well-known new tracking error measure [21].

Lemma 6.1 Let $e(t) = h * r$, where $h = L^{-1}(H(s))$ and $H(s)$ is an $n \times n$ strictly proper, exponentially stable, transfer function. Then $r \in L_n^2 \Rightarrow e \in L_n^2 \cap L_n^\infty$, $\dot{e} \in L_n^2$, e is continuous and $e \to 0$ as $t \to \infty$. If, in addition, $r \to 0$ as $t \to \infty$, then $\dot{e} \to 0$ [22].

Therefore, to study the stability of e and $\dot{e}$, we need to study only the properties of r for $q \in A$.

Let $(\hat{.})$ be the estimate of $(.)$, and define $(\tilde{.}) = (.) - (\hat{.})$. Let $\hat{D}_N(q)$, $\hat{C}_N(q, \dot{q})$, and $\hat{G}_N(q)$ correspond to the estimates of $D_N(q)$, $C_N(q, \dot{q})$, and $G_N(q)$, respectively, obtained by substituting the true parameter matrix $\{\Omega\}$ and vector $\{\theta_0\}$ by the estimated parameter matrix $\{\hat{\Omega}\}$ and vector $\{\hat{\theta}_0\}$. From the properties listed for the NN model (5.14), we have

$$\hat{D}_N(q)\ddot{q}_r + \hat{C}_N(q, \dot{q})\dot{q}_r + \hat{G}_N(q) = [\{\hat{\Omega}\}^T \bullet \{\Xi\}]z + [\{\hat{\theta}_0\}^T \bullet \{\xi_0\}] \tag{6.4}$$

where

$$z = \ddot{q}_r - \frac{1}{\sigma^2}[q - \mu]^T \dot{q}\dot{q}_r - \frac{1}{\sigma^2}\dot{q}[q - \mu]^T \dot{q}_r + \otimes[q - \mu]\dot{q}^T \dot{q}_r \frac{1}{\sigma^2} \tag{6.5}$$

Since $\dot{q} = \dot{q}_r - r(t)$, $\ddot{q} = \ddot{q}_r - \dot{r}(t)$, we have

$$D(q)\ddot{q} + C(q, \dot{q})\dot{q} + G(q) = D(q)\ddot{q}_r + C(q, \dot{q})\dot{q}_r + G(q) - D(q)\dot{r} - C(q, \dot{q})r \tag{6.6}$$

Consider the general controller of the form, for $q \in A_d$,

$$\tau = \hat{D}_N(q)\ddot{q}_r + \hat{C}_N(q, \dot{q})\dot{q}_r + \hat{G}(q) + Kr + u_{sl} \tag{6.7}$$

$$= \{\hat{\Omega}\}^T \bullet \{\Xi\}z + \{\hat{\theta}_0\}^T \bullet \{\xi_0\} + Kr + u_{sl} \tag{6.8}$$

where the sliding mode control component u_{sl} is defined as

$$u_{sl} := F \bullet \mathrm{sgn}(r) = \begin{bmatrix} f_1 \, \mathrm{sgn}(r_1) \\ f_2 \, \mathrm{sgn}(r_2) \\ \cdots \\ f_n \, \mathrm{sgn}(r_n) \end{bmatrix}, \qquad F = \begin{bmatrix} f_1 \\ f_2 \\ \cdots \\ f_n \end{bmatrix}, \mathrm{sgn}(r) = \begin{bmatrix} \mathrm{sgn}(r_1) \\ \mathrm{sgn}(r_2) \\ \cdots \\ \mathrm{sgn}(r_n) \end{bmatrix} \qquad (6.9)$$

Depending on whether q is in the compact set A or not, the sliding mode control u_{sl} is switched from one stage to another as is shown later in **Theorem 6.1**. Hence, in this chapter, it is referred to as a two-stage controller.

By combining Equations (6.6) and (6.7), we have an error Equation of the form

$$D(q)\dot{r} + C(q,\dot{q})r + Kr + F \bullet \mathrm{sgn}(r)$$

$$= D(q)\ddot{q}_r + C(q,\dot{q})\dot{q}_r + G(q) - \hat{D}_N(q)\ddot{q}_r - \hat{C}_N(q,\dot{q})\dot{q}_r - \hat{G}_N(q) \quad (6.10)$$

Recalling Equation (5.31), the right-hand side of Equation (6.10), for $q \in A$, can be written as

$$D(q)\ddot{q}_r + C(q,\dot{q})\dot{q}_r + G(q) - \hat{D}(q)\ddot{q}_r - \hat{C}(q,\dot{q})\dot{q}_r - \hat{G}(q)$$

$$= D_N(q)\ddot{q}_r + C_N(q,\dot{q})\dot{q}_r + G_N(q) - \hat{D}(q)\ddot{q}_r - \hat{C}(q,\dot{q})\dot{q}_r - \hat{G}(q) + E$$

$$= -[\{\tilde{\Omega}\}^T \bullet \{\Xi\}]z - [\{\tilde{\theta}_0\}^T \bullet \{\xi_0\}] + E \qquad (6.11)$$

where $E = E_D\ddot{q}_r + E_C\dot{q}_r + E_G$ with E_D, E_C, and E_G being the bounded NN approximation errors.

In general, Equation (6.10) can be written as

$$D(q)\dot{r} + C(q,\dot{q})r + Kr + F \bullet \mathrm{sgn}(r)$$

$$= D(q)\ddot{q}_r + C(q,\dot{q})\dot{q}_r + G(q) - \hat{D}_N(q)\ddot{q}_r - \hat{C}_N(q,\dot{q})\dot{q}_r - \hat{G}_N(q) + E$$

$$= -[\{\tilde{\Omega}\}^T \bullet \{\Xi\}]z - [\{\tilde{\theta}_0\}^T \bullet \{\xi_0\}] + E + M \qquad (6.12)$$

with $M = 0$ for $q \in A$, and $M \neq 0$ as indicated by (5.33).

The closed-loop stability properties of the system (6.10) and (6.12) are stated in the following theorem:

Theorem 6.1 Considering the mappings (6.10) and (6.12), under the assumptions $K^T > 0$, $K_i = K_i^T \geq 0$, we have the two stages

(1)   When $q \notin A$, if

$$\dot{\hat{\omega}}_i = 0 \qquad (6.13)$$

$$\dot{\hat{\theta}}_{i0} = 0 \qquad (6.14)$$

$$F > \left| D(q)\ddot{q}_r + C(q,\dot{q})\dot{q}_r + G(q) - \{\hat{\Omega}\}^T \bullet \{\Xi\}z - \{\hat{\theta}_0\}^T \bullet \{\xi_0\} \right| \quad (6.15)$$

then, region A is attractive, hence, invariant; all trajectories starting inside $A(t = 0)$ remain inside A for all $t \geq 0$.

(2)   In addition, when $q \in A$, if

$$\dot{\hat{\omega}}_i = \Gamma_i \bullet \{\xi_i\} z r_i \tag{6.16}$$

$$\dot{\hat{\theta}}_{i0} = \Gamma_{i0}\xi_{i0}r_i \tag{6.17}$$

$$F \geq |E| \tag{6.18}$$

where Γ_i and Γ_{i0} are dimensionally compatible symmetric positive-definite matrices, and z is defined by (6.5), then

(a) $\hat{\omega}_i$ and $\hat{\theta}_{i0} \in L^\infty$;

(b) $r \to 0$ as $t \to \infty$.

Thus, from Lemma 6.1, we have $e \in L_n^2 \cap L_n^\infty$, e is continuous, and $\dot{e} \to 0$ as $t \to \infty$.

Proof:

Define the non-negative function V as

$$V = \frac{1}{2}r^T D(q)r + \frac{1}{2}\sum_{i=1}^n \tilde{\omega}_i^T \Gamma_i^{-1}\tilde{\omega}_i + \frac{1}{2}\sum_{i=1}^n \tilde{\theta}_{i0}^T \Gamma_{i0}^{-1}\tilde{\theta}_{i0} \tag{6.19}$$

Its time derivative is given by

$$\dot{V} = r^T \left[D(q)\dot{r} + C(q,\dot{q})r \right] + \sum_{i=1}^n \tilde{\omega}_i^T \Gamma_i^{-1}\dot{\tilde{\omega}}_i + \sum_{i=1}^n \tilde{\theta}_{i0}^T \Gamma_{i0}^{-1}\dot{\tilde{\theta}}_{i0}$$

where the skew-symmetric property of $\dot{D}(q) - 2C(q,\dot{q})$ has been used.

Substituting the error Equation (6.10), we have

$$\dot{V} = -r^T Kr - r^T F \bullet \text{sgn}(r) + r^T \left[D(q)\ddot{q}_r + C(q,\dot{q})\dot{q}_r + G(q) \right.$$

$$\left. -\hat{D}_N(q)\ddot{q}_r - \hat{C}_N(q,\dot{q})\dot{q}_r - \hat{G}_N(q) \right] + \sum_{i=1}^n \tilde{\omega}_i^T \Gamma_i^{-1}\dot{\tilde{\omega}}_i + \sum_{i=1}^n \tilde{\theta}_{i0}^T \Gamma_{i0}^{-1}\dot{\tilde{\theta}}_{i0} \tag{6.20}$$

When $q \notin A$, the parameter adaptations (6.13)-(6.14) are inactive. Since $\{\Omega\}$ and $\{\theta_0\}$ are constant matrices, we have

$$\dot{\tilde{\omega}}_i = 0 \tag{6.21}$$

$$\dot{\hat{\theta}}_{i0} = 0 \tag{6.22}$$

which leads to

$$\dot{V} = -r^T Kr - r^T [F \bullet \text{sgn}(r) + D(q)\ddot{q}_r + C(q,\dot{q})\dot{q}_r + G(q)$$

$$-\{\hat{\Omega}\}^T \bullet \{\Xi\}z - \{\hat{\theta}_0\}^T \bullet \{\xi_0\}] \tag{6.23}$$

From Equation (6.15) in Theorem 6.1,

$$F > \left| D(q)\ddot{q}_r + C(q,\dot{q})\dot{q}_r + G(q) - \{\hat{\Omega}\}^T \bullet \{\Xi\}z - \{\hat{\theta}_0\}^T \bullet \{\xi_0\} \right| \tag{6.24}$$

Therefore

$$\dot{V} \le -r^T Kr - \varepsilon|r|, \quad \varepsilon > 0 \tag{6.25}$$

which guarantees that region A is attractive, hence invariant; all trajectories starting inside $A(t = 0)$ remain inside A for all $t \ge 0$. In finite time, t_0, the control will drive the system into A.

When $q(t_0) \in A$, from Equation (6.11), we have

$$\dot{V} = -r^T Kr - r^T F \bullet \text{sgn}(r) + r^T \left[\{\hat{\Omega}\}^T \bullet \{\Xi\}z + \{\hat{\theta}_0\}^T \bullet \{\xi_0\} + E \right]$$

$$+ \sum_{i=1}^{n} \tilde{\omega}_i^T \Gamma_i^{-1} \dot{\tilde{\omega}}_i + \sum_{i=1}^{n} \tilde{\theta}_{i0}^T \Gamma_i^{-1} \dot{\tilde{\theta}}_{i0}$$

$$= -r^T Kr - r^T F \bullet \text{sgn}(r) + \begin{bmatrix} r_1 & r_2 & \cdots & r_n \end{bmatrix} \begin{bmatrix} \{\tilde{\omega}_1\}^T \bullet \{\xi_1\}z \\ \{\tilde{\omega}_2\}^T \bullet \{\xi_2\}z \\ \cdots \\ \{\tilde{\omega}_n\}^T \bullet \{\xi_n\}z \end{bmatrix}$$

$$+ \begin{bmatrix} r_1 & r_2 & \cdots & r_n \end{bmatrix} \begin{bmatrix} \tilde{\omega}_{10}^T \xi_{10} \\ \tilde{\omega}_{20}^T \xi_{20} \\ \cdots \\ \tilde{\omega}_{n0}^T \xi_{n0} \end{bmatrix} + \sum_{i=1}^{n} \tilde{\omega}_i^T \Gamma_i^{-1} \dot{\tilde{\omega}}_i + \sum_{i=1}^{n} \tilde{\theta}_{i0}^T \Gamma_i^{-1} \dot{\tilde{\theta}}_{i0} + r^T E$$

$$= -r^T Kr - r^T F \bullet \text{sgn}(r) + \sum_{i=1}^{n} \{\tilde{\omega}_i\}^T \bullet \{\xi_i\} z r_i + \sum_{i=1}^{n} \tilde{\omega}_{i0}^T \xi_{i0} r_i$$

$$+ \sum_{i=1}^{n} \tilde{\omega}_i^T \Gamma_i^{-1} \dot{\tilde{\omega}}_i + \sum_{i=1}^{n} \tilde{\theta}_{i0}^T \Gamma_i^{-1} \dot{\tilde{\theta}}_{i0} + r^T E$$

Substituting into the above Equation the parameter adaptation algorithms (6.16)-(6.17), (noting that the ideal neural network weights ω_i and θ_{i0} are constants),

$$\dot{\tilde{\omega}}_i = -\Gamma_i \bullet \{\xi_i\} z r_i \tag{6.26}$$

$$\dot{\tilde{\theta}}_{i0} = -\Gamma_{i0}\xi_{i0} r_i \tag{6.27}$$

we have

$$\dot{V} = -r^T K r - r^T \left(F \bullet \text{sgn}(r) + E\right) \tag{6.28}$$

From Theorem 6.1, we have

$$F \geq |E| \tag{6.29}$$

Therefore, V is a Lyapunov function and

$$\lambda_{\min}(K)\int_0^t r^T r d\tau \leq \int_0^t r^T K r d\tau \leq V(0) \tag{6.30}$$

(a) Since

$$\dot{V} \leq -r^T K r \leq 0 \tag{6.31}$$

it follows that $0 \leq V(t) \leq V(0)$, $\forall\ t \geq 0$. Hence $V(t) \in L^\infty \Rightarrow \int_0^t r(\tau)d\tau$, $\tilde{\omega}_i$ and $\tilde{\theta}_{t0} \in L^\infty$, i.e., $\hat{\omega}_i$ and $\hat{\theta}_{i0} \in L^\infty$.

(b) Since $V(0)$ and $\lambda_{\min}(K)$ are positive constants, it follows that $r \in L_n^2$. Consequently, from Lemma 6.1, $e \in L_n^2 \cap L_n^\infty$, e is continuous and goes to 0 as $t \rightarrow \infty$ and $\dot{e} \in L_n^2$. By noting that $r \in L_2^n$, q_d, $\dot{q}_d$, $\ddot{q}_d \in L_n^\infty$, E are bounded error tolerances, and $\{\Xi\}$ and $\{\xi_0\}$ are bounded basis functions, we obtain $\dot{r} \in L_\infty^n$ from Equation (6.10). Since $\dot{r} \in L_\infty^n$, r is uniformly continuous. The proof is completed using the implication:

$$r \text{ uniformly continuously, and } r \in L_2^n \Rightarrow r \rightarrow 0 \text{ as } t \rightarrow \infty \Rightarrow \dot{e} \rightarrow 0.$$

QED.

Remarks:

(1)   If $q(t=0) \notin A$, the error of neural network approximation may be large. In this stage, the NN approximations, $\{\hat{\Omega}\}^T \bullet \{\Xi\} z - \{\hat{\theta}_0\}^T \bullet \{\xi_0\}$, will, at best, approximately cancel the nonlinear couplings, $D(q)\ddot{q}_r + G(q) + C(q,\dot{q})\dot{q}_r$.+ Therefore, a relatively large sliding mode gain (6.15) is required to drive the system into region A and guarantee that A be attractive. The presence of the sliding mode term will guarantee the system reaches A in a finite time interval. This is referred to as the first stage of

(2) the controller. Since NN approximation is not valid for $q(t = 0) \notin A$, parameter adaptation is not activated as shown in Equations (6.13)-(6.14).

Since A is invariant, all trajectories starting inside $A(t = 0)$ remain inside A for all $t \geq 0$. For $q \in A$, the NN approximations, $[\{\hat{\Omega}\}^T \bullet \{\Xi\}]z - [\{\hat{\theta}_0\}^T \bullet \{\xi_0\}]$ can "accurately" approximate the nonlinear couplings, $D(q)\ddot{q}_r + C(q,\dot{q})\dot{q}_r + G(q)$ with a small tolerance of E. In this stage (referred to as the second stage), the parameter adaptation is activated to solve problems of the unknown neural network weights as shown by Equations (6.16)-(6.17). Subsequently, a relatively small sliding mode gain (6.18) is enough to guarantee robust closed-loop stability.

(3) If the compact set A includes all the possible configurations of the robot, i.e., $A = A_d$, then the proposed two-stage controller becomes a single stage controller.

(4) If the nonlinear functions $D(q)$ and $G(q)$ are in the functional range of the neural network emulators $D_N(q)$ and $G_N(q)$, i.e., $E_D = 0$ and $E_G = 0$, and subsequently, $E = 0$ in (5.17), the sliding mode control part can be dropped by letting $F = 0$ while the closed-loop system is still stable. If the nonlinear functions $D(q)$ and $G(q)$ are not in the NN functional ranges, the existence of $F > 0$ will inevitably introduce chattering in the control signal. To solve this problem, a boundary layer can be introduced as shown in [21].

7 Case Study

Similar controllers have been successfully implemented in controlling a multi-axis mechanical system which is highly nonlinear and exhibits strong couplings among axes [23,24]. Intensive simulation results are presented here to show the effectiveness of the controller for robots as well. For the simulation studies, we consider a planar two-link manipulator, whose dynamics are described by

$$D(q)\ddot{q} + C(q,\dot{q})\dot{q} + G(q) = \tau \tag{7.1}$$

where

$$D(q) = \begin{bmatrix} m_1 + m_2 + 2m_3 \cos q_2 & m_2 + m_3 \cos q_2 \\ m_2 + m_3 \cos q_2 & m_2 \end{bmatrix}$$

$$C(q,\dot{q}) = \begin{bmatrix} -m_3 \dot{q}_2 \sin q_2 & -m_3(\dot{q}_1 + \dot{q}_2)\sin q_2 \\ m_3 \dot{q}_1 \sin q_2 & 0 \end{bmatrix}$$

$$G(q) = \begin{bmatrix} m_4 \cos q_1 + m_5 \cos(q_1 + q_2) \\ m_5 \cos(q_1 + q_2) \end{bmatrix}$$

The parameters of interest, m_is, are defined by $M = P + p_l L$ with

$$M = [m_1, m_2, m_3, m_4, m_5]^T, \quad P = [p_1, p_2, p_3, p_4, p_5]^T$$

$$L = [L_1^2, L_2^2, L_1 L_2, L_1, L_2]^T$$

where p_l is the payload, L_1 and L_2 are the lengths of links 1 and 2, respectively, and P is the parameter of the robot itself.

7.1 Trajectory Planning

The desired trajectory for each axis is expressed as a Hermite polynomial of the third degree in t with continuous bounded position, velocity, and bounded acceleration. The general expression for the desired position trajectory is

$$q_d(t) = q_0 + \left(-2\frac{t^3}{t_d^3} + 3\frac{t^2}{t_d^2} \right)(q_f - q_0) \qquad (7.2)$$

where q_0 and q_f are the initial and final positions of the arm, and t_d represents the time at which the desired arm trajectory reaches the desired final position. In simulation tests, the following values were chosen for the desired trajectories:

$$t_d = 1.0 \text{ sec}, \quad q_d(0) = [0.0, 0.0]^T \text{ rad}, \quad q_d(t_d) = [1.0, 2.0]^T \text{ rad}$$

7.2 Simulation Settings

For the elements of $D(q)$ and $G(q)$, 200-node static networks were chosen, with $\mu = 0$ and $\sigma^2 = 10.0$. The parameter matrices $\{\hat{\Omega}\}(0)$, $\{\hat{\theta}_0\}(0)$ are initialized to zero by assuming that no knowledge about the system is available.

The true parameters of the robot used for the simulation are

$$P = [1.66, 0.42, 0.63, 3.75, 1.25]^T \text{ kgm}^2$$

Suppose the robot initially rests at

$$q_0 = [0.09, -0.09]^T \text{ rad}, \quad \dot{q}_0 = [0.0, 0.0]^T \text{ rad}$$

and the gains for the controller were chosen as

$$K = \text{diag}[10.0], \quad \Lambda = \text{diag}[5.0] \qquad (7.3)$$

In order to test load disturbance rejection of the proposed controller, a payload $p_l =$ 4.0 kg was added at time $t = 3.0$ sec.

7.3 Non-Adaptive Control

Let us first investigate the control performance when the weight adaptations of the neural network controller were not activated. In this case, the position, velocity tracking performances of the robot and the corresponding control signals are shown in Figures 2, 3, and 4, respectively. It can be seen that the non-adaptive neural network control has a large tracking error, and cannot handle changes in the system such as modeling errors.

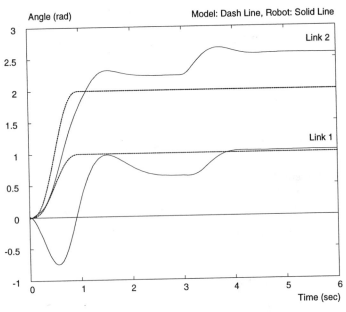

Figure 2: Position tracking without adaptation.

7.4 Adaptive Control

Several cases studied with the proposed adaptive controller include (i) slow adaptation without sliding mode control, (ii) slow adaptation with sliding mode control, and (iii) fast adaptation. First, the adaptation algorithms (6.16)-(6.17) were activated with relatively small gains $\Gamma_i = \text{diag}[0.01]$, $\Gamma_{i0} = \text{diag}[2.0]$. The position and velocity tracking performance of the robot and the corresponding control inputs are shown in Figures 5, 6, and 7, respectively. It can be seen that the tracking error is much smaller than in the non-adaptive case   because of the "learning" mechanism.

Second, the sliding mode term is included in the controller with $F = 5.0$. The position and velocity tracking performance of the robot and the corresponding control inputs are shown in Figures 8, 9, and 10. It can be seen that the tracking performance of the system further improved because of the introduction of sliding mode control term.

Finally, the adaptation algorithms (6.16)-(6.17) were activated with relatively large gains $\Gamma_i = \text{diag}[0.1]$, $\Gamma_{i0} = \text{diag}[10.0]$ without sliding mode control. The position and velocity tracking performance of the robot and the corresponding control inputs are shown in Figures 11, 12, and 13, respectively. It can be seen that the tracking error of the system becomes even smaller. It was found that the tracking performance of the system becomes better for relatively faster adaptation. It verified that the neural networks and adaptive mechanisms can control the system satisfactorily if the sampling interval is small enough. The simulation results thus demonstrate that the proposed adaptive neural network control can effectively control an unknown nonlinear dynamic system with load disturbances. Different tracking performance can be achieved by adjusting the parameter adaptation gains and other factors, such as the size of the networks.

8 Summary

In this chapter, the control problems of model mis-matching were investigated for the static neural network model of robots [1,2]. Static NNs $D_N(q)$ and $G_N(q)$ are used only to model $D(q)$ and $G(q)$, respectively, and $C_N(q,\dot{q})$ for $C(q,\dot{q})$ can be constructed directly from $D_N(q)$. The size of the resulting model of robots is smaller than the dynamic ones. This is very desirable for real-time implementation in view of low computational power needed. The well-known structural properties of robots, such as linear-in-the-parameters dynamics, were examined for the neural network models. Subsequently, a two-stage adaptive controller was suggested based on the neural network model. It has been shown that the closed-loop system can be made asymptotically stable, by appropriately adjusting the weights of the neural networks on line and introducing the sliding mode term. Similar controllers have been successfully implemented in controlling a free gyro stabilized mirror system [23,24]. Intensive simulations were carried out to show the effectiveness and feasibility of the proposed controllers.

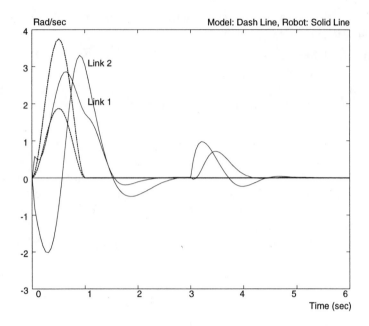

Figure 3: Velocity tracking without adaptation.

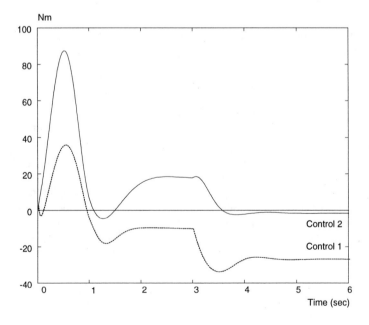

Figure 4: Control variations without adaptation.

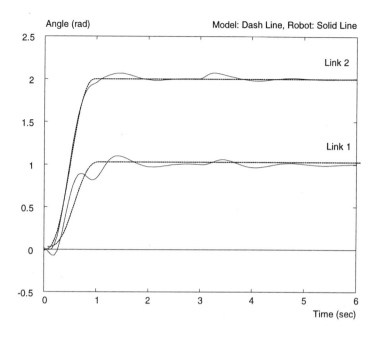

Figure 5: Position tracking with slow adaptation.

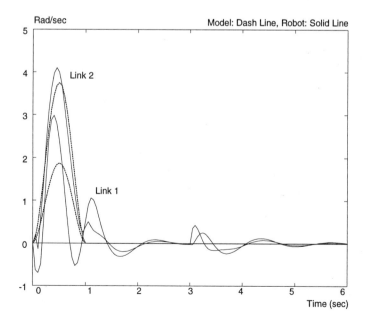

Figure 6: Velocity tracking with slow adaptation.

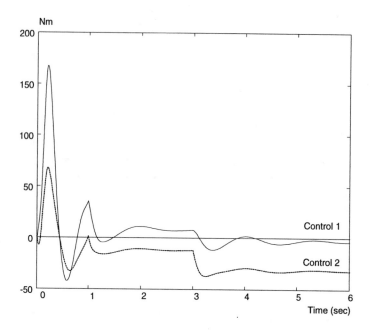

Figure 7: Control variations with slow adaptation.

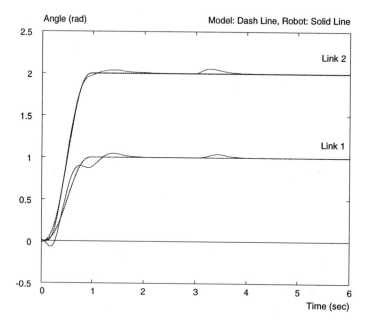

Figure 8: Position tracking with slow adaptation and sliding mode.

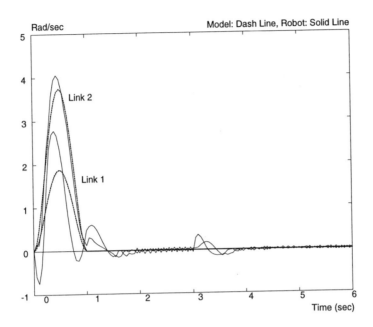

Figure 9: Velocity tracking with slow adaptation and sliding mode.

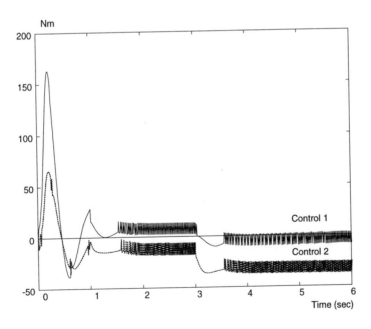

Figure 10: Control variations with slow adaptation and sliding mode.

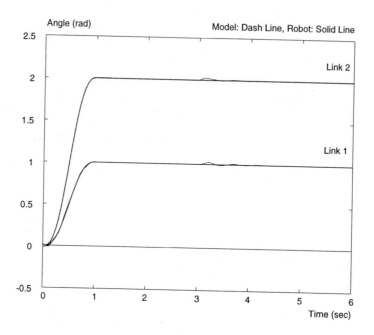

Figure 11: Position tracking with fast adaptation.

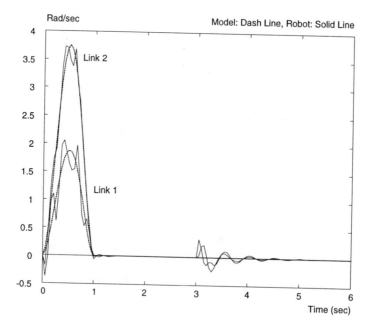

Figure 12: Velocity tracking with fast adaptation.

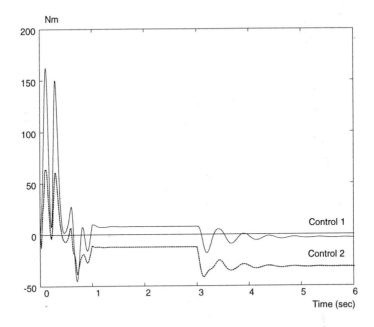

Figure 13: Control variations with fast adaptation.

References

[1] Ge, S.S., Wang, Z.-L., and Chen, Z.-J. (1994), "Adaptive static neural network control of robots," *IEEE Intl. Conf. on Industrial Technology*, 5-9, Dec., Guangzhou, P.R. China, pp. 240-244.

[2] Ge, S.S. (1996), "Robust adaptive control of robots based on static neural networks," Proceedings of 1996 Triennial World Congress (IFAC), June 30-July 5, San Francisco, California, USA, pp. 139-144.

[3] Miller, W.T., Glanz, F.H. and Kraft, I.G. (1987), "Application of a general learning algorithm to the control of a robotic manipulator," *Int. J. Robotics Res.*, Vol. 6, No. 2, pp. 84-98.

[4] Miyamoto, H., Kawato, M., Setoyama, T., and Suzuki, R. (1988), "Feedback error learning neural network for trajectory control of a robotic manipulator," *Neural Networks*, Vol. 1, pp. 251-265.

[5] Ozaki, T., Suzuki, T., Furuhashi, T., Okuma, S., and Uchikawa, Y. (1991), "Trajectory control of robotic manipulators using neural networks," *IEEE Trans. on Industrial Electronics*, Vol. 38, No. 3, pp. 195-202.

[6] Yabuta, T. and Yamada, T. (1993), "Force control using neural networks," *Advanced Robotics*, Vol. 7, No. 4, pp. 395-408.

[7] Ge, S.S. (1996), "Robust adaptive NN feedback linearization control of nonlinear systems," *Int. J. of Systems Science*, Vol. 27, No. 12, pp. 1327-1338.

[8] Ge, S.S. and Hang, C.C. (1995), "Direct adaptive neural network control of robots," *Int. J. of Systems Science*, Vol.27, No. 6, pp. 533-542.

[9] Ge, S.S. and Lee, T.H. (1997), "Robust adaptive neural network control of a class of nonlinear mechanical systems," *Journal of Systems and Control Engineering*, Proceedings of Institution of Mechanical Engineers, Vol. 211, No. 3, pp. 171-181, 1997.

[10] Lewis, F.L., Liu, K., and Yesildirek, A., (1995), "Neural Net Robot Controller with Guaranteed Tracking Performance," *IEEE Trans. on Neural Networks*, Vol. 6, No. 3, pp. 703-715.

[11] Lewis, F.L., Yesildirek, A. and Liu, K. (1996), "Multilayer neural-net robot controller with guaranteed tracking performance," *IEEE Trans. on Neural Networks*, Vol. 7, No. 2, pp. 388-399.

[12] Sanner, R.M. and Slotine, J.J.E. (1992), "Gaussian networks for direct adaptive control," *IEEE Trans. on Neural Networks,* Vol. 3, No. 6, pp. 837-863.

[13] Tzirkel-Hancock, E. and Fallside, F. (1991), *A Direct Control Method for a Class of Nonlinear Systems Using Neural Networks*, Cambridge University, CUED/F-INFENG/TR65.

[14] Runmelhart, D.E. and McClelland, J.L. (1986), *Parallel Distributed Processing*, Cambridge, MA: MIT Press.

[15] Agaian, S.S. (1985), *Hadamard Matrices and Their Applications*, Springer-Verlag, Berlin Heidelberg.

[16] Cameron, P.J. and Van Lint, J.H. (1991), *Designs, Graphs, Codes and Their Links*, Cambridge University Press, Cambridge.

[17] Poggio, T. and Girosi, F. (1989), "A theory of networks for approximation and learning," *Artificial Intelligence Lab. Memo*, No. 1140, MIT, Cambridge, MA, July.

[18] Girosi, F. and Poggio, T. (1989), "Networks and the best approximation property," *Artificial Intelligence Lab. Memo*, No. 1164, MIT, Cambridge, MA, Oct.

[19] Poggio, T. and Girosi, F. (1990), "Networks for approximation and learning," *Proc. of IEEE*, Vol. 78, No. 9, pp. 1481-1497.

[20] Spong, M.W. and Vidyasagar, M. (1988), *Robot Dynamics and Control*, John Wiley & Sons, Inc., New York.

[21] Slotine, J.J.E. and Li, W. (1991), *Applied Nonlinear Control*, Prentice Hall, Englewood Cliffs.

[22] Desoer, C.A. and Vidyasagar, M. (1975), *Feedback Systems: Input-Output Properties*, Academic Press, New York.

[23] Ge, S.S., Lee, T.H., and Harris, C.J. (1998) *Adaptive Neural Network Control of Robotic Manipulators*, World Scientific, Singapore.

[24] Ge, S.S., Lee, T.H., and Zhao, Q. (1997), "Real-time neural network control of a free gyro stabilized mirror system," *Proceedings of American Control Conference,* Albuquerque, NM, U.S.A., pp. 1076-1080.

8 | ERROR CORRECTION USING FUZZY LOGIC IN VEHICLE LOAD MEASUREMENT

L. Su
Info-Telematics R&D Division
Yazaki Corporation
Mishuku, 1500, Susono City, Shizuoka Prefecture, 410-11
Japan

C. Komata
In-Vehicle Integration Systems R&D Division
Yazaki Corporation
Ohoka, 2771, Numazu City, Shizuoka Prefecture, 410
Japan

Vehicle load measurement is an important technology in ITS (Intelligent Transport Systems). For public authorities, load measurement using bypass stations or other facilities has been proposed to enable warning of overloading, enforcement of regulations, and possible linkage with traffic management systems like electronic tolling. The vehicle load indicator by Yazaki Corporation is an in-vehicle device for load measurement.

In this chapter, we introduce an error correction technique using fuzzy logic for accuracy improvement in vehicle load measurement.

1 Introduction

A number of in-vehicle systems have been realized for implementing Intelligent Transport Systems (ITS). Vehicle load measurement is considered a very important tool by public authorities and private users for enforcing regulations and other possible linkage with traffic management systems. Weighing station bypass [1] and an automatic measuring system for heavy freight vehicles [2] are examples in public use.

Conversely, from the viewpoint of a combination of active safety control of braking and speed into an in-vehicle networking system, in-vehicle load measurement is more desirable. Also, there is a need for assisting the vehicle driver to check the loading states beforehand to avoid the accident caused by overloading. An vehicle load indicator by Yazaki Corporation [3] is an in-vehicle device for this purpose.

Based on the structure of in-vehicle load indicator, many measurement methods have been proposed, which differ from current sensor types, the sensor installation, and the sensor output and signal processing. In this chapter, we consider the vehicle load indicator using magnetostrictive gage sensors [3]. This load indicator provides a basic function of measuring a vehicle load by summing all respective outputs from magnetostrictive gage sensors embedded in the shackle pins which are used to fix the laminated leaf springs to the brackets on the deck frame, as shown in Figure 1.

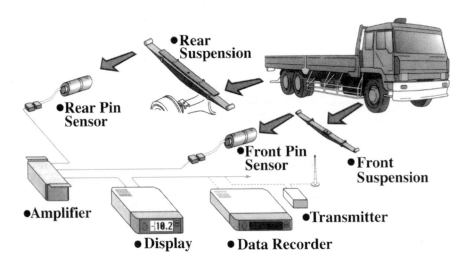

Figure 1: Overview of vehicle load indicator.

In the setup stage, the sensor outputs have to be pre-adjusted under a particular condition. For example, we use a 10-ton standard load (for a model truck of 3 axles with maximum authorized payload: 10 ton). We assume that the load distributes uniformly on the deck, then measure the weight changes on the axles, and adjust the sensor outputs with reference to the weight changes.

Due to physical constraints for sensor installation and methodology for sensor adjustment, accuracy is still a difficult obstacle in practical application of vehicle load indicator. Usually, the real load measuring condition is critical and differs from the setup condition. This fact has a mixed influence on the accuracy of vehicle load measurement due to the changes of sensor characteristics, load distribution, and so on. It is not easy to represent these inaccuracies in vehicle load measurement in a mathematical model.

We used fuzzy logic for improving the accuracy of vehicle load measurement. In this study, we introduce two new parameters, i.e., the deck inclinations in right-left and front-back directions, to describe the load distribution. Our error correction technique is to treat these parameters using fuzzy logic to estimate the error contained in a load measurement.

Finally, we present a discussion on vehicle load measurement and the newly introduced inclination parameters for ITS use, including the simulation results based on loading tests.

2 Vehicle Load Indicator

As shown in Figure1, the in-vehicle load indicator generally consists of two parts, i.e., load sensing and data processing. For the sake of safety regulation by law, the place for installing sensors is restricted.

In this chapter, we consider the construction of magnetostrictive gage sensors and assume that the vehicle symbol 146 \f "Times New Roman" \s 10'}s deck is held on the deck frame which is supported by axles in front, middle, and back via laminated leaf springs. The sensors are embedded in the shackle pins (see Figure 2) which are used to fix the laminated leaf springs to the brackets on the deck frame, and installed symmetrically on the right and left sides of the front, middle, and back axles.

Figure 3 shows the block diagram of the load sensing unit. Due to the load change, an exciting current signal from the oscillator flows to the primary coil of the transformer which produces an excitation voltage on the secondary coil. The detector changes the excitation voltage into a DC voltage and the V/F circuit converts the DC voltage to a frequency signal as the external sensing output with respect to the load change, in a linear sense.

We represent the sensor outputs as W_1, W_3, W_5 and W_2, W_4, W_6, by left and right in the front, middle and back, respectively (see Figure 4).

The vehicle load is then calculated by summing the outputs from all sensors, i.e.,

$$W = \text{sum} (W_i), \quad I = 1, 2, \dots \tag{1-a}$$

and

$$W_i = \kappa_i \, \varepsilon_i + \alpha_i \tag{1-b}$$

where $W_i(0)$ is a function of the load distributed around i-th sensor, κ_i is the gain factor determined during pre-adjustment, ε_i is the sensing output and α_i is the sensor hysteresis factor.

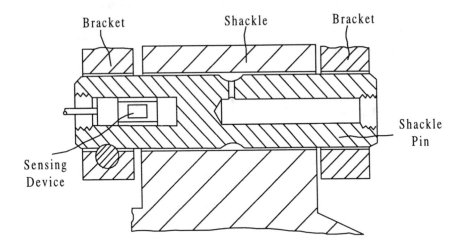

Figure 2: Sensor installation.

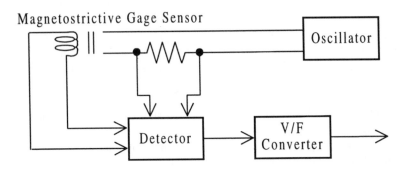

Figure 3: Block diagram of load sensing.

For the calculation of gain κ_i in (1-b), we assume that the vehicle load distributes uniformly on the deck. For example, we use a 10-ton standard load (for a model truck with 3 axles and 10-ton maximum authorized payload), measure the weight changes on the axles, and determine the sensor output gains with reference to the weight changes compared with the case of an empty load.

In other words, we assumed that a real load distributes uniformly on the deck and the weights on the axles change linearly. Obviously, such requirements for real load measurement are too critical in practical application.

The gap between the real and desirable measuring conditions contributes to the inaccuracy in load measurement. Our approach to accuracy improvement is to first quantitatively evaluate the gap of measuring conditions, i.e., the changes of loading states with a diversity of uncertainties, and then build up a fuzzy rule base to estimate the error contained in a real load measurement.

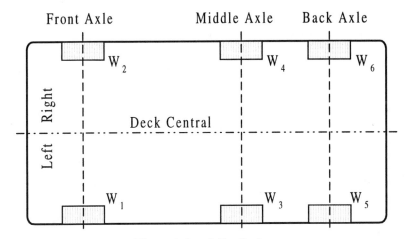

Figure 4: Load distribution.

3 Describing Vehicle Loading States

To evaluate change of measuring conditions, i.e., the change of loading states with a diversity of uncertainties, we introduce two new parameters concerned with the load distribution features.

From the sensor outputs W_i ($I = 1,2,...$), we can definitely determine other loading features, e.g., deck inclinations.

First, we introduce the deck inclination in the right-left direction (parameter δ). For the front axle, we define the inclination parameter δ_F as

$$\delta_F = (W_1 - W_2)/(W_1 + W_2). \tag{2-a}$$

For the middle and back axles, the inclination parameters δ_M and δ_B can be defined similarly.

Therefore, the deck inclination in the right-left direction is given as

$$\delta = Q_F \delta_F + Q_M \delta_M + Q_B \delta_B, \tag{2-b}$$

where Q_F, Q_M, and Q_B are weight coefficients determined by the deck frame structure.

On the other hand, to represent the inclination along front-back direction (parameter γ), we simply define

$$\gamma = [(W_3+W_4+W_5+W_6)-(W_1+W_2)]/(W_1+W_2+W_3+W_4+W_5+W_6), \qquad (3)$$

where W_i ($i=1,2,...,n$) are briefly divided into two groups by front and back, i.e., (W_1+W_2) and $(W_3+W_4+W_5+W_6)$, and $\gamma = 0$ if (3) has a zero denominator.

Given for the model truck with three axles, the above definitions can be easily extended to other types of freight vehicles.

Therefore, we can represent the changes of loading states quantitatively using these deck inclination parameters, which can be used for other purposes like error correction.

4 Fuzzy Reasoning

Using inclination parameters, the influence on load measurement accuracy can be shown using a data table obtained from previous loading tests. The next step is a data analysis to determine a basic input/output relation of the affect on accuracy and what can be used for building a fuzzy inference engine.

Based on the above analysis, we proposed a technique of treating a load measurement and its loading states, i.e., the deck inclination parameters, as input variables for the estimation of the measurement error ΔW using fuzzy logic as shown in Figure 5.

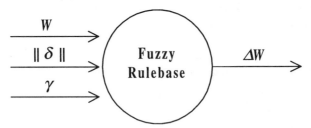

Figure 5: Fuzzy reasoning diagram.

Therefore, the load calculation of (1) is revised as

$$W_{new} = W - \Delta W \qquad (4)$$

where the inclination parameters δ and γ are used to represent the changes of loading states, while fuzzy reasoning is adopted to realize the input-output relation to the error.

For the construction of fuzzy rule base, the following steps are generally repeated in simulations.

STEP 1: Data analysis
　　　　Determine the basic input/output relation and
　　　　make a reference data table from data collected in loading tests,
　　　　which is used as desirable input/output for rule base tuning.

STEP 2: Variable analysis
　　　　Determine the details for simulation such as membership functions
　　　　and inference method.

STEP 3: Fuzzy rule base generation
　　　　Automatic fuzzy rule base generation using neural network by TIL
　　　　Shell, a tool of Togai InfraLogic, Inc. [5]
　　　　or using other methods.

STEP 4: Rule base tuning
　　　　Evaluate the rule base generated in STEP 3 using loading tests.
　....

Readers are directed to refer to the literature on fuzzy logic [6] to [10] for detailed information.

5 Simulation Experiment

In our simulation, a model truck with three axles and 10-ton maximum authorized payload is considered as a plant, in which four sensors have been installed on the right and left sides of the front and back axles, respectively. This setup uses $Q_F = 0.3$, $Q_M = 0$ and $Q_B = 0.7$ in (2-b).

To obtain the reference I/O data table, i.e., the desirable input/output reference model for fuzzy reasoning, loading tests were repeated, with known loads from 0 ton to 16 tons (load on and off) and artificial inclinations in the front-back and right-left directions.

Finally, about 3,000 data samples have been collected as the reference I/O data table for rule base generation.

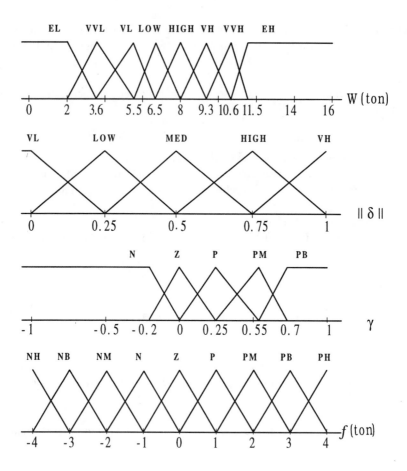

Figure 6: Membership functions.

Throughout the simulation, we use the TIL Shell, a development tool by Togai InfraLogic, Inc. [5] for an automatic generation of rules, and then evaluate the rule base using loading tests. Figure 6 shows the membership functions for variables used in fuzzy reasoning.

Based on iterative simulation experiments, a set of fuzzy rules has been determined as follows.

Applying the rules of Table 1 to all test data, by max-dot centroid inference defuzzification method (refer to [4] [5]), it is confirmed that the ratio of errors in norm form, i.e., the ratio of norm of errors after using fuzzy to without using fuzzy, has been reduced to 40%. It appears that the average error of approximately 3 tons in real loading tests was reduced to less than 0.5 ton.

In real loading tests with known loads and artificial inclinations, the effect of fuzzy logic based rules can be clearly observed. Figure7 is just a numerical example showing the effectiveness in accuracy improvement using fuzzy logic compared with the conventional method, in which known loads from 0 ton to 16 tons were used on and off with a step of 1 ton, and inclinations in the right-left direction were added artificially.

Table 1: A set of fuzzy rules.

W	$\|\| \delta \|\|$	γ symbol 224 \f "Wingdings" \s 10→"	ΔW
EL	VL or LOW	N or P	Z
EL	VL	Z or PM or PB	P
EL	LOW	PM	P
EL	MED or HIGH	P or PB	P
EL	MED or HIGH	PM	Z
EL	VH	P	P
VVL	VL	N	Z
VVL	VL	Z or P or PM or PB	P
VVL	LOW	N or Z	Z
VVL	LOW or MED	P or PM or PB	P
VVL	MED	Z	Z
VVL	HIGH	Z or P	Z
VVL	HIGH	PM	P
VVL	VH	N or Z	Z
VVL	VH	P	P
VVL	VH	PM	PM
VL	VL or LOW	N or Z or P or PM or	P
VL	MED or HIGH	PB	Z
VL	MED or HIGH	Z	P
VL	VH	P or PM	N
VL	VH	N	Z
VL	VH	Z	PM
		P	
LOW	VL	N	Z
LOW	VL	Z or P or PM or PB	P
LOW	LOW	N or P	P
LOW	LOW	PM	Z
LOW	LOW	PB	N
LOW	MED	N	N
LOW	MED	Z	Z
LOW	MED	P	P
LOW	HIGH	Z or P	Z
LOW	HIGH	PM	P
LOW	VH	N or Z	N
LOW	VH	P	PB

HIGH	VL	Z	P
HIGH	VL	N or P or PM or PB	Z
HIGH	LOW	PM	Z
HIGH	LOW or MED	Z or P	Z
HIGH	HIGH or VH	N or Z	NM
HIGH	HIGH	P	Z
HIGH	HIGH	PM	P
HIGH	VH	P	PM
VH	VL	N	P
VH	VL	Z or P	Z
VH	VL	PM or PB	N
VH	LOW	Z	N
VH	LOW	P	Z
VH	MED	N	NM
VH	MED	Z	N
VH	MED	P	Z
VH	HIGH	N	NM
VH	HIGH	Z	NB
VH	HIGH	P	Z
VH	HIGH	PM	P
VH	VH	Z	PM
VH	VH	P	PB
VVH	VL	N	N
VVH	VL	Z or P	Z
VVH	VL	PM	NM
VVH	VL	PB	NB
VVH	LOW	Z or P	N
VVH	MED	Z	N
VVH	MED or HIGH	P	Z
VVH	VH	Z	PB
EH	VL	N or Z	N
EH	VL	P	NM
EH	VL	PM	NB
EH	VL	PB	NH
EH	LOW	N	Z
EH	LOW	Z or P	NM
EH	LOW	PB	NH
EH	MED	Z	P
EH	MED	P	NM
EH	HIGH	P	Z
EH	VH	Z	PM
EH	VH	P	PB

6 Summary

In-vehicle load measurement is an important factor in ITS technology, which can be possibly linked to various uses in real applications related to ITS.

Basically, the vehicle load indicator by Yazaki Corporation was developed for freight vehicle use as a warning of overloading due to its accuracy limit. Based on a combination with proposed inclination parameters, the warning functions are expected to include increased driver awareness of the driving situation.

The advanced use of vehicle load measurement is deemed necessary for the active control of distance, speed, and braking, in-vehicle networking systems, fleet management within a logistic system, and other purposes like linkage to electronic tolling. However, there is no doubt that high accuracy and reliability of vehicle load measurement is the key for high level extensions.

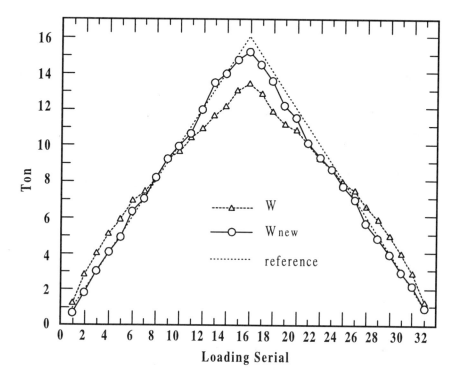

Figure 7: Loading test with artificial inclinations (right-left).

Generally speaking, sensor performance (accuracy, stability, non linearity) and methodology for sensor adjustment and data processing are the main difficulties in in-vehicle load measurement. Using knowledge-based techniques like neural network, fuzzy logic, and genetic algorithms, the influence on accuracy by the methodology could be reduced to a satisfactory level.   Sensor performance improvement, especially in stability and precision, is a big question, because of the serious sensing environment with heavy load, vibration, and a diversity of uncertainties.

Future work is necessary on the fuzzy rule base optimization and the analysis of deck frame structures for better performance for extensive applications, including endless effort for a stable and precise sensing device.

References

[1]   HTMS, U.S.A. (1995), Weighing Station Bypass. *Traffic Technology International*, Winter'95, pp. 40-44. U.K. and International Press.

[2]   Karuo, S. and Koyasu, Y. (1995), Automatic Measuring System for Heavy Freight Vehicles. The 2nd World Congress on Intelligent Transport System, '95 Yokohama, Japan, pp. 479-503.

[3]   Su, L. (1996), Accuracy Improvement of Vehicle Load Measurement Using Fuzzy Logic, *Methodologies for the Conception, Design, and Application of Intelligent Systems*, edited by Yamakawa, T. and Matsumoto, G., World Scientific, pp. 887-890, Proceedings of the 4th International Conference on Soft Computing, IIZUKA'96, Japan.

[4]   Mamdani, E. H. (1974), Applications of Fuzzy Algorithms for Control of Simple Dynamic Plant. *Proc. IEE*, 121, 12, pp. 1585-1588.

[5]   Togai InfraLogic, Inc. (1993), TIL Shell for Windows 3.0.0.

[6]   Zimmermann, H. J. (1985), Fuzzy Set Theory and Its Application, Kluwer Academic Publishers.

[7]   Klir, G. and Folger, T. (1988), Fuzzy Sets, Uncertainty and Information, Prentice- Hall International, Inc.

[8]   Kosko, B. (1992), Neural Networks and Fuzzy Systems, Prentice-Hall International, Inc.

[9]   Jain, L. C. (Editor) (1997), Soft Computing Techniques in Knowledge-Based Intelligent Engineering Systems, Springer-Verlag, Germany.

[10] Sato, M., Sato, Y., and Jain, L. C. (1997), Fuzzy Clustering Models and Applications, Springer-Verlag, Germany.

9 | INTELLIGENT CONTROL OF AIR CONDITIONING SYSTEMS

T. Iokibe
System Development Section
System Technology Division
Meidensha Corporation
36-2, Nihonbashi Hakozakicho Chuo-ku, Tokyo 103 Japan

T. Tobi
R & D of Dept.1
Technical Research Laboratory
Dai-dan Co., LTD.
390 Kita-Nagai, Miyoshi-Machi, Iruma-Gun, Saitama 354, Japan

S. Araki
Intelligent Agent Team
Central Research Laboratories
Matsushita Electric Industrial Co., LTD.
3-4 Hikaridai, Seika-cho, Souraku-gun, Kyoto 619-02, Japan

In 1989, Mitsubishi Heavy Industries, LTD. developed and brought to market the world's first intelligent air conditioning system that employed fuzzy control (named "Beaver Air Conditioner Warp 65"). By the beginning of 1990s, many intelligent air conditioning systems that employ fuzzy theory and neural nets appeared on the market.

In this chapter, we give three practical applications on the intelligent control of air conditioning systems. In Section 1, Iokibe et al. show the air conditioning system applied to the synthetic fiber plant. At synthetic fiber plants, the atmosphere in the production process needs to be kept constant for stabilizing and enhancing product quality. Therefore, temperature, humidity, and internal pressure are controlled with a large-scale air conditioner. Since the conventional control system is dependent on control of one input to one output, the atmosphere is controllable to a constant level, but energy conservation is difficult to be taken into account. The author et al.

developed a fuzzy-controlled air conditioning system for energy conservation, which is capable of implementing a control of multiple inputs to multiple outputs, introduced it into the drawing and twisting process in a synthetic fiber plant, and obtained an energy conservation effect of about 26% in the autumn. In Section 2, Tobi et al. show the air conditioning system applied to the clean room. The clean room for semiconductor production requires constant maintenance against uncertain changes of sensors and actuators caused by any environmental changes. The maintenance activities, by trial and error after installation of a control system, need substantial engineering effort. Therefore, the author et al. are attempting to make the systems learn. The results of the experiments are very promising. In Section 3, Araki et al. introduce an approach for a new air conditioning system with "eyes," which is a pyroelectric infrared rays detector to take thermal images in a room, for detecting the occupant condition. In order to achieve a more comfortable thermal environment, it is necessary to detect not only physical environmental conditions (air temperature, humidity, etc.) but also the information about occupants in a room (the number, positions, etc.). They developed a segmentation method of a thermal image using a fuzzy clustering algorithm to identify the number of occupants, and positions of occupants are located by estimating the distance between the infrared rays detector and each occupant using fuzzy if-then rules. Room temperature and wind direction can be controlled according to the identified number and positions of occupants. For example, it is possible to prevent the cool air from reaching occupants when an air conditioner starts to overheat.

1 Fuzzy Controlled Air Conditioning System for Energy Conservation Applied to the Synthetic Fiber Plant

1.1 Introduction

The atmosphere in spinning and drawing/twisting rooms is one of the factors determining the product quality of the polyester filament which is an excellent material for fashion garments. To keep the ambient conditions constant, a large-scale air conditioner is installed in each synthetic fiber plant and used for control of the temperature, humidity, and internal pressure. The conventional method employs an independent PID control of temperature and humidity, and a computer control of air mixture ratio. However, this method is not efficient from the viewpoint of energy conservation although a constant atmosphere control is achievable.

We developed a fuzzy-controlled air conditioning system for energy conservation and introduced it into the same process as the one using the conventional method for comparison. It was confirmed that the newly developed system provided a much higher energy conservation effect.

This section begins with the role and problems of the air conditioning system in a synthetic fiber plant, explains the configuration and function of the fuzzy air conditioning system for solving the above problems, and finally reports the effect obtained by introduction of this system.

1.2   Role and Problems of Air Conditioning Equipment in Synthetic Fiber Plant

Polyester filament as a synthetic fiber is generally produced in the processes shown in Figure 1. Uniformity is one of the important elements of these processes. For maintaining uniformity, there are conditions to be met such as minimizing a variation in the analytical value of raw material quality, properly controlling the temperature, pressure, flow rate, and level in each process, and keeping the atmosphere constant in the spinning and drawing/twisting rooms.

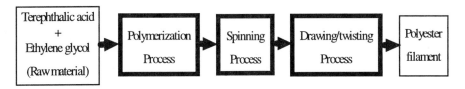

Figure 1: Synthetic fiber production process.

1.2.1 Role of Air conditioners in Spinning and Drawing/Twisting Processes

The target of the application of fuzzy control is the large-scale air conditioners which control the atmosphere in the drawing/twisting process. The space to be air conditioned is 40,000 m³ and there are 4 air conditioners provided (3 units for routine operation and 1 unit on standby). The role played by these air conditioners is as follows.

(1) Aging of yarn yet to be drawn
For ensuring product quality, aging is required between the spinning process and drawing/twisting process under even atmospheric conditions for the determined period of time.

(2) Prevention of yarn from changing in physical properties before drawing
The air conditioners prevent yarn from changing its physical properties by securing appropriate temperature and humidity conditions.

(3) Maintaining level of dyeing
Level dyeing of yarn can be maintained under temperature deviation ±1°C,

humidity deviation ±5%, and constant control of internal pressure.

(4) Prevention of occurrence of yarn cut and fluff

Fluctuation of the room temperature or humidity will increase occurrence of defects.

(5) Purification of room atmosphere

The oil mist and monomer generated at heat setting of filament is partially exhausted and replaced with the outdoor air for purification.

1.2.2   Problems of Air conditioners in Spinning and Drawing/Twisting Processes

Because the target of air conditioning is a big space of 40,000 m^3 as mentioned before, electric power, water, and steam are consumed in a large volume. This is a negative factor from the viewpoint of manufacturing cost. As a countermeasure, the energy conservation operation using the outdoor air is effective. For this operation, the following control is conceivable.

(1) Volume reduction of cold water to be used

In a hot climate, cold water is prepared with a refrigerator. When reducing this load of cooling by taking the outdoor air, the energy conservation will become effective.

(2) Volume reduction of steam to be used

In a cold climate, the volume of steam to be used can be reduced by taking an appropriate amount of the outdoor air, so the effect of energy conservation is expectable.

However, the conventional computer control is dependent on one input to one output, with the temperature of mixing air taken as an input parameter and the mixing ratio of outdoor air and return air taken as an output parameter. This method entails a wide variation and the expected result cannot be obtained.

1.3   Fuzzy-Controlled Air Conditioning System for Energy Conservation

Fuzzy control has been employed for solving the problems of the conventional air conditioning control system as mentioned in Section 1.2 and for obtaining effective energy conservation throughout the year.

The relevant fuzzy control system is intended to reduce the volume of steam and cold water to be used through mixing the outdoor air and return air properly based on a fuzzy inference.

Also provided for energy conservation operation throughout the year is the function for automatically judging the current season by a fuzzy inference which uses the temperature change of industrial water.

As mentioned before, this control system targets 4 air conditioners. The entire air conditioning system is shown in Figure 2 and the control flow of one air conditioner is detailed in Figure 3. In this system, 3 air conditioners are controlled normally. Of these conditioners, the central unit is operated in energy-conserving mode and the units at both sides are under energy conservation operation and internal pressure-controlled operation.

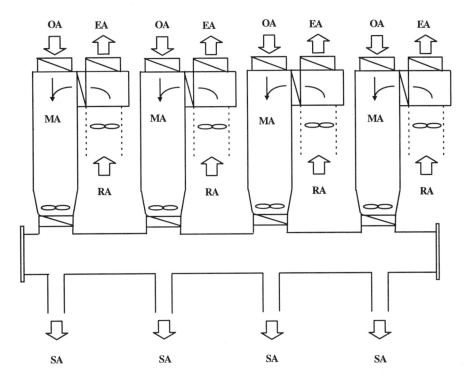

OA: outdoor air,   EA: exhaoust air,   MA: mixing air,   RA: retuen air,   SA: supply air

Figure 2: Diagram of air conditioning system.

The outdoor air and return air are mixed in the following process as shown in the stoichiometric chart of Figure 4.
(1) Mixing air (MA) is obtained by mixing the outdoor air (OA) and return air (RA).
(2) From the mixing air (MA), point Q is obtained with an air washer.
(3) The temperature of supply air (SA) is raised from point Q with a fan.

For point Q to be appropriate, proper mixing air (MA) is indispensable.

The fuzzy system obtains the proper mixing air (MA) by

(1) acquiring the detected data of outdoor air temperature/humidity, return air temperature/humidity, outdoor air flow, return air flow, and industrial water temperature,
(2) comparing the above data with the standard values for judgment, and
(3) obtaining the mixing air temperature, outdoor air damper, return air damper, and exhaust damper opening by fuzzy inference.

Table 1 lists the fuzzy inference rules of the target mixing air temperature which are used in the relevant control, and Figure 5 shows the membership functions used for the same purpose. The names of the variables in Table 1 are

TK:	Temperature of industrial water
TA:	Target temperature of mixing air

Table 1: Fuzzy inference rules of target mixing air temperature (example)

IF	TK is LL	THEN	TA is H	
IF	TK is L	THEN	TA is O	
IF	TK is O	THEN	TA is L	
IF	TK is H	THEN	TA is LL	
DEFUZZY TA				
END				

Table 2 lists the fuzzy rules for damper control and Figure 6 shows the membership functions. The names of the variables in Table 2 are as follows.

TE:	Temperature deviation
TED:	Rate of change in mixing air temperature
PE:	Internal pressure
PED:	Rate of change in internal pressure
DUEA:	Manipulated variable of exhaust damper
DUOAM:	Manipulated variable of outdoor air damper

1.4 Results

The newly developed fuzzy control system began operation in KANEBO's Hokuriku synthetic fiber factory in September, 1991. Operation of this system involved modification of the existing equipment, including installation of sensors. The result of operation is shown in Figure 7. It is evident from this figure that an energy conservation of about 26% in terms of cooling load (USRT: ton of refrigeration) could be obtained in a time period from the first day of September to the last day of November when comparing year 1991 under fuzzy control and 1990 under conventional control.

1.5 Future Directions

Low-energy operation of air conditioners through intake of the outdoor air, could be obtained by using a fuzzy control as shown in Figure 7. However, maintaining the energy conserving effect throughout a year, under different meteorological conditions and application to processes having different target values, remains to be investigated.

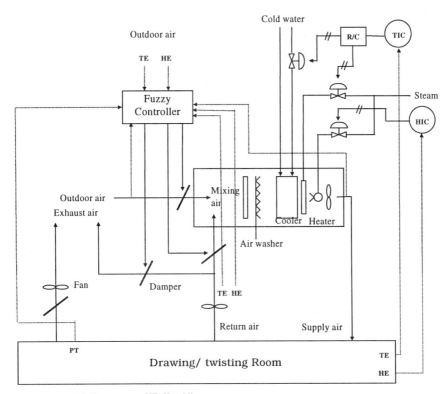

PT: Pressure, TE: Temperature, HE: Humidity

Figure 3: Air conditioner control flow (1 unit).

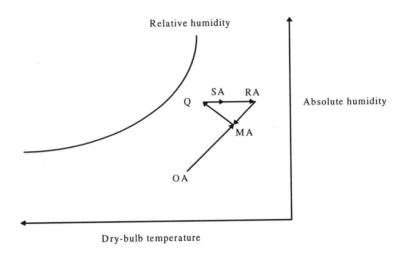

Figure 4: Stoichiometric chart.

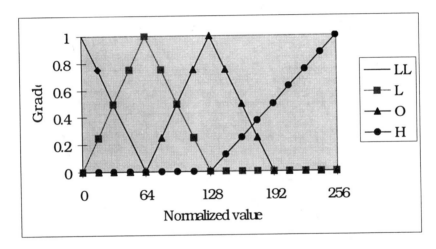

Figure 5: Membership functions of industrial water for target mixing air
           temperature.

1.6 Acknowledgment

The authors appreciate Kazuo Satake, Toshimi Ishiguro with Kanebo LTD for
offering helpful advice, and Toshio Karakama of Meidensha Corporation for
developing the system.

2 A Learning Type Fuzzy Logic Control for Stabilizing Temperature and Humidity in a Clean Room

2.1 Introduction

Recent industrial growth has highlighted the need for accurate control methods for an air conditioning system, especially in facilities such as biohazard and clean room for semiconductor production. Constant maintenance of the air conditioning system is important because the performances of sensors and actuators are changed by various environmental changes.

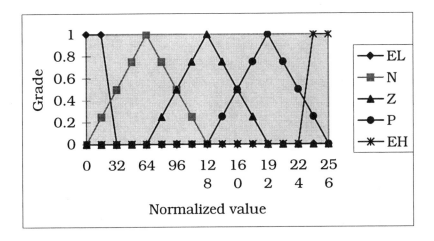

Figure 6: Membership functions of internal pressure for damper control.

Table 2: Fuzzy inference rules of damper control (example)

IF	PE is NM THEN			
	IF	TE is N and TED is N	THEN	DUOAM is N
	IF	TE is N and TED is Z	THEN	DUOAM is NS
	IF	TE is N and TED is P	THEN	DUOAM is NS
	IF	TE is Z and TED is N	THEN	DUOAM is NS
	IF	TE is Z and TED is Z	THEN	DUOAM is Z
	IF	TE is Z and TED is P	THEN	DUOAM is PS
	IF	TE is P and TED is N	THEN	DUOAM is PS
	IF	TE is P and TED is Z	THEN	DUOAM is PS
	IF	TE is P and TED is P	THEN	DUOAM is P
	IF	PE is N and PED is N	THEN	DUEA is N
	IF	PE is N and PED is Z	THEN	DUEA is NS
	IF	PE is N and PED is P	THEN	DUEA is NS

```
         IF      PE is Z and PED is N      THEN    DUEA is NS
         IF      PE is Z and PED is Z      THEN    DUEA is Z
         IF      PE is Z and PED is P      THEN    DUEA is PS
         IF      PE is P and PED is N      THEN    DUEA is PS
         IF      PE is P and PED is Z      THEN    DUEA is PS
         IF      PE is P and PED is PTHEN  DUEA is P
ENDIF
IF       PE is EL  THEN    DUEA is S
IF       PE is EH  THEN    DUEA is O
DEFUZZY DUEA
DEFUZZY DUOAM
END
```

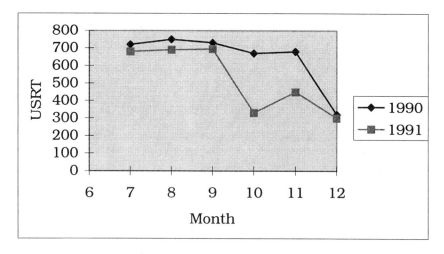

Figure 7: Result of operation.

However, it is time consuming to execute the maintenance activities by trial and error after installing a control system. For this reason, we have attempted to develop a learning type fuzzy logic control for stabilizing temperature and humidity in a clean room. In this chapter, the feasibility of a hierarchical fuzzy model [1] with learning for controlling temperature and humidity in the clean room is discussed, and the results of simulation studies are reviewed. The effectiveness of the present method is also confirmed by experiments.

2.2 Ordinary Fuzzy Logic Control System

The structure of a fuzzy logic control system for stabilizing temperature and humidity is illustrated in Figure 8.

In this control system, there are two steps of fuzzy inference, (1) the temperature request inference and humidity request inference and (2) integrated inference of economical allotment to three final controlling elements used in the conclusion of the first step. These steps of fuzzy inference are organized as follows:

[Step 1]   *Temperature request and humidity request inference.*

A Temperature Request Value (TRV) and a Humidity Request Value (HRV) are inferred. The TRV is inferred from the error of measured temperature and desired temperature, and from its deviation. The HRV is inferred from its foreseen error and deviation, because relative humidity is influenced by temperature.

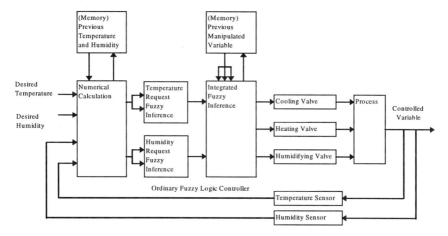

Figure 8: Diagram of the structure of the ordinary fuzzy logic control system.

[Step 2]   *Integrated inference.*

In this step, temperature and humidity are not only integrated, but an economical operation is taken into consideration. We represent this inference as the following fuzzy rule:

Rule i :   If x_1 is A_{i1},, x_5 is A_{i5} Then y_1 is B_{i1},, y_3 is B_{i3},

where x_1 is the last manipulated variable for the cooling valve, x_2 is the last manipulated variable for the heating valve, x_3 is the last manipulated variable for the humidifying valve, x_4 is the temperature request (decided in the first step of fuzzy inference), and x_5 is the humidity request (decided in the first step of fuzzy inference). y_1, y_2, and y_3 are the change in manipulated variable for the cooling, heating, and humidifying valve, respectively. Needless to say, A_{i1}, A_{i2}, A_{i3}, A_{i4}, A_{i5}, B_{i1}, as well as B_{i2} and B_{i3} are fuzzy sets that express an ambiguous condition. Readers are referred to the paper by T. Tobi et al. [2] for a more detailed description and discussion.

2.3 A Learning Type Fuzzy Logic Control System

The structure of a learning type fuzzy logic control system is shown in Figure 9. The learning of this fuzzy logic controller is achieved by using a set of learning time series data of the controlled variable and the manipulated variable. In practice, these learning data sets are obtained from the practical system.

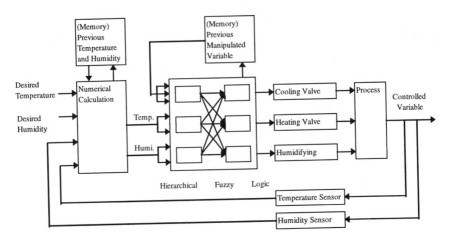

Figure 9: Diagram of the structure of a learning type fuzzy logic
control system.

2.3.1   Structure of a Hierarchical Fuzzy Model

Using fuzzy models, a hierarchical fuzzy model is shown in Figure 10. A hierarchical fuzzy model is divided into hidden and output layers, each of which is then divided into three models, the first model, the second model, and the third model, giving six models in all. These models are either two-input and single-output system or three-input and single-output system. Each model is implemented by Neuro-Fuzzy method [3] based on the simplified fuzzy inference [4], i.e., a method to tune the fuzzy rules by a delta-rule [5]. In the simplified fuzzy inference, the consequent part of the fuzzy rule is expressed by a real number.

The real numbers in the consequent parts are adjusted by learning so that they are optimized, after they are initialized randomly.

In Figure 10, note that $x_{ij}*$ is the j-th object value to the i-th input variable of the model in the input layer; x_i^1, the i-th input variable of the model in the input layer; $w_{u,k}^1$, the real number in the consequent part of the k-th rule of the u-th model, in the hidden layer; x_u^2, the output of the u-th model in the hidden layer, i.e., the input of the u-th model in the output layer; $w_{u,k}^2$, the real number in the consequent part of the k-th rule of the u-th model, in the output layer; x_u^3, the output of the u-th model in the output layer; $y_{uj}*$, the j-th observed output value corresponding to the j-th object's input value $x_{ij}*$, i.e., the desired output of the u-th model in the output layer.

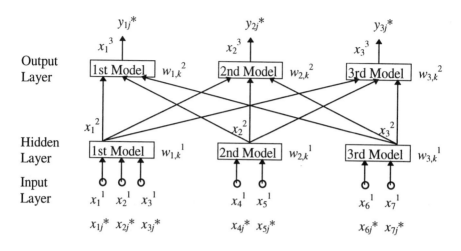

Figure 10: Seven-input and three-output hierarchical fuzzy model.

In Figure 11, an example of membership functions characterizing fuzzy sets as triangles and trapezoids is shown.

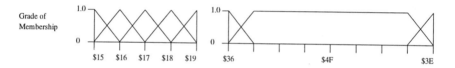

Figure 11: One example of definition of membership functions.

In these diagrams, the horizontal axis represents the entire finite set and the vertical axis is the grade.

The First Model in the Hidden Layer

$_\bullet CCV(\bullet\text{-}1)$, $_\bullet HCV(\bullet\text{-}1)$ and $_\bullet HUV(\bullet\text{-}1)$ were used as the input variable. Where $_\bullet CCV(\bullet\text{-}1)$ is the last manipulated variable for the cooling valve at discrete time $(\bullet\text{-}1)$, $_\bullet HCV(\bullet\text{-}1)$ is the last manipulated variable for the heating valve, and $_\bullet HUV(\bullet\text{-}1)$ the last manipulated variable for the humidifying valve. The number of fuzzy rules was set to be 27. In addition, the labels of the indices which modify the membership functions in the antecedent part of the fuzzy rule are expressed in hexadecimal in Table 3.

The Second and the Third Model in the Hidden Layer

In the second model, $e_{DB}(\bullet)$ and $de_{DB}(\bullet)$ were used as the input variables. Where $e_{DB}(\bullet)$ is the temperature error at discrete time $\bullet$, $de_{DB}(\bullet)$ is the change in temperature error. On the other hand, $e_{RH}(\bullet)$ and $de_{RH}(\bullet)$ were used as the input variables to the third model. Where e_{RH} (•) is the foreseen humidity error, de_{RH} (•) is the change in foreseen humidity error. The number of fuzzy rules was set to be 25. In addition, the labels of the indices which modify the membership functions in the antecedent part of the fuzzy rule are expressed in hexadecimal in Table 4.

Table 3: Hexadecimal representation of the membership functions

Rule	Antecedents			Rule	Antecedents			Rule	Antecedents		
	1	2	3		1	2	3		1	2	3
1	$36	$36	$36	2	$4F	$36	$36	3	$3E	$36	$36
4	$36	$4F	$36	5	$4F	$4F	$36	6	$3E	$4F	$36
7	$36	$3E	$36	8	$4F	$3E	$36	9	$3E	$3E	$36
10	$36	$36	$4F	11	$4F	$36	$4F	12	$3E	$36	$4F
13	$36	$4F	$4F	14	$4F	$4F	$4F	15	$3E	$4F	$4F
16	$36	$3E	$4F	17	$4F	$3E	$4F	18	$3E	$3E	$4F
19	$36	$36	$3E	20	$4F	$36	$3E	21	$3E	$36	$3E
22	$36	$4F	$3E	23	$4F	$4F	$3E	24	$3E	$4F	$3E
25	$36	$3E	$3E	26	$4F	$3E	$3E	27	$3E	$3E	$3E

Table 4: Hexadecimal representation of the membership functions

Rule	Antecedents		Rule	Antecedents		Rule	Antecedents		Rule	Antecedents		Rule	Antecedents	
	1	2		1	2		1	2		1	2		1	2
1	$15	$15	2	$16	$15	3	$17	$15	4	$18	$15	5	$19	$15
6	$15	$16	7	$16	$16	8	$17	$16	9	$18	$16	10	$19	$16
11	$15	$17	12	$16	$17	13	$17	$17	14	$18	$17	15	$19	$17
16	$15	$18	17	$16	$18	18	$17	$18	19	$18	$18	20	$19	$18
21	$15	$19	22	$16	$19	23	$17	$19	24	$18	$19	25	$19	$19

Each of the Models in the Output Layer

An output of the model in the hidden layer is employed as an input of the model in the output layer. This is illustrated in Figure 10. Consequently, the input-output mapping needs corresponding by normalizing input and output values in interval [0, 1]. $d•_{CCV}$ (•), $d•_{HCV}$ (•), and $d•_{HUV}$ (•) were used as output variables. Where $d•_{CCV}$ (•) is the change in manipulated variable for the cooling valve, $d•_{HCV}$ (•) is the change in manipulated variable for the heating valve, and $d•_{HUV}$ (•) is the change in manipulated variable for the humidifying valve. With respect to the practical manipulated variable, $•_{CCV}$ (•) = $•_{CCV}$ (•-1) + $d•_{CCV}$ (•) was used as the manipulated variable for the cooling valve, $•_{HCV}$ (•) = $•_{HCV}$ (•-1) + $d•_{HCV}$ (•) as the manipulated variable for the heating valve, and $•_{HUV}$ (•) = $•_{HUV}$ (•-1) + $d•_{HUV}$ (•) as the manipulated variable for the humidifying valve. The number of fuzzy rules was set to be 75. In addition, the labels of the indices which modify the membership functions are expressed in hexadecimal in Table 5.

Table 5: Hexadecimal representation of the membership functions

Rule	Antecedents			Rule	Antecedents			Rule	Antecedents			Rule	Antecedents			Rule	Antecedents		
	1	2	3		1	2	3		1	2	3		1	2	3		1	2	3
1	$36	$15	$15	2	$4F	$15	$15	3	$3E	$15	$15	4	$36	$16	$15	5	$4F	$16	$15
6	$3E	$16	$15	7	$36	$17	$15	8	$4F	$17	$15	9	$3E	$17	$15	10	$36	$18	$15
11	$4F	$18	$15	12	$3E	$18	$15	13	$36	$19	$15	14	$4F	$19	$15	15	$3E	$19	$15
16	$36	$15	$16	17	$4F	$15	$16	18	$3E	$15	$16	19	$36	$16	$16	20	$4F	$16	$16
21	$3E	$16	$16	22	$36	$17	$16	23	$4F	$17	$16	24	$3E	$17	$16	25	$36	$18	$16
26	$4F	$18	$16	27	$3E	$18	$16	28	$36	$19	$16	29	$4F	$19	$16	30	$3E	$19	$16
31	$36	$15	$17	32	$4F	$15	$17	33	$3E	$15	$17	34	$36	$16	$17	35	$4F	$16	$17
36	$3E	$16	$17	37	$36	$17	$17	38	$4F	$17	$17	39	$3E	$17	$17	40	$36	$18	$17
41	$4F	$18	$17	42	$3E	$18	$17	43	$36	$19	$17	44	$4F	$19	$17	45	$3E	$19	$17
46	$36	$15	$18	47	$4F	$15	$18	48	$3E	$15	$18	49	$36	$16	$18	50	$4F	$16	$18
51	$3E	$16	$18	52	$36	$17	$18	53	$4F	$17	$18	54	$3E	$17	$18	55	$36	$18	$18
56	$4F	$18	$18	57	$3E	$18	$18	58	$36	$19	$18	59	$4F	$19	$18	60	$3E	$19	$18
61	$36	$15	$19	62	$4F	$15	$19	63	$3E	$15	$19	64	$36	$16	$19	65	$4F	$16	$19
66	$3E	$16	$19	67	$36	$17	$19	68	$4F	$17	$19	69	$3E	$17	$19	70	$36	$18	$19
71	$4F	$18	$19	72	$3E	$18	$19	73	$36	$19	$19	74	$4F	$19	$19	75	$3E	$19	$19

2.3.2 Succession of the Ordinary Fuzzy Model

Traditionally, experts based their decisions to adjust or not to adjust depending upon how well the air conditioning system was running. In the ordinary fuzzy model, the control algorithm based on fuzzy production rules uses the expert's common sense and experiences. In order to succeed the control algorithm of the ordinary fuzzy model, a hierarchical fuzzy model was implemented as follows:

[Step 1]   *Production of the input values of a learning data set.*
Define a set of the input value $X = \{x_{1j}*,, x_{7j}*| j = 1, 16,875\}$, the sequence of random numbers X is generated from a computer system.

[Step 2]   *Acquisition of the learning data set.*
Define a set of the observed value $Y = \{y_{1j}*, y_{3j}*| j = 1, 16,875\}$ as a result of the ordinary fuzzy model reasoned from X, and a set of the input/output data $T = \{x_{1j}*, x_{7j}*, y_{1j}*, y_{3j}*| j = 1, 16,875\}$ is used for the learning of a hierarchical fuzzy model.

[Step 3]   *Learning of a hierarchical fuzzy model.*
Learning was completed after 50 iterations (one iteration is defined as applying a learning data set to the hierarchical fuzzy model).

2.4 Simulation Experiments

A lot of work has been done to investigate the performance of a hierarchical fuzzy model and the ordinary fuzzy model by computer simulation. Figure 12 shows the results under summer conditions. The desired values were changed every 30 minutes.

The upper graph shows the atmospheric temperature (DB) and the relative humidity (RH) plotted against time. The graph at the bottom shows the percentage of the manipulated variable for the each valve, cooling valve (CCV), heating valve (HCV), and humidifying valve (HUV). In this simulation, summer conditions mean the state of 33.2 outside temperature (OADB) and 62.6% outside humidity (OARH), and the initial state of space is the same. In this case, the set point of temperature is 23, and the set point of relative humidity is 50%. After changing the desired temperature to 25, the offset temperature of the hierarchical fuzzy model was -0.1. On the other hand, the offset humidity was -0.4% after changing the desired humidity to 60%.

To meet changes in heat load, the ordinary fuzzy logic control system infers optimal and economical manipulated variables according to the fuzzy production rules. It is shown from the manipulated variables that the control system is economical because all of the control valves did not open at the same time. However, in the case of a hierarchical fuzzy model, the cooling valve was somewhat opened. In short, it is not economical.

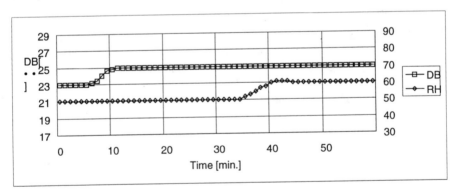

2.5 Practical Results

We have briefly presented the results of practical use of a learning type fuzzy logic control system for a clean room. Control data from the March 20, 1995 run is presented in Figure 13.

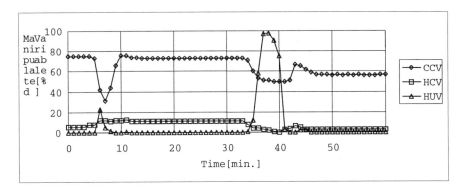

(a) The ordinary fuzzy model.

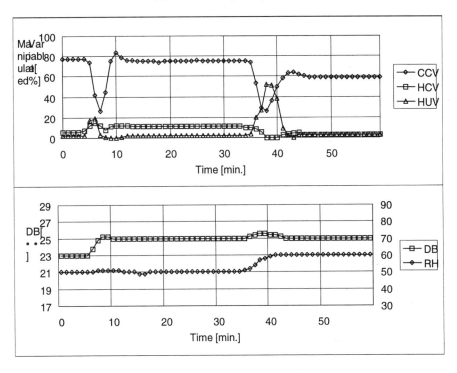

(b) A hierarchical fuzzy model.

Figure 12: One example of step response curve.

In this control system, the temperature set point is 24 and the humidity set point is 50%. The clean room is rectangular with dimensions 6.5W4.5D2.7H[m³]. Then, the temperature and humidity set point was changed from 2450% to 2555% at 13:03:29, from 2555% to 2450% at 15:05:42. As the result of practical use, a learning type fuzzy logic control system has been able to achieve the state of both constant temperature and humidity, and the biggest temperature error is about -0.4, and the biggest humidity error is about -3.8%.

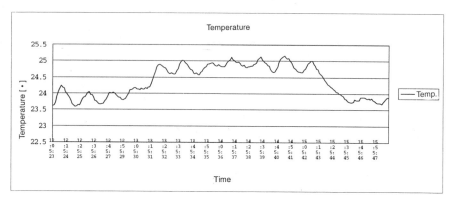

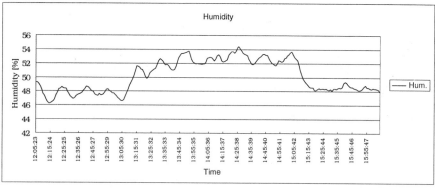

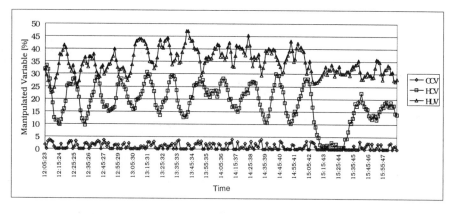

Figure 13: Results of practical use for a clean room.

2.6 Conclusion

In this study, we focused on a self-learning fuzzy logic control for stabilizing temperature and humidity in a clean room, and our goal is to contribute to the establishment of high efficiency and economical control. We performed the feasibility of a seven-input and three-output three-layered hierarchical fuzzy model for a learning controller of temperature and humidity in the clean room. Also, we reported the results of simulation studies and the quantitative results of the experiments. Favorable results were nearly obtained, and we were able to confirm the validity of a hierarchical fuzzy model.

3 Occupant Condition Detecting Algorithm for Air Conditioning Systems

3.1 Introduction

PMV (Predicted Mean Vote) index [6] is well known as a quantitative index for thermal sensation of human registered in ISO. PMV index value is derived from six factors. Four factors, temperature, air velocity, radiation, and humidity are the features of the thermal environment around occupants. Two factors, clothing and metabolic rate are the personal information of occupants. If the PMV value is equal to 0, occupants feel comfortable. If it has a positive or a negative value, they feel hot or cool, respectively. If PMV value is within the interval [-0.5, 0.5], it is said that at least 80% occupants feel comfortable.   For this reason, recent air conditioning systems in consumer electronics are controlled based on the PMV index.   However, such systems cannot always produce a comfortable thermal environment for all occupants in an air conditioned room since conventional sensors and algorithm installed in the systems cannot detect PMV factors for each occupant.

In this section, we introduce an approach for a new air conditioning system with "eyes" which is a pyroelectric infrared rays detector [7] to take thermal images in a room to detect the occupant condition. We developed a segmentation method of a thermal image using a fuzzy clustering algorithm to identify the number and positions of occupants [8][9]. Room temperature can be controlled according to the identified number. The proposed algorithm consists of three stages. The first stage is to distinguish occupants from background in an image using the Fuzzy C-Means algorithm (FCM). The second stage is to distinguish each occupant by locating local temperature peaks in the image. The third stage is a region growing algorithm for more accurate segmentation of occupants based on the membership values determined by the FCM and the number of located peaks. We also developed a method of locating occupants in a room. Wind direction can be controlled if their positions are identified.   Positions of occupants are located by reasoning the distance between the infrared rays detector and each occupant using fuzzy if-then rules constructed by iterative rule generation method [10].

3.2 Structure of the Pyroelectric Infrared Rays Detector

Schematic structure of the developed sensor is shown in Figure 14. The sensor consists of a 1-dimensional linear pyroelectric array detector, a horizontal scanning mechanical body, a silicon Infrared Rays (IR) lens, and a mechanical chopper. The array detector, which has 8 elements, and the IR lens are united to be a revolving part scanning horizontally. The part is located in the center of a cylindrical chopper. The chopper has 3 windows, so the chopper opens and closes at 60-degree intervals. It opens and closes three times for one rotation. The revolving part and the chopper are linearly rotated at a constant ratio. A 2-dimensional thermal image is produced by a horizontal scanning of the vertical array detector. Figure 15(a) shows a visual image of a room scene. The corresponding thermal image is shown in Figure 15(b). The thermal image is composed of 8x20 elements. Each element has a value of temperature expressed by 8bit as shown in Figure 15(c).

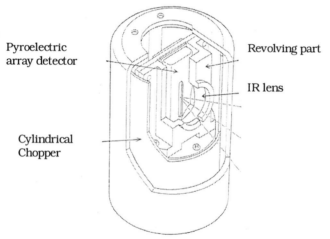

Figure 14: Schematic structure of the infrared rays detector.

3.3 Segmentation of Occupants from Thermal Images

A segmentation algorithm to identify the number of occupants is described here. The algorithm consists of three procedures: removal of background, identification of each occupant, and region growing. The principle of our approach is based on the feature of thermal images that the temperature of occupants is higher than that of background and each occupant has a local temperature peak in an image.

3.3.1 Removing Background Using the Fuzzy C-Means Algorithm

The fuzzy c-means algorithm (FCM) is applied to remove background or to distinguish occupants from background in a thermal image. FCM is applied to a data set of temperature of 8x20 elements of an image. Figure 16 shows an example of clustering result. The elements, which belong to the cluster whose weighted mean temperature is the highest with high membership value, are selected as the component of regions representing occupants. As shown in Figure 16, selected elements depend on the threshold level of membership value. Therefore, at first, we set the threshold level at 0.5 to roughly distinguish occupants from background. Figure 17 shows the initial selected elements as the components occupants.

(a)

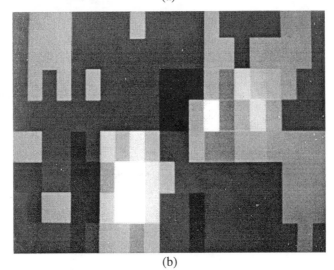

(b)

search step

	1	2	3	4	5	6	7	8	9	10	11	12	13	14	15	16	17	18	19	20
1	81	84	82	64	68	74	75	80	74	72	73	69	85	75	76	77	77	80	83	75
2	78	82	82	71	73	71	63	73	72	62	72	66	78	78	69	83	86	84	83	76
3	80	72	82	74	82	73	70	71	74	62	55	59	96	101	111	132	118	84	71	68
4	71	75	77	65	71	75	72	72	66	42	38	80	164	162	131	113	86	74	68	66
5	78	76	69	55	63	120	151	135	93	66	66	79	107	108	98	93	91	81	82	80
6	53	69	77	32	50	149	215	207	133	84	73	70	70	67	63	65	73	78	84	78
7	54	78	87	51	47	121	186	197	138	71	71	58	66	70	72	65	69	82	89	85
8	62	72	76	57	65	79	92	102	86	64	65	56	64	67	64	72	73	76	82	73

element number

(c)

Figure 15: (a) An example of a visual image of a room scene. (b) The corresponding thermal image. (c) Sensor output values of the thermal image.

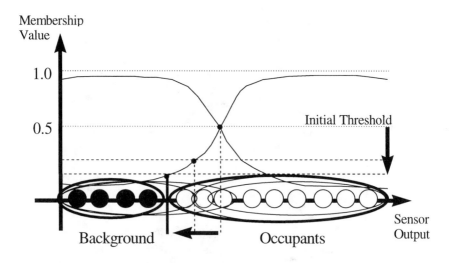

Figure 16: An example of clustering result.

3.3.2 Identifying the Number of Occupants

In order to distinguish each occupant, local temperature peaks in an image are located from among elements selected as occupants. A particular element searches an element which has maximum temperature in its 8-neighbors. If the temperature of the searched element is higher than the particular one, the searched one is taken as the parent of the particular one. The result of peak location is shown in Figure 18. An arrow linked between two elements point to the parent. Elements with no parent are local peaks of temperature. A distinct peak indicated by a half-toning circle is regarded as the representative element of an occupant. Each of the peaks is given a distinct label such as a different natural number. Every element selected as a component of occupants is given the same label as its parent. As shown in Figure 19, the region composed of the elements which have same label represents one occupant.

search step

	1	2	3	4	5	6	7	8	9	10	11	12	13	14	15	16	17	18	19	20
1	81	84	82	64	68	74	75	80	74	72	73	69	85	75	76	77	77	80	83	75
2	78	82	82	71	73	71	63	73	72	62	72	66	78	78	69	83	86	84	83	76
3	80	72	82	74	82	73	70	71	74	62	55	59	96	101	111	(132)	(118)	84	71	68
4	71	75	77	65	71	75	72	72	66	42	38	80	(164)	(162)	(131)	113	86	74	68	66
5	78	76	69	55	63	(120)	(151)	(135)	93	66	66	79	107	108	98	93	91	81	82	80
6	53	69	77	32	50	(149)	(215)	(207)	(133)	84	73	70	70	67	63	65	73	78	84	78
7	54	78	87	51	47	(121)	(186)	(197)	(138)	71	71	58	66	70	72	65	69	82	89	85
8	62	72	76	57	65	79	92	102	86	64	65	56	64	67	64	72	73	76	82	73

element number

Figure 17: Initial selected elements as components of occupants.

3.3.3 Region Growing Algorithm

The region growing algorithm settles the regions of occupants in an image. It is performed by lowering the threshold level of the membership value of the element to be regarded as occupants under the restriction that the number of occupants determined by locating peaks is not changed. Figure 20 shows the result of region growing. Half-toning elements are newly selected as components of occupants.

3.4 Method to Locate Occupants

A method to find the positions of occupants is explained below. We can control the wind direction according to their positions; for example, it is possible to prevent the cool wind from reaching occupants when an air conditioner start to overheat.

3.4.1 Estimating the Distance between the Sensor and Occupants

Positions of occupants in a room is easily calculated if the distance Lso between the infrared rays sensor and occupants can be estimated. In general, the output of sensor becomes low and the position of occupant in an image rises as the distance Lso becomes long. On the contrary, as the distance Lso becomes short, the output becomes high and the position descends to the lower part of the image. However, it is difficult to express this relationship in a simple model. We apply the iterative rule generation method [5] to identify the relationship among the distance Lso, the output of local temperature peak, and its element number using fuzzy if-then rules. The method consists of parameter adjustment of membership functions based on a steepest descent method and new rule generation procedures, which are repeated until the reasoning error is reduced to the desirable accuracy. Some rules are generated by creating a new membership function in the antecedent part when the reasoning error can no longer be reduced by adjusting parameters of current rules. New membership functions are created one by one so as to partition the domain with the highest reasoning error in input space, which is fuzzy partitioned by antecedent membership functions. The method can automatically determine the number of rules necessary to identify the input-output relationship of an objective system.

3.4.2 Experimental Results

Figure 21 shows a room layout for the experiment to acquire fuzzy if-then rules for estimating the distance Lso between the sensor and each occupant. The sensor is set 2.3m high from the floor and its decline angle is 28 degrees. First, we got 4 thermal images of one occupant who is sitting at every position from A to L. Two images at one position are for identification of fuzzy if-then rules, and the rest are for evaluation of the rules. Next, we segmented a region of an occupant from all images. The output of local temperature peak and its element number are collected from each image. They are recorded with the distance Lso measured, in practice, as two data sets for identification and evaluation. We generated fuzzy if-then rules which average reasoning error is less than 0.1m using the data set for identification. Nine rules are generated in total. Reasoning results of the data set for evaluation using the generated rules are shown in Table 6. Average reasoning error for the evaluation data is also less than 0.1m. Some experiments including this result show that the generated rules can estimate the distance Lso with desirable accuracy.

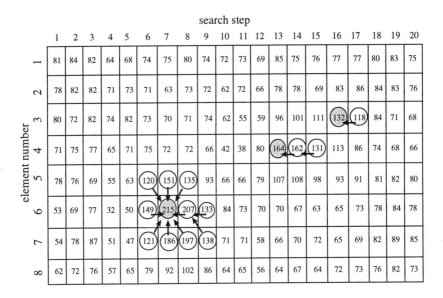

Figure 18: Peak location.

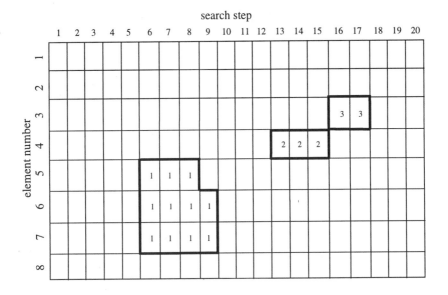

Figure 19: Initial label.

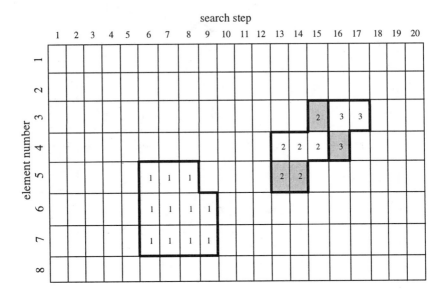

Figure 20: Result of region growth.

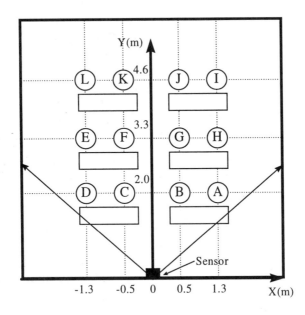

Figure 21: Room layout for experiment.

3.5 Conclusion

We have introduced a segmentation method of thermal images to identify the number and positions of occupants in a room.   We evaluated the proposed algorithm in a more practical environment, such as an office and a living room.

Table 6: Reasoning result of distance

Position	Peak Temp.	Element No.	Reasoning result	Real distance
A	182	6	2.58	2.63
A	188	6	2.52	2.63
B	203	6	2.35	2.34
B	193	6	2.48	2.34
C	207	6	2.36	2.34
C	201	6	2.41	2.34
D	209	6	2.34	2.63
D	184	6	2.56	2.63
E	173	4	3.57	3.72
E	161	4	3.68	3.72
F	175	4	3.55	3.52
F	176	4	3.55	3.52
G	168	4	3.61	3.52
G	161	4	3.68	3.52
H	167	4	3.62	3.72
H	167	4	3.62	3.72
I	129	3	4.86	4.91
I	131	3	4.84	4.91
J	128	3	4.87	4.76
J	138	3	4.78	4.76
K	131	3	4.84	4.76
K	125	3	4.89	4.76
L	136	3	4.80	4.91
L	121	3	4.92	4.91

References

[1]   H. Ichihashi, (1991), Iterative Fuzzy Modeling and a Hierarchical Network, *Proc. of Int'l. Fuzzy System Association Fourth World Congress*, pp. 49-52.

[2]   T. Tobi, T. Hanafusa, (1991), A Practical Application of Fuzzy Control for an Air-Conditioning System, *Int'l. Journal of Approximate Reasoning*, 5, 3, pp. 331-348.

[3]   H. Ichihashi, (1995), Efficient Algorithms for Acquiring Fuzzy Rules from Examples, *Theoretical Aspects of Fuzzy Control* (H. T. Nguyen, M. Sugeno, R. Tong, and R. Yager Eds.), John Wiley & Sons, pp. 261-280.

[4]   M. Braae, D.A. Rutherford, (1978), Fuzzy Relations in a Control Setting, *Kybernetes*, 7, pp. 185-188.

[5] D. E. Rumelhart, J. L. McClelland, and the PDP Research Group, (1986), *Parallel Distributed Processing*, MIT Press.

[6] P. O. Fanger, *Thermal Comfort*, Danish Tech. Press, 1970.

[7] N. Yoshiike, K. Arita, and K. Morinaka, (1993) Human information sensor, *Proceedings of the 7th International Conference on Solid-State Sensor and Actuators*, pp. 1015-1018.

[8] S. Araki, H. Nomura, and N. Wakami, (1993) Segmentation of thermal images using the fuzzy c-means algorithm, *Proceedings of the IEEE Second International Conference on Fuzzy Systems*, pp. 719-724.

[9] N. Wakami, S. Araki, and H. Nomura, (1993) Recent applications of fuzzy logic to home appliances, *Proceedings of the International Conference on Industrial Electronics, Control, and Instrumentation*, pp. 155-160.

[10] S. Araki, H. Nomura, I. Hayashi and N. Wakami, (1991) A self-generating method of fuzzy inference rule, *Proceedings of the International Fuzzy Engineering Symposium*, pp. 1047-1058.

10 INTELLIGENT AUTOMATION SYSTEMS AT PETROLEUM PLANTS IN TRANSIENT STATE

T. Tani, T. Kobayashi, and K. Nakane
System Technology Development Center
Manufacturing Department
Idemitsu Kosan Co., Ltd.
Japan

This chapter describes the applications of artificial intelligence (AI) techniques to the automation of petroleum refinery plants. It begins with a discussion of the history of automated control in the refining industry in Japan. Then, three applications of AI techniques are described, with particular focus given to feed oil switching, start-up plant, and other transient states that have been considered difficult to automate. These are (a) PID controller supervised by neuro-fuzzy hierarchical system in feed oil switching, (b) Fuzzy control system in pump start up, and (c) Fuzzy-PID hybrid control system in feed property changing.

Like other equipment-intensive industries, the petroleum refining industry has actively adopted automation technology in order to improve product quality, stabilize operations, boost productivity, and save labor. The history of refining automation began in the 1930s with the adoption of automatic control using pneumatic analog control devices. Many advances followed, with automatic control becoming particularly widespread in the 1980s following the introduction of distributed control systems (DCS).

A 1990 survey in Japan found that most process control methods used in plants employed proportional integral derivative (PID) control methods based on classical control theory. The exact breakdown was 84.5% for PID control, 6.8% for advanced PID control, 1.5% for modern control, 0.6% for AI control, and 6.6% for manual control [1].

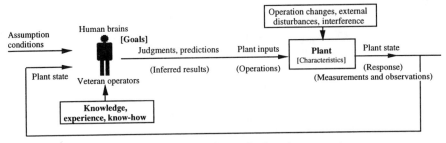

(a) The adjustment of operation by veteran operators

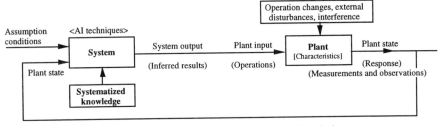

(b) The automation of operation using AI techniques

Figure 1: Automation of plant operation using AI techniques.

There are several reasons for the petroleum refining industry not becoming completely automated despite its long history of adopting plant automation techniques. It is difficult to automate procedures such as feed oil switching, operating mode switching, and other transient states. It is especially hard to automate shutdown and start-up operations. Other reasons include the complexity of many processes, the difficulty of measuring some important process variables, and

- Interference (external disturbance) at the operating points is too large.
- Dead times (i.e., time delays) are too long.
- Response is nonlinear.
- Properties of the feed (i.e., the component blends) change.
- Not enough measuring points.
- Difficult to collect and summarize knowledge about process operation.

Even under these circumstances, expert operators are able to keep temperatures stable and achieve their other goals during plant operation by observing the basic operating conditions of the plant and applying their knowledge, experience, and know-how to analyze the plant's operation and to make appropriate judgments and predictions. The knowledge, experience, and know-how of expert on-site operators are the basis for automation using AI techniques (fuzzy logic, neural networks, etc.) [2]-[9]. This automation process is diagrammed conceptually in Figure 1.

1 PID Controller Using Neuro-Fuzzy Hierarchical System in Feed Oil Switching

1.1 Introduction

This section shows a practical control system using neural networks and fuzzy control in a transient state of a petroleum plant [10]. The neuro-fuzzy hierarchical control system plays the role as a supervisor of PID controller. Neural networks effectively simulate an expert operator's procedure to control the plant and estimate the rough control target. The control system keeps the receiver level within a certain range using this rough target. Fuzzy control rules of operator's knowledge compensate for the rough control target to make the receiver level stable even if the process is noisy. This PID and neuro-fuzzy hierarchical control system is applied to the receiver level control of a desulfurizing plant in the transient state of feed oil switching. It can control the level effectively, and thus results in reduced boiler consumption.

1.2 Process Description

Crude oil is distilled into products, such as liquefied petroleum gases (LPG), naphtha, kerosene, and light gas oil (LGO). Each product has a sulfur content. A desulfurizing plant desulfurizes kerosene and LGO to improve their quality as shown in Figure 2. Either kerosene or LGO is mixed with hydrogen gas to promote reaction, and heated to reaction temperature. The heated oil/hydrogen gas is channeled to a reactor for desulfurization. The oil/hydrogen gas from the reactor is separated into oil and hydrogen gas in a separator. Through a flash drum, the oil/gas is fed to a stripper to remove the gas dissolved in the oil. The gas is removed from the reflux receiver and the oil at the bottom of the stripper goes to a product tank.

1.3 Control Problems in Feed Oil Switching

We must maintain the receiver level at a stable level and control the reflux flow amount to a minimum to save energy as shown in Figure 2. The plant desulferizes LGO or kerosene alternately. We have already automated this plant in steady state.

Feed oil is switched quickly from LGO to kerosene or vice versa. In feed oil switching from LGO to kerosene, for example, kerosene flows rapidly into the stripper and evaporates because of the difference in the boiling temperature of kerosene and LGO. An expert, therefore, decreases the reboiler temperature beforehand so as not to increase the receiver level by quick evaporation. This rapid changing of receiver level prevents the perfect automation of the plant in feed oil switching.

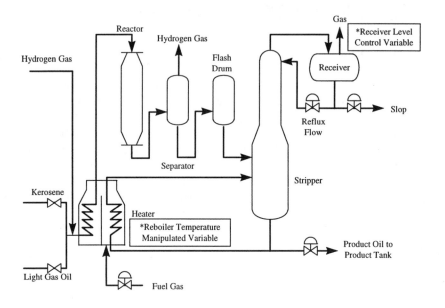

Figure 2: Process flow diagram of desulfurizing plant.

In this process, several problems are encountered in stabilizing the receiver level in feed oil switching as follows:

(1) There are two ways to control the receiver level; one is to change the reflux flow amount and the other is to change the reboiler temperature. These two variables interact strongly with each other.

(2) The response of the receiver level by the reboiler temperature is nonlinear with a long dead time .

(3) We use a center-wall type heater, which is separated into two parts by the wall. One part is used as a heater of the reactor and the other as a heater (reboiler) of stripper.

(4) The properties of feed oil vary continuously for several hours after feed oil switching and one cannot precisely observe this phenomena.

(5) This plant does not have enough inventory to absorb rapid changes at the separator, the flush drum, and the receiver of the stripper during feed oil switching.

1.4 Neuro-Fuzzy Hierarchical Control System

We observed the procedures used by an expert operator. He knows the changing profiles of reboiler temperature from experience. He calculates a rough target of the reboiler temperature using this changing profile and the current reboiler temperature. He then corrects the rough target by observing the receiver level, reflux flow amount, and pressure of stripper (see Figure 3).

We designed a neuro-fuzzy hierarchical control system which can replace the manual operation by an expert operator to keep the receiver level stable in feed oil switching [8][11]. The hierarchical control system consists of two functions, that is, a prediction function (neural network) and a correction function (fuzzy controller) and they vary frequently under the control target of PID controller, as shown in Figure 4.

1.4.1 Prediction Function

The prediction function block is a neural network for predicting the changing profile of the reboiler temperature for a given flow of charge.

(1) Structure of neural networks
A neural network with feedback can identify a nonlinear system with a long dead time. It has been reported that a neural network where several outputs go back to inputs, using dead time units or differential units, is useful [11]. However, this type of neural network is too complex for our purpose. Therefore, we use a simple multilayered neural network with time as an input variable.

We use a three-layered model which uses backpropagation algorithm [12]. We use two kinds of neural networks to identify the operator's procedure for the different feed oil switching, i.e., from kerosene to LGO and vice versa. Our neural network, in the case of kerosene to LGO, is shown in Figure 5. The input layer has six units, the hidden layer has 16, and the output layer has one, reboiler temperature (MV: Manipulated Variable). We defined the input variables by interviewing expert operators.

(2) Output of trained neural networks
The changing profiles of reboiler temperature in feed oil switching are obtained using the trained neural network. Input variables of the charge flow amount, the tower-top temperature of the stripper, the reflux flow amount, the slop flow amount, and the product flow amount are fixed to specified target values and the time range is over a several hour period. Figure 6 shows the reboiler temperature changing profile using a trained neural network.

Since we require a rapid response by a distributed control system (DCS), the changing profile of the reboiler temperature is expressed by a logistic function,

$$y_1(t) = k/(1+exp(a)\ exp(bt)), \tag{10.1}$$

where k stands for the temperature target and a and b are parameters which are defined by the charge flow amount. Those parameters are determined by multiple regression analysis.

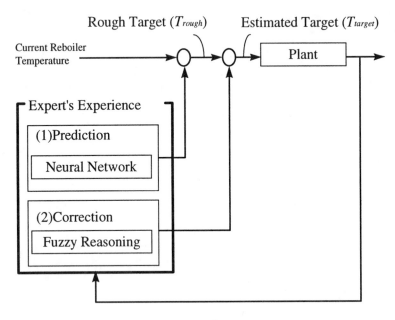

Figure 3: Procedure of expert operator.

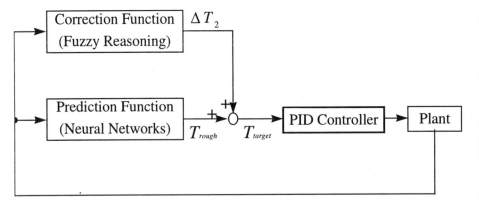

Figure 4: Structure of neuro-fuzzy hierarchical control system.

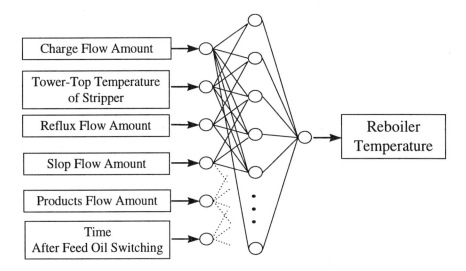

Figure 5: Neural network for feed oil switching in case of kerosene to LGO.

1.4.2 Correction Function

The correction function is implemented as a fuzzy controller for compensating the estimated rough reboiler temperature. We want to know how to adjust reboiler temperatures when the level makes it unstable.

The system observes a current reflux flow amount and the pressure of the receiver, and calculates the compensating value for the reboiler temperature to keep the receiver level stable. We use the fuzzy reasoning of Takagi-Sugeno type [13] whose consequence is a linear function.

The control rules of experienced operators to stabilize the receiver level are as follows:

$$\text{Rule 1: If } DF \text{ is N then } y_2^1 = \alpha_1 DF + \chi_1,$$

$$\text{Rule 2: If } DF \text{ is Z then } y_2^2 = \alpha_2 DF + \chi_2,$$

$$\text{Rule 3: If } DF \text{ is P and } DP \text{ is N then } y_2^3 = \alpha_3 DF + \beta_3 PR + \chi_3, \qquad (10.2)$$

$$\text{Rule 4: If } DF \text{ is P and } DP \text{ is Z then } y_2^4 = \alpha_4 DF + \beta_4 PR + \chi_4,$$

$$\text{Rule 5: If } DF \text{ is P and } DP \text{ is P then } y_2^5 = \alpha_5 RE + \beta_5 PR + \chi_5,$$

where y_2^i is the compensating value for the rough reboiler temperature of rule i; DF is the difference between the current reflux flow amount and a specified target; DP is the change-rate of the pressure of the receiver; PR is the pressure of the receiver; N, Z, and P are fuzzy sets, which mean negative, zero (near target), positive, respectively; and α_i, β_i, and χ_i are constants calculated by multiple regression analysis ($i = 1, ..., 5$).

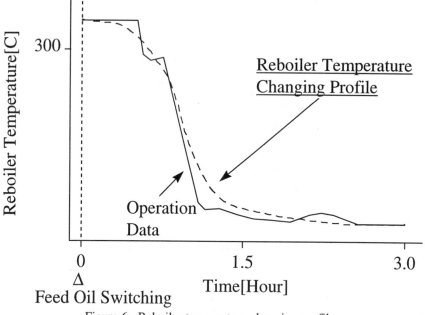

Figure 6: Reboiler temperature changing profile.

When the values of *DF*, *DP* and *PR* are given, output value (y_2) of fuzzy reasoning is calculated as

$$y_2 = \sum_{i=1}^{5} h^i y_2^i \Big/ \sum_{i=1}^{5} h^i ,$$
(10.3)

where h^i is the degree of fitness in the premise part for Rule i ($i = 1, ..., 5$) calculated by the minimum of each premise.

1.5 Control Algorithm

The system has several steps to calculate the reboiler temperature target as follows (see Figure 7).

Step 1: The changing profile of reboiler temperature is defined by the current charge flow amount using Equation (10.1).

Step 2: The difference of the temperature (ΔT_1) in time t and $t+\Delta t$ is calculated by the Equation (10.1),

$$\Delta T_1 = y_1(t+\Delta t) - y_1(t),$$
(10.4)

where Δt is the time difference.

Step 3: The rough target of the reboiler temperature (T_{rough}) is calculated by adding ΔT to the current reboiler temperature T,

$$T_{rough} = T + \Delta T_1,$$ (10.5)

Step 4: The system observes the current reflux flow amount ($CREF(t)$) and the current pressure of the receiver ($PR(t)$) and calculates DF, DP, PR for fuzzy rules, Section 1.4.2, as follows:

$$DF = CREF(t) - TGT,$$ (10.6)
$$DP = PR(t) - PR(t - \Delta t),$$ (10.7)
$$PR = PR(t),$$ (10.8)

where TGT is a specified target of the reflux flow amount and Δt is time difference. The system calculates a compensating value (ΔT_2) for the reboiler temperature at time t to keep the reflux flow amount at a specified target by fuzzy rules.

Step 5: The reboiler temperature target (T_{target}) at time $t + \Delta t$ is calculated by adding a compensation value to the rough target,

$$T_{target} = T_{rough} + \Delta T_2,$$ (10.9)

The system gives the control target T_{target} to PID controller.

Step 6: Go to Step 2 until the transient state ends.

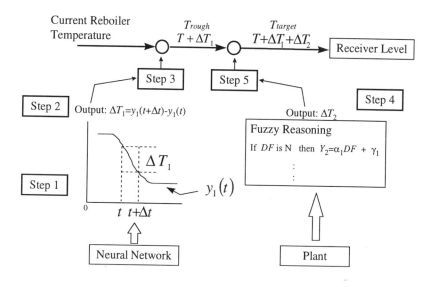

Figure 7: Control algorithm.

1.6 Results

We have applied the PID and neuro-fuzzy hierarchical control system to a commercial desulfurizing plant at Idemitsu Aichi Refinery. Figure 8 shows the result of a field test with the control system of feed oil switching. On the other hand, Figure 9 shows the result of manual control of feed oil switching. The receiver level ranges smoothly from 58% to 63% by the control system, while it ranges rapidly from 49% to 84% by manual control. The reflux flow amount ranges from 700 kl/day to 1050 kl/day by the control system, while it ranges from 780 kl/day to over 1200 kl/day by manual control.

The system shows the advantages of: (1) keeping the receiver level stable; (2) reducing energy consumption by a reflux flow amount minimum; and (3) reducing operator's intervention during feed oil switching .

1.7 Conclusion

We have shown a PID and neuro-fuzzy hierarchical control system which can maintain the stable receiver level of petroleum plants in transient state of feed oil switching. The neuro-fuzzy hybrid control system, which acts as the supervisor of PID controller, consists of two functions; one estimates the rough target of reboiler temperature using a neural network, the other compensates the rough target to maintain the receiver level stable using fuzzy reasoning. The hierarchical control system shows its effectiveness at the commercial plant of the desulfurizing process.

2 Fuzzy Control System in Pump Start-up

2.1 Introduction

This section shows the use of fuzzy logic in a pump start-up operation [1]. Automation of pump start-up is achieved by replacing the expert operator's knowledge with fuzzy control rules. The results show this system's effectiveness in a commercial plant. The system reduces the time for pump start-up operation by 50 percent compared with a ramp algorithm controller.

We developed a practical automatic operation system that can automate the procedures and sequences in start-up, shut-down, and feed oil-switching in a commercial plant [15]. However, it is difficult to automate valve control in pump start-up operation using conventional techniques, such as ramp algorithm and PID algorithm, because of operating characteristics. On the other hand, an operator with his/her know-how and experience can control the valve manually without any pump trouble. We apply the operator's expertise to automate the procedure of pump start-up. Since it is difficult to express this know-how using a mathematical model, we

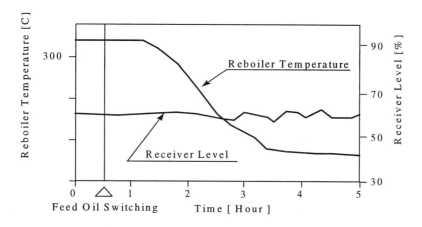

Figure 8: Control result of hierarchical control system in feed oil switching.

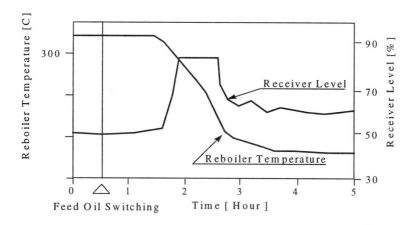

Figure 9: Control result of manual control in feed oil switching.

2.2 Outline of Pump Start-up Operation

Figure 10 shows the flow diagram of the plant discussed in this section. A pump transfers oil from a feed tank to a vessel. The discharge pressure sensor measures a discharge pressure of the pump. The flow sensor measures a pump discharge flow rate. In a steady state, the flow valve controls a pump discharge flow rate, and the discharge valve is fully opened. During start-up of a pump, the discharge valve is manually opened to control a discharge pressure, and the flow valve is slightly opened.

Manual start-up procedure of a pump is as follows:

Step 0: Stop the pump.

Step 1: Shut the discharge valve and the flow valve.

Step 2: Fill the pump with oil to eliminate gas in the pump (it is difficult to eliminate gas completely by this operation.)

Step 3: Open the flow valve slightly.

Step 4: Start the pump.

Step 5: Open the discharge valve gradually. To open the discharge valve quickly causes a drop in a discharge pressure of pump. If the discharge pressure reaches a certain value (P_{low}:P_{low} is a minimum operation value of discharge pressure). When the discharge pressure is below P_{low}, we must stop the pump, and this procedure must be restarted from Step 0.

Step 6: Start steady control. The pump discharge flow rate is controlled automatically by the flow valve.

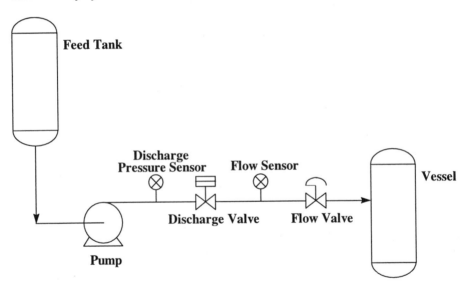

Figure 10: Process flow diagram of plant.

Our purpose of automating a pump start-up is

(1) To reduce operating time of the above procedure.

(2) To keep the discharge pressure above P_{low}. However, it is difficult to automate Step 5 due to the following reasons:

  (a) To reduce operating time, it is necessary to increase the opening rate of discharge valve. However, quick opening of discharge valve causes cavitation of the pump (cavitation is caused by the vacuum generated inside the inlet nozzle of running centrifugal pump). Cavitation causes the

the inlet nozzle of running centrifugal pump). Cavitation causes the discharge pressure to drop abruptly, as shown in Figure 11(a).

(b) To keep the discharge pressure, it is required to suppress the opening rate of discharge valve. However, to suppress the opening rate takes a long operation time, as shown in Figure 11(b).

The above two conditions should be compromised. An operator with his/her know-how and experience can open the discharge valve smoothly by manual operation to prevent cavitation. We attempted to apply the operator's expertise in order to automate the procedure of Step 5. Since this know-how is difficult to express in a mathematical model, we replaced it with fuzzy control rules.

2.3 Problems to Automate by Conventional Controller

2.3.1 Ramp controller

A discharge valve openings and a discharge pressure of pump start-up by ramp controller is shown in Figure 12. A ramp controller opens the discharge valve at a constant rate, regardless of discharge pressure. Therefore, we set the ramp rate small to prevent cavitation. This takes a long operation time.

2.3.2 PID controller

Using PID controller, it is difficult to control pump start-up by fixed PID parameters. When PID controller with fixed parameters is applied to discharge pressure control, we assume the result as shown in Figure 13.

Using PID controller with fixed parameters, the discharge valve will be opened abruptly when pump starts, because the discharge pressure becomes very high. After that, PID controller adjusts the discharge valve only to keep the pressure constant. The curve of PID controller is completely different from that of ideal operation. In order to fit the curve of PID controller to that of ideal operation, PID parameters must be adjusted every moment, which is difficult to do.

2.4 Fuzzy Controller

2.4.1 Input Variable and Output Variable

Three input variables and one output variable are determined by observing an expert operator's procedure. Input variables are the discharge valve openings, discharge pressure, and change-rate of discharge pressure, and output variable is the change-rate of discharge valve openings as shown in Figure 14. To apply the fuzzy controller to all pumps that have different specifications, we normalized the values of inputs and outputs from zero to one.

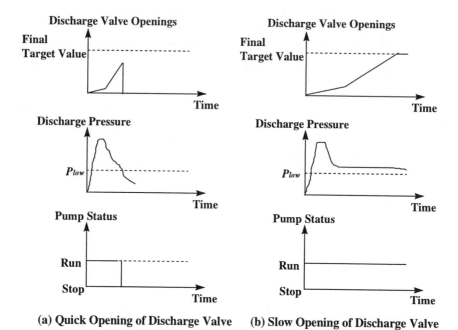

(a) Quick Opening of Discharge Valve   (b) Slow Opening of Discharge Valve

Figure 11: Pump start-up by ramp algorithm.

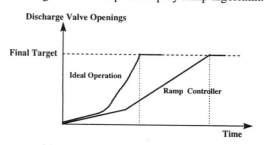

(a) Discharge Valve Openings by Ramp Controller

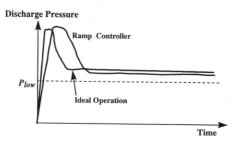

(b) Discharge Pressure by Ramp Controller

Figure 12: Problem by ramp controller.

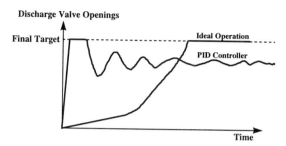

(a) Discharge Valve Openings by PID Controller

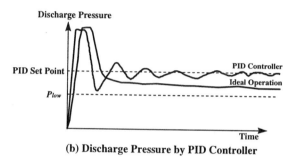

(b) Discharge Pressure by PID Controller

Figure 13: Problem by PID controller.

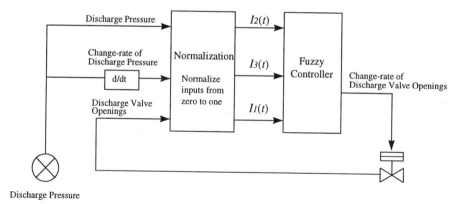

Figure 14: Input and output variables of fuzzy controller.

A discharge valve openings at time t is normalized as

$$I_1(t) = VO(t) / 100, \tag{10.10}$$

where $I_1(t)$ is the normalized discharge valve openings, and $VO(t)$ is the discharge valve openings (0 - 100 %) at time t. The discharge pressure at time t is normalized as

$$I_2(t) = (P_1(t) - Pmin) / (Pmax - Pmin), \qquad (10.11)$$

where $I_2(t)$ is the normalized discharge pressure, $P_1(t)$ is the discharge pressure at time t, $Pmax$ is maximum value, and $Pmin$ is minimum value of the discharge pressure. The change-rate of discharge pressure at time t is normalized as

$$I_3(t) = (P_1(t) - P_1(t\text{-}1) / (Pmax - Pmin), \qquad (10.12)$$

where $I_3(t)$ is normalized change-rate of discharge pressure, $P_1(t)$ and $P_1(t\text{-}1)$ are the discharge pressure at time t and t-1, respectively. Output of fuzzy controller at time t, i.e., the change-rate of discharge valve openings is normalized as

$$O_1(t) = Fuz(t) / Vmax, \qquad (10.13)$$

where $O_1(t)$ is the normalized change-rate of discharge valve openings, $Fuz(t)$ is the result of fuzzy reasoning at time t, and $Vmax$ is maximum opening rate of the discharge valve in a single control cycle. We use an area method in fuzzy reasoning [16].

2.4.2 Fuzzy Control Rules

Fuzzy control rules are divided into three stages (initial opening stage, main opening stage, and final opening stage) based on normalized discharge valve openigs as shown in Table 1. These membership functions are shown in Figure 15. The properties of each stage are as follows:

(1) Initial opening stage (Rules 1–3 in Table 1)
The purpose of this stage is to open a discharge valve at small valve openings while keeping a discharge pressure with some margin above P_{low}. If the discharge valve is opened abruptly at this stage, cavitation occurs and the discharge pressure drops. Fuzzy control rules are summarized as follows:

- If the discharge pressure stays above P_{low} with some margin, the discharge valve should be opened gradually.
- If the discharge pressure stays near P_{low}, the discharge valve openings should be reduced.

(2) Main opening stage (Rules 4–12 in Table 1)
The purpose of this stage is to increase the discharge valve openings in order to reduce the operation time. However, the opening discharge valve abruptly causes cavitation and results in discharge pressure drops. On the other hand, opening the discharge valve at small rate takes a long operation time. Therefore, suitable control

for the discharge valve is required. Fuzzy control rules are summarized as follows:

- If the discharge pressure stays above P_{low} with some margin and the discharge pressure tends to increase or stay the same, opening rate of the discharge valve should be increased gradually.
- If the discharge pressure stays above P_{low} with some margin and the discharge pressure tends to decrease, the discharge valve should be opened at a small rate.
- If the discharge pressure stays near P_{low} and the discharge pressure tends to increase or stay the same, the discharge valve should be opened at a small rate.
- If the discharge pressure stays near P_{low} and the discharge pressure tends to decrease, the discharge valve openings should be decreased.

Table 1: Fuzzy control rules.

No.	$I_1(t)$	$I_2(t)$	$I_3(t)$	$O_1(t)$
1	Small	Low	--	NB
2	Small	Middle	--	PS
3	Small	High	--	PS
4	Middle	Low	Negative	NB
5	Middle	Low	Zero	PS
6	Middle	Low	Positive	PM1
7	Middle	Middle	Negative	PM1
8	Middle	Middle	Zero	PM3
9	Middle	Middle	Positive	PB
10	Middle	High	Negative	PM2
11	Middle	High	Zero	PB
12	Middle	High	Positive	PB
13	Large	Low	Negative	NS
14	Large	Low	Zero	PM2
15	Large	Low	Positive	PM3
16	Large	Middle	--	PM
17	Large	High	--	PB

No. : Rule Number
$I_1(t)$: Normalized Discharge Valve Openings
$I_2(t)$: Normalized Discharge Pressure
$I_3(t)$: Normalized Change-rate of Discharge Pressure
$O_1(t)$: Normalized Change-rate of Discharge Valve Openings

(3) Final opening stage (Rules 13–17 in Table 1)
The purpose of this stage is to open the discharge valve to the final target and reduce operation time. The discharge pressure remains almost constant in this stage. Fuzzy control rules are summarized as follows:

- If the discharge pressure stays above P_{low} with some margin, the discharge valve should be opened at a high rate.
- If the discharge pressure stays near P_{low} and change-rate of discharge pressure tends to increase or stay the same, opening rate of a discharge valve should be decreased slightly.

• If the discharge pressure stays near P_{low} and change-rate of discharge pressure tends to decrease, the discharge valve openings should be decreased.

2.5 Results

The fuzzy controller is applied to a commercial plant at Idemitsu Chiba Refinery. The result of on-line testing is shown in Figure 16. Our findings are

(1)   The trajectory of discharge valve openings by fuzzy controller is almost the same as the trajectory by ideal operation.
(2)   Fuzzy controller keeps the discharge pressure of a pump almost constant. (This means fuzzy controller opened the discharge valve to the final target without cavitation.)

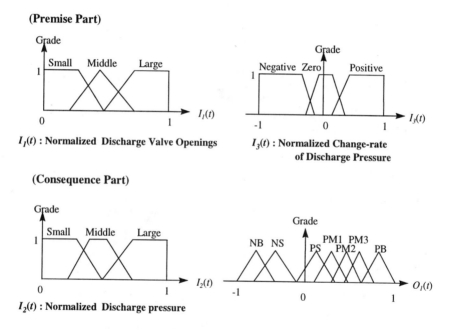

Figure 15: Membership functions.

We compare the discharge valve openings by the fuzzy controller, ramp controller, and operators in Figure 17. The solid line in the figure shows the trajectory of discharge valve openings by the fuzzy controller, the dotted line by the ramp controller, curve 1 shows the best case by an expert operator, and curve 2 shows the case by a novice operator. Time T_1, T_2, T_3, and T_4 are the operating times to open the discharge valve to the final target by an expert operator, a fuzzy controller, a novice operator, and a ramp algorithm controller, respectively. Our findings are

(1) Fuzzy controller reduces the time for pump start-up by about 50 percent compared with a ramp controller.
(2) Fuzzy controller acts similar to an average expert operator.

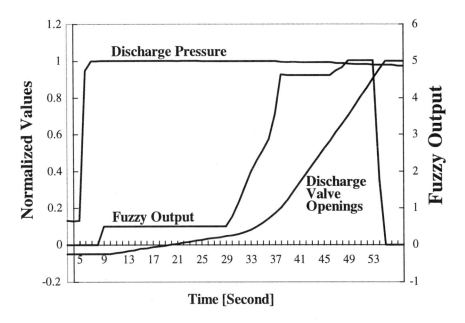

Figure 16: Control results of on-line testing.

2.6 Conclusion

We used a fuzzy controller to control a discharge valve. The discharge valve is difficult to control using conventional techniques. We replaced the expert operator's know-how and experience with fuzzy control rules. The fuzzy controller acts similar to an expert operator and the automation of pump start-up is achieved.

3 Fuzzy-PID Hybrid Control System in Feed Property Changing

3.1 Introduction

This section presents a practical control system using fuzzy and conventional PID controller, where the output of fuzzy controller compensates for the output of PID controller directly [7][17]. In most fuzzy and PID hybrid control systems, the fuzzy reasoning adjusts three parameters of the PID controller or the fuzzy controller combines some different PID controllers [18]-[21]. This control system is applied for controlling the tower-top temperature in a petroleum plant at the Idemitsu Hokkaido refinery. Evidently, the control system can control the temperature effectively in both steady state and in transient state.

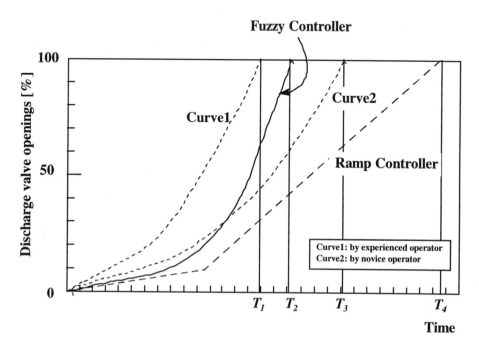

Figure 17: Comparison with another controller.

3.2 Process Description

Figure 18 shows the process of a naphtha desulfurizing plant which desulfurizes sour liquefied petroleum gas (LPG) and sour naphtha in order to improve their quality. The sour LPG and sour naphtha are mixed with hydrogen gas and heated by a heater, then desulfurized at a reactor. The outlet from the reactor is a mixture of LPG, naphtha, and sour gas. This mixture is then fed to the stripper to separate it into product LPG and product naphtha. The product LPG is extracted from the top of the stripper, and the product naphtha, from the bottom. They are separated by the difference in their boiling points, so the stripper is always heated by the reboiler. To keep the product quality high, we have to keep the tower-top temperature stable.

3.3 Control Problems

This plant has several characteristics which keep the product quality high as follows:

(1) When the reboiler temperature is too high, the product LPG becomes poor because of the pentane (C_5) impurity in LPG. And when the temperature is too low, the product naphtha quality is reduced because of the remaining butane (C_4) which makes the VP (Vapor Pressure) high. Therefore, the reboiler temperature must be controlled suitably for the product specification.

(2) Two ways to control the tower-top temperature are by the reflux flow amount and by the reboiler temperature. Therefore, to minimize the reflux flow amount and thus save energy, we must control the tower-top temperature by the reboiler temperature.

(3) Since it takes too much time to sense the VP of product naphtha, it is difficult to determine the reboiler temperature.

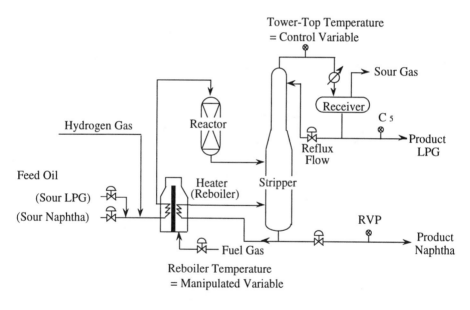

Figure 18: Process flow diagram of naphtha desulfurizing plant.

Therefore, the control variable (CV) is the tower-top temperature and manipulated variable (MV) is the reboiler temperature. We have the following difficulties in controlling the tower-top temperature:

(1) Feed oils, sour LPG, and sour naphtha come from the crude unit. The feed amount and the mixture rate of feed oils usually vary every moment. The feed properties of this naphtha desulfurizing plant varies greatly when the crude oil is switched. We termed this transient state "feed property changing."

(2) The center-wall type heater is separated into two parts by a wall. One part is used for reactor heater and the other for stripper heater (reboiler).

(3) The behavior of the tower-top temperature affected by the reboiler has a nonlinear response with a long dead time.

3.4 Fuzzy-PID Hybrid Control System

An expert operator can decide the tower-top temperature of the stripper by experience and thus keep the product quality high by using PID controller. We developed a fuzzy-PID hybrid control system by replacing operator's procedure, which consists of two controllers, PID controller and fuzzy controller (Figure 19) [5].

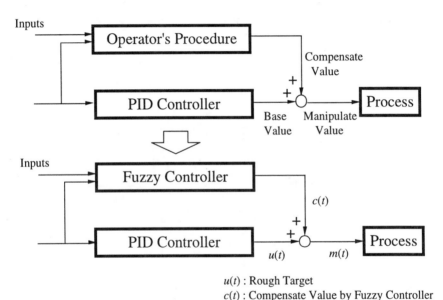

$u(t)$: Rough Target
$c(t)$: Compensate Value by Fuzzy Controller
$m(t)$: Compensated PID Output

Figure 19: Configuration of fuzzy-PID hybrid control system.

A rough target of the tower-top temperature $u(t)$ is calculated by the PID controller, based on the closed-loop error $e(t) = y_r - y(t)$, where y_r is a prescribed target of the tower-top temperature and $y(t)$ is a process variable of the tower-top temperature. This PID controller has the following standard form:

$$u(t) = K_c[e(t) + T_d\frac{de(t)}{dt} + \frac{1}{T_i}\int e(t)dt],$$ (10.14)

where K_c, T_d, T_i are the proportional gain, the derivative time, and the integral time of the controller, respectively.

The calculated rough target $u(t)$ is compensated for by the output $c(t)$ of the fuzzy controller, where we use a simplified fuzzy reasoning method [18][22][23], which has a crisp value but which is not fuzzy set for the consequence of fuzzy rules. The fuzzy control rules and these membership functions are shown in Figure 20. The fuzzy rules for this method are of the following form:

$$R^i : if\ TT\ is\ A^i{}_1\ and\ \Delta TT\ is\ A^i{}_2\ then\ Y^i = b^i, (i = 1, \cdots, 15) \quad (10.15)$$

where TT, ΔTT are input variables for i-th rules, Y^i is the output variable for the i-th rule, A^i_j is a fuzzy set, and b^i is a crisp value; TT means the difference between target and actual tower-top temperatures, ΔTT means the change-rate of tower-top temperature.

Given inputs ($TT = x_1$, $\Delta TT = x_2$) at time t, the degree of matching in the premise for the i-th rule w^i is calculated as follows:

$$w^i = A^i{}_1(x_1) \times A^i{}_2(x_2),\ (i = 1, \cdots, 15). \quad (10.16)$$

Then the output of $c(t)$ is inferred by taking the weighted average of the Y^i's,

$$c(t) = \sum_{i=1}^{15} (w^i b^i) \Big/ \sum_{i=1}^{15} w^i. \quad (10.17)$$

Now, the tower-top temperature is controlled by the compensated PID output $m(t)$, which is calculated as

$$m(t) = u(t) + c(t). \quad (10.18)$$

This control system has the following characteristics:

(1) With PID and fuzzy controllers, the process of nonlinear response with a long dead time can be controlled unlike with the conventional PID controller.
(2) The fuzzy reasoning compensates for the PID controller output directly. Therefore, the control rules are simpler than those which adjust three parameters of the PID controller. We can easily tune the parameters of membership functions.

3.5 Parameter Tuning

Most applications of the fuzzy controller have difficulty tuning membership functions on fuzzy rules. Some tuning methods for parameters of the membership functions are proposed [24][25], but we cannot get good tuning data because of noise in the process, interaction between variables, and nonlinear response with a long dead time. Therefore, we tuned the PID controller and the fuzzy controller manually [17].

○ Control Rule Table

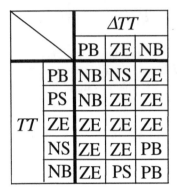

		ΔTT		
		PB	ZE	NB
	PB	NB	NS	ZE
	PS	NB	ZE	ZE
TT	ZE	ZE	ZE	ZE
	NS	ZE	ZE	PB
	NB	ZE	PS	PB

PB : Positive Big
PS : Positive Small
ZE : Zero
NS : Negative Small
NB : Negative Big
p_i : Parameter

○ Membership Functions in Premise
   TT : Difference between Target
   and Actual Tower-Top Temperature

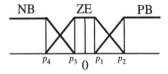

ΔTT : Change -rate of Tower-Top Temperature

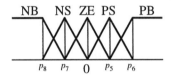

○ Membership Functions in Consequence

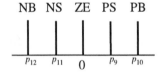

Figure 20: Fuzzy control rule table and membership functions.

The formulated 15 control rules which keep the tower-top temperature stable (Figure 20), contain 12 parameters ($p_1, p_2, ..., p_{12}$) which define membership functions. The steps of tuning the parameters are as follows:

Step 0: Set the PID parameters to correspond to those of a system in steady state by a conventional method.

Step 1: Set the cycle of fuzzy reasoning equal to that of an expert operator's compensation for the PID output.

Step 2: Tune the parameters for the consequence of fuzzy rules.

Step 3: Start to control the process with the fuzzy-PID hybrid control system and tune the parameters of membership functions of the premise of fuzzy rules, step-by-step. We move the parameters of the membership functions closer to zero and stop moving them before the process becomes unstable (Figure 21). We tune the parameter of the TT first, then the parameters of ΔTT

next. If we cannot achieve the aim of control ability, go to Step 4.

p_b, p_j: parameters of Membership

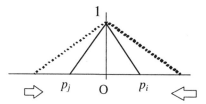

Figure 21: Step of parameter tuning.

Step 4:   Shorten the cycle of the fuzzy reasoning until it satisfies the needed control ability. This step corresponds to that of the increase in frequency of operator's intervention. Note that the more frequently the system is compensated, the smaller the value of consequence of the fuzzy rules. If we cannot achieve the aim of control ability, go to Step 5.

Step 5:   Set the T_i parameter of the PID controller smaller until the control ability that we need is satisfied.

3.6 Results

We apply the control system to the naphtha desulfurizing plant at Idemitsu Hokkaido Refinery. The target of control range in tower-top temperature is 4°C. Table 2 shows the control range of the tower-top temperature by each control technique.

Table 2: Control ranges of tower-top temperature

	Steady State	Feed Property Changing
Fuzzy-PID Hybrid Control System	1.5 °C	3 °C
PID Controller	10 °C	Unstable
Manual Operation	0.7- 11.5 °C	7.5 °C

In steady state, the result of using a fuzzy-PID hybrid control system shows that the tower-top temperature is controlled within 1.5 °C (Figure 22). Inversely, for result of the PID controller, we could not produce stable control i, and it ranged at 10 °C (Figure 23). By manual operation, it controlled within 0.7 °C usually; however, it ranged at 11.5 °C in the worst case (Figure 24).

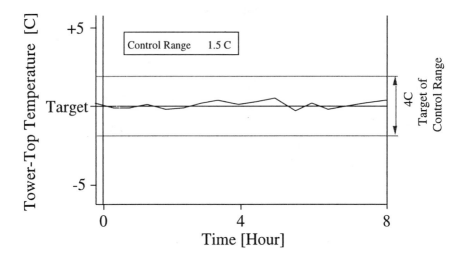

Figure 22: Control result of fuzzy-PID hybrid control system in steady state.

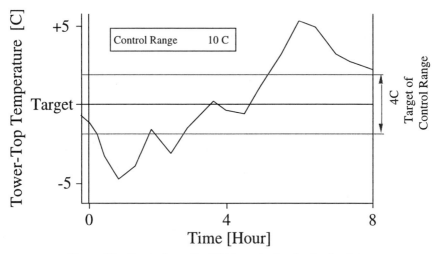

Figure 23: Control result of PID controller in steady state.

In feed property changing, the result of using the fuzzy-PID hybrid control system indicates that the tower-top temperature is controlled stable within 3 °C (Figure 25). Inversely, for result of manual operation, the tower-top temperature changed unstable within 7.5 °C (Figure 26). That means the fuzzy-PID hybrid control system can control the plant in both steady state and in transient state.

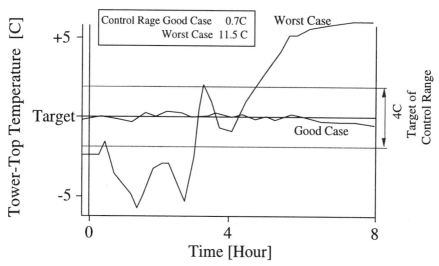

Figure 24: Control result of manual operation in steady state.

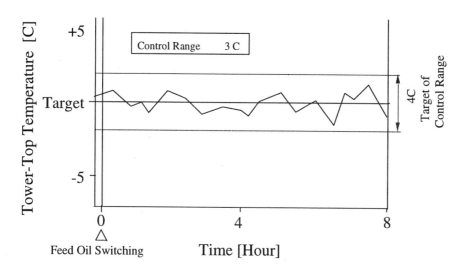

Figure 25: Control result of fuzzy-PID hybrid control system in feed property
changing.

3.7   Conclusion

We have shown a practical fuzzy-PID hybrid control system in this section. The
advantage of this method is as follows:

(1) The fuzzy controller compensates for the output of the PID controller directly, so we can use fuzzy rules to facilitate control of the tuning of parameters for fuzzy membership functions.
(2) The hybrid system can control the process which has a nonlinear response with a long dead time, unlike that of a conventional PID controller.

Our technique shows its effectiveness at a commercial plant both in steady state and in transient state, such as feed property changing.

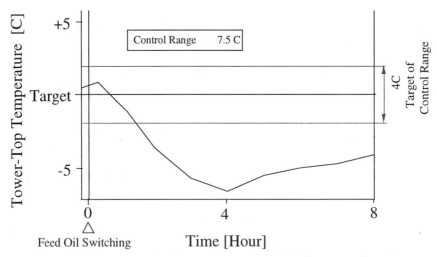

Figure 26: Control result of manual operation in feed property changing.

References

[1] K.Hiroi (1990): Concepts of Automatic Control and the Basics of PID Control Technology, *Instrumentation and Control Engineering*, Vol. 33, No. 9, pp. 91–96 (in Japanese).

[2] M.Sugeno (Ed.) (1985): *Industrial Applications of Fuzzy Control*, Amsterdam, North-Holland.

[3] T.Terano, K.Asai, and M.Sugeno (1987): *Fuzzy Theory and Its Applications*, Ohm-sha (in Japanese).

[4] W.H.Bare, R.J.Mulholl, and S.S.Sofer (1988): Design of a Self-tuning Expert Controller for a Gasoline Refinery Catalytic Reformer, *Proc. American Control Conf.*, Vol. 1, pp. 247-253.

[5] M.Kishimoto, T.Yoshida, and M.M.Young (1989): Application of the Fuzzy Expert System to Fermentation Processes, *Preprint of IFAC workshop on PCPI*, pp. 2-38.

[6] N.V.Bhat, P.A.Minderman, Jr., T.McAvoy, and N.S.Wang (1990): Modeling Chemical Process Systems via Neural Computation, *IEEE Control System Magazine*, Vol. 10, No. 3, pp. 24-30.

[7] T.Tani and K.Tanaka (1991): A Design of Fuzzy-PID Combination Control System and Application to Heater Output Temperature Control, *Trans. Society of Instrument and Control Engineers (SICE) of Japan*, Vol. 27, No. 11, pp. 1274-1280 (in Japanese).

[8] T.Tani, T.Sato, M.Umano, and K.Tanaka (1993): Application of Neural Network to Tank Level Control of Petrochemical Plants, *Proc. Int. Conf. on Industrial Electronics, Control, Instrumentation and Automatic (IEEE-IECON '93)*, Maui, Hawaii, U.S.A., pp. 321-326.

[9] Y.Suzuki (ed.) (1994): *Applications of Neuro/Fuzzy/AI System*, in Neuro/Fuzzy/AI Handbook, N.Kamata (ed.), the Society of Instrument and Control Engineers, Ohm-sha, pp. 863-998 (in Japanese).

[10] T.Tani, T.Sasada, M.Utashiro, M.Umano, and K.Tanaka (1995): "Neuro-fuzzy Hybrid Control System of Nonlinear Process in Petroleum Plant," *Proc. IEEE Int. Conf. on Neural Networks (ICCN '93)*, Perth, Western Australia, pp. 2501–250.

[11] T.Fukuda, T.Shibata, M.Tokita, and T.Mitsuoka (1990): Neural Network Application for Robotic Motion Control: Adaptation and Learning, *Proc. Int. Joint Conf. Neural Network (IJCNN '90)*, San Diego, CA. U.S.A., Vol. 3, pp. 447-451.

[12] D.E.Rumelhart, G.E.Hinton, and R.J.Williams (1986): Learning Internal Representations by Error Propagation, *Parallel Distributed Processing*, Vol. 1, MIT Press, Cambridge, pp. 318-362.

[13] T.Takagi and M.Sugeno (1985): Fuzzy Identification of Systems and Its Applications to Modeling and Control, *IEEE Trans. Systems Man & Cybernetics*, SMC-15, pp. 116-132.

[14] T.Tani, S.Mishima, K.Nakane, and T.Asaao (1997): Fuzzy Control System Applied to Pump Star in Petroleum Plant, *Proc. KES Int. Conf. on Conventional and Knowledge-Based Intelligent Electronic Systems (KES'97)*, Adelaid, Australia, pp. 174-180.

[15] T.Tani and K.Nakane (1996): Automation in Transition Stage Utilizing AI Techniques, *Proc. General Session of Honeywell Users Group Symposium*, Phoenix, Arizona, U.S.A.

[16] L.Zheng (1992): A Practical Guide to Tune of PI Like Fuzzy Controllers, *Proc. IEEE Int. Conf. on Fuzzy Systems (FUZZ-IEEE '92)*, San Diego (U.S.A.), pp. 633-640.

[17] T.Tani, M.Utashiro, M.Umano, and K.Tanaka (1994): Application of Practical Fuzzy-PID Hybrid Control System to Petrochemical Plant, *Proc. IEEE Int. Conf. on Fuzzy Systems (FUZZ-IEEE '94)*, Orlando, Florida, U.S.A., pp. 1211-1216.

[18] H.Ichihasi and H.Tanaka (1988): PID and Fuzzy Hybrid Controller, *4th Fuzzy System Symposium*, pp. 97-102 (in Japanese).

[19] N. Matsunaga and S.Kawaji (1991): Fuzzy Hybrid Control for DC Servomotor, *Trans. IEE of Japan*, Vol. 111- D, No. 3, pp. 195-200 (in Japanese).

[20] S.H.He, S.H.Tan, F.Z.Xu, and P.Z.Wang (1993): PID Self-Tuning Control Using a Fuzzy Adaptive Mechanism, *IEEE Int. Conf. on Fuzzy Systems (FUZZ-IEEE'93)*, California, pp. 708-713.

[21] A.Rueda and W.Pedrycz (1993): A Design Method for a Class of Fuzzy Hierarchical Controllers, *IEEE Int. Conf. on Fuzzy Systems (FUZZ-IEEE'93)*, California, pp. 196-199.

[22] M.Maeda and S.Murakami (1988): Self-Tuning Fuzzy Controller, *Trans. SICE of Japan*, Vol. 24, No. 2, pp. 191-197 (in Japanese).

[23] M.Mizumoto (1990): Fuzzy Controls by Product-Sum-Gravity Method, *Advancement of Fuzzy Theory and System in China and Japan* (Ed. by X.H.Liu and M.Mizumoto). International Academic Publishers, Vol. c1.1, No. 1.4.

[24] H.Nomura, I.Hayashi, and N.Wakami (1992): A Self-Tuning Method of Fuzzy Reasoning by Delta Rule and Its Application to a Moving Obstacle Avoidance, *Journal of Japan Society for Fuzzy Theory and Systems*, Vol. 4, No. 2, pp. 379-388 (in Japanese).

[25] L.Zheng (1993): A Practical Computer-Aided Tuning Technique for Fuzzy Control, IEEE Int. Conf. on Fuzzy Systems (*FUZZ-IEEE'93*), California, pp. 702-707.

11 | INTELLIGENT CONTROL FOR ULTRASONIC MOTOR DRIVE

Faa-Jeng Lin and Rong-Jong Wai
Department of Electrical Engineering
Chung Yuan Christian University
Chung Li 32023, Taiwan

This chapter proposes three intelligent controllers, the fuzzy model-following controller, the neural network model-following controller and the fuzzy neural network (FNN) model-following controller to control the traveling-wave type ultrasonic motor (USM) drive. Since the lumped dynamic model of the USM are difficult to obtain and the parameters of the motor are time varying, the intelligent controllers are very suitable for the control of the USM drive. The theory of the intelligent controllers is described in detail. The effectiveness of the proposed controllers is demonstrated by some experimental results.

1 Introduction

The newly developed traveling-wave type ultrasonic motor (USM) has many excellent performance features such as high-retention torque, high torque at low speed, silence, compact in size, and no electromagnetic interferences. It has been used in many practical applications [1]. The driving principles of the USM are based on high-frequency mechanical vibrations and frictional force. Therefore, its mathematical model is difficult to derive and a lumped motor model of the USM is presently unavailable. Moreover, the control characteristic of the USM is complicated and highly nonlinear. The control characteristic and motor parameters are time-varying due to increases in temperature and changes in motor drive operating conditions, such as drive frequency, source voltage and load [2,3]. Since the two-phase construction of the USM are coupled mechanically and the reaction from the electrical to the mechanical part are unbalanced for the two phases, the equivalent two-phase loads of the rotor are unbalanced and the equivalent resistor values are varied for different rotating directions, rotor speeds, load torque, applied voltages, and static pressure force between the stator and the rotor.

A driving circuit for balanced two-phase sinusoid output voltages with 90 degree phase difference under variable frequency control to effectively drive the USM, is

0-8493-9805-3/99/$0.00+$.50

designed and implemented in this chapter [4]. The proposed circuit is a two-phase chopper-inverter driving circuit. Each phase contains a high-frequency boost-chopper DC-DC converter and a half-bridge series-resonant inverter with capacitor-parallel load. The boost chopper is implemented to control the two-phase output voltages with constant magnitude and the switching frequency of the resonant inverter, which is adjusted by the control algorithm. It provides variable frequency control of the output voltages.

Fuzzy control technique [5-8] has been successfully applied to the control of motor drives in recent years [9,10]. The main advantages of fuzzy controllers are: (a) they are constructed on the basis of intuition and operating experience concerning the plant to be controlled without using mathematical derivation; (b) they possess some degree of robust control capability; however, under typical circumstances, the response trajectory of a fuzzy controller will change significantly when changes occur in the system parameters. Model reference adaptive control [11], which also has been applied to the control of motor drives [12-15], is an effective control technique to deal with time-varying control characteristics and parameter variations. Moreover, the desired tracking performance can be specified in the reference model. A fuzzy adaptive model-following mechanism for position control of the USM drive was described in [14]. A simple linear position control loop is designed and augmented by the model-following error-driven fuzzy adaptive mechanism to reduce the influence of external disturbances and parameter variations.

In recent years researchers have used the neural network in the control field to deal with nonlinearities and uncertainties of the control system [16-21]. The neural network applications for control can be classified as supervised control, inverse control, or neural adaptive control [17]. Moreover, the neural networks can be mainly classified as general learning (off-line) and special learning (on-line) [17,22]. In general learning, the connective weights of a back-propagation neural controller are trained off-line; therefore, a tremendous amount of training data must be used to include all possible operating conditions and the training time undertaken is long. On the other hand, since the weights cannot be changed on-line during real-time control, the neural network based on off-line learning does not have the property of true adaptive control and is, therefore, difficult to apply to a wide range of real-time control problems [23]. The special learning method can overcome the problem in general learning. The connective weights of the neural network are trained during on-line control, but some prior knowledge such as the sensitivity derivatives or the Jacobian of the system must be available in order to apply the traditional backpropagation algorithm [17]. Since the dynamic characteristics of the USM are difficult to obtain and the motor parameters are time varying, an on-line trained neural network model-following controller was proposed to control the rotor position of the USM in [21]. In the proposed neural network controller, the error between the states of the plant and the reference model is used to train the connective weights of the neural network controller on-line. Therefore, when parameter variations and external disturbances of the USM servo drive occur, using the model-following error driven approach, robust control performance can be obtained. Moreover, prior off-line learning of the neural network is not necessary and the Jacobian of the system is replaced by a sign function to increase the learning speed of the connective weights.

Fuzzy neural network (FNN) systems, which have the advantages of both fuzzy systems [5-10] and neural networks [16-23], have been used to deal with nonlinearities and uncertainties of the control systems [24-31]. For instance, in Chapter 3 of Wang [7], the adaptive fuzzy systems, or the FNNs, are introduced as identifiers for nonlinear dynamic systems based on backpropagation algorithm; Chen and Teng [27] proposed a model reference control structure using FNN controller, which is trained on-line using FNN identifier with adaptive learning rates; Zhang and Morris [28] described a technique for the modeling of nonlinear systems using an FNN topology; and Jang and Sun [29] reviewed the fundamental and advanced developments in neuro-fuzzy synergisms for modeling and control based on adaptive network. Moreover, according to the structure of the control systems, the applications of FNN for the control systems can be classified as supervised control, inverse control, neural adaptive control, etc. [29]. The central part of a learning rule for FNN concerns how to recursively obtain a gradient vector in which each element in the learning algorithm is defined as the derivative of an energy function with respect to a parameter of the network. This is done by means of the chain rule and the method is generally referred to as the backpropagation learning rule, because the gradient vector is calculated in the direction opposite to the flow of the output of each node [24-31]. According to the learning methods, the FNN can be mainly classified as general learning (off-line) and special learning (on-line). In general learning the connective weights and membership functions of FNN are trained off-line; therefore, a tremendous amount of training data must be used and the training time undertaken is long. On the other hand, since the parameters of the network cannot be changed on-line during real-time control, the network based on general learning does not have the property of true adaptive control and is, therefore, difficult to apply to a wide range of real-time control problems [27]. The special learning method can overcome the problem in general learning. The connective weights and membership functions of FNN are trained during on-line control, but some prior knowledge such as the sensitivity derivatives or the Jacobian of the system must be available in order to apply the traditional backpropagation algorithm. An on-line trained FNN model-following controller was proposed to control the USM drive in [31]. A delta adaptation law is proposed for the on-line training of the FNN. In [31], first, a proportional-integral (PI) position controller is adopted to control the position of the USM. Then, the FNN model-following controller is added to reduce the influence of parameter uncertainties. Moreover, to test the performance of the FNN, the inner loop PI position controller is omitted using only the FNN model-following controller to control the position of the USM.

This chapter, dealing with the problem of intelligent control of ultrasonic motor drive, is organized as follows. In Section 2, the equivalent circuit model of the USM is discussed and the proposed driving circuit for the USM is introduced.

Sections 3, 4 and 5 present three intelligent control algorithms, the fuzzy model-following control, the neural network model-following control and the FNN model-following control. In Section 6, a PC-based control system is devised to test the position control characteristic of the USM drive system. The experimental results of the three intelligent controllers are implemented and compared with a proportional-integral (PI) position controller in Section 7.

2 Ultrasonic Motor Drive

The traveling-wave type ultrasonic motor (USM) has many excellent performance features such as high-retention torque, high torque at low speed, silence, compact in size, and no electromagnetic interferences. It has been used in many practical applications [1].

2.1 The Equivalent Model of the USM

The driving principle of the USM is based on high-frequency mechanical vibration and the vibration force is generated by the piezoelectric ceramic located in the stator [1]. A single-phase equivalent circuit model of the USM driven by a half-bridge series-resonant inverter is shown in Figure 1(a). The definition of the parameters shown in Figure 1 are given as

A :   force factor
K :   spring constant of the ceramic-and-metal assembly of the stator
m :   mass of the ceramic-and-metal assembly of the stator
C_d :   blocking capacitor, due to dielectric properties of piezoelectric ceramic
L_m :   equivalent inductor, representing the mass effect of the stator assembly
C_m :   equivalent capacitor, representing the spring effect of the stator assembly
r_o :   equivalent resistor, representing mechanical losses that occur within the stator assembly
R_F :   equivalent resistor, representing sliding losses between the stator and the rotor
R_r :   equivalent resistor, representing mechanical losses resulting from the movement of the rotor
R_L :   equivalent load

The electrical load of the inverter is the polarized piezoelectric ceramic, which is the exciting source of the ultrasonic vibration and the mechanical vibration system, which is primarily composed of the elastic body (stator) and the rotor.

During the conduction of the T-switch or the D-diode, the circuit can be reduced to the specifications in Figure 1(b), where

$$L_m = m / A^2 \tag{11.1}$$

$$C_m = A^2 / K \tag{11.2}$$

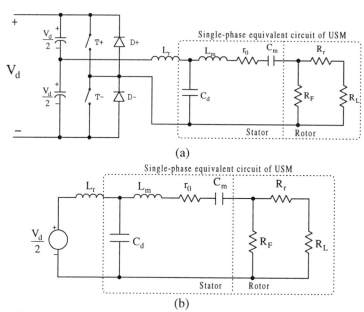

(a)

(b)

Figure 1: Single-phase equivalent circuit model of USM: (a) driving by a half-
bridge series resonant inverter; (b) reduced circuit during T- or D-
conduction [4].

In order to achieve high efficiencies, the USM should be driven at the approximate
frequency which will create resonance between C_m and L_m in the equivalent
circuit [1]. The mechanical resonance frequency is

$$f_m = 1 / 2\pi\sqrt{1 / C_m L_m} = 1 / 2\pi\sqrt{k / m} \tag{11.3}$$

and the electrical resonant frequency of the series-resonant inverter is

$$f_0 = 1 / 2\pi\sqrt{1 / C_d L_r} \tag{11.4}$$

The motor parameters of the USM are all dependent on the operating temperature
and the time of usage. In addition, since the two-phase construction of the USM are
coupled mechanically and the reaction from the electrical to the mechanical part is
unbalanced for the two phases, the equivalent two-phase loads of the rotor, R_F, R_r
and R_L are unbalanced and the equivalent resistor values are varied for different
rotating directions, rotor speeds, load torque, applied voltages and static pressure
force between the stator and the rotor. According to the above description, the Q-

factor [32] of the resonant inverter circuit of the A and B phases are not equal, moreover, they are time-varying and load-condition dependent.

2.2 The Driving Circuit for the USM

To drive the USM effectively, a two-phase chopper-inverter driving circuit is proposed in Figure 2 [4]. The proposed driving circuit can provide balanced two-phase sinusoid output voltage under variable frequency control. The two chopper-inverter driving circuits are identical and each is composed of one boost chopper cascading with one half-bridge series resonant capacitor-parallel load inverter. The boost choppers are adopted here to provide variable DC voltages, V_{dA} and V_{dB}, for the half-bridge series-loaded resonant inverters in order to maintain the two-phase sinusoid output voltages V_A and V_B at a constant peak value, V^*, under the variable output frequency control. The rotor position is measured using an optical encoder with 1000 pulses per revolution and the position message is sent to the control computer. The control input to the drive circuit, u_p, is calculated by the control algorithm in the control computer. Using the control input, the inverter frequency is regulated by means of a voltage controlled oscillator (VCO).

The split-phase circuit is designed for the two-phase inverter to provide two-phase output voltages, V_A and V_B, with a phase difference of 90 degrees and the rotating direction (CW or CCW) can be controlled by letting V_A lead or V_B lead. The L_A and L_B inductance are inserted in series with the load for each inverter, respectively, in order to become resonant with the inherent parasitic capacitances, C_{d1} and C_{d2}, of the USM. The resonant frequency of the L-C tank, f_0, is 35 kHz and the mechanical resonant frequency of the USM (USR-60), f_m, is from 39 kHz to 40 kHz in the ultrasonic frequency range. The switching frequency, f_s, of the inverter, which is designed to vary between 41 kHz and 43 kHz, should be designed higher than the resonant frequency of the mechanical vibration [2,3]. The resonant frequency of the L-C tank is designed to be lower than the resonant frequency of the mechanical vibration. Since the revolution speed of the rotor is closely related to the switching frequency of the inverter [2,3], i.e., the driving frequency of the USM, the switching frequency should be controlled only according to the control input resulting from the closed-loop control algorithm. The function of the choppers, which are implemented with MOSFETs, are to regulate output voltages of the inverters with PWM direct-duty cycle control. The resonant inverters, using parasitic impedance of the USM to provide sinusoidal voltage sources, are also implemented with MOSFETs.

3 Fuzzy Model-Following Control

The control algorithm for a fuzzy controller is formulated intuitively. Basically, a fuzzy controller is an error-driven control-signal generating mechanism. In a conventional fuzzy controller, the error between the set point and the plant output is used to drive the fuzzy controller. Though the fuzzy control law has inherent robust control characteristics, its tracking response trajectories may change significantly due to parameter variations in the plant. To enhance the tracking control performance, a reference model can be adopted to generate the desired response trajectory. The error between the outputs of the reference model and the plant is used to drive the fuzzy controller. To design the fuzzy adaptive controller, the model following dynamic is analyzed.

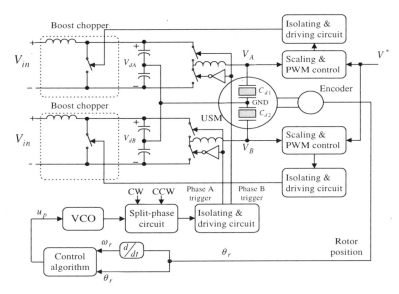

Figure 2: Schematic block diagram of the two-phase chopper-inverter USM driving circuit [4].

Dynamic Signal Analysis

The dynamic signal analysis of the rotor position tracking response is implemented to acquire some information concerning the plant before designing a fuzzy controller. A step command is issued at the beginning of the time axis, as shown in Figure 3 [14]. Figure 3 also shows the desired rotor position step-tracking response of the fuzzy adaptive controller and the output of the reference model.

The error e and error change Δe are defined as

$$e(k) \underline{\underline{\Delta}} \theta_r^*(k) - \theta_r(k)$$
(11.5)
$$\Delta e(k) = e(k) - e(k-1)$$
(11.6)

where

$\theta_r^*(k)$:   the response of the reference model at kth sampling interval,

$\theta_r(k)$:   the rotor position signal at kth sampling interval,

$e(k)$:   the error signal at kth sampling interval,

$\Delta e(k)$:   the error change signal at kth sampling interval,

and the indices shown in Figure 3 denote that

$C_1, C_2, C_3, \cdots, C_6$:   reference crossover points

$P_1, P_2, P_3, \cdots, P_6$:   reference extreme points

$A_1, A_2, A_3, \cdots, A_{12}$:   reference ranges.

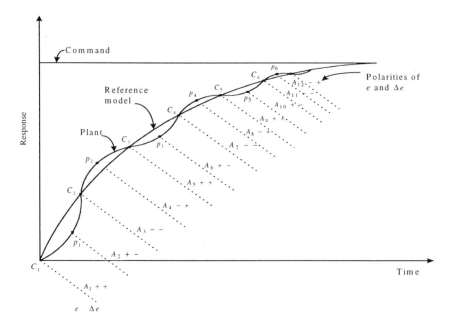

Figure 3: Dynamic signal analysis of the desired model-following response [14].

For convenience in analysis, the polarities of e and Δe at each range are also indicated in Figure 3. From Figure 3 one can observe the following properties at these points and ranges [15]:

$C_1 (e < 0 \rightarrow e > 0)$ and $\Delta e >>> 0$

$C_2 (e > 0 \rightarrow e < 0)$ and $V_{dB} <<< 0$

$$C_3 (e< 0 \rightarrow e > 0) \text{ and } \Delta e >> 0$$
$$C_4 (e> 0 \rightarrow e < 0) \text{ and } \Delta e << 0$$
$$C_5 (e< 0 \rightarrow e > 0) \text{ and } \Delta e > 0$$
$$C_6 : (e> 0 \rightarrow e < 0) \text{ and } \Delta e < 0$$
$$p_1 \Delta e \cong 0 \text{ and } e>>>0$$
$$p_2 \Delta e \cong 0 \text{ and } e<<<0$$
$$p_3 \Delta e \cong 0 \text{ and } e>>0$$
$$p_4 \Delta e \cong 0 \text{ and } e<<0$$
$$p_5 \Delta e \cong 0 \text{ and } e>0$$
$$p_6 \Delta e \cong 0 \text{ and } e<0 \tag{11.7}$$

and

Areas A_1, A_5, A_9 e = "+" and Δe = "+", the error is positive and increased

Areas A_2, A_6, A_{10} e = "+" and Δe = "−", the error is positive and decreased

Areas A_3, A_7, A_{11} e = "−" and Δe = "−", the error is negative and increased

Areas A_4, A_8, A_{12} e = "−" and Δe = "+", the error is negative and decreased. (11.8)

The linguistic control rules are constructed on the basis of the above analysis. Next, the fuzzy numbers, membership functions and quantization levels of the proposed fuzzy adaptive controller are chosen as follows:

Fuzzy Numbers

Positive Big: PB,	Negative Small: NS,
Positive Medium: PM,	Negative Medium: NM,
Positive Small: PS,	Negative Big: NB,
Zero: ZE	

Membership Functions

Depending on the application and the preference of the user, a number of membership functions can be selected. In this study, the trapezoidal-shaped functions shown in Figure 4 [14] are chosen, due to the best control performance for the step-command tracking position control of the USM and simplicity.

Quantization Levels

In the proposed controller, the position model following error and the error change are chosen as the input signals for the fuzzy controller. The quantization levels of these input variables are listed in Table 1 [14]. To increase the sensitivities, $e(k)$ and $\Delta e(k)$ are appropriately scaled by the factors G_e and G_d, respectively. It is known that the dynamic behavior of a fuzzy control system is highly dependent on these scaling factors. Factors G_e and G_d are determined by observing the experimental results to minimize the model-following error using the phase-plane trajectory of the error and error derivative during the transient condition of the step-command tracking. Accordingly, G_e and G_d are both selected to be 30 in this study.

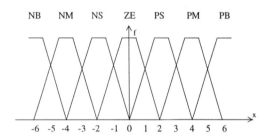

Figure 4: The membership functions [14].

Using the dynamic signal analysis (Figure 3), one can easily formulate the fuzzy control rules on the basis of intuition and experience.

Construction of Fuzzy Control Rules:

For convenience, the indices shown in Figure 3 are listed in Table 2 [14]. The control rules for the fuzzy adaptive model following position control of the USM, which are the "IF-THEN" forms [5], are formulated and listed, respectively, in Table 3 [14] and Table 4 [14] for reference points and reference areas, respectively. The CI, which is shown in Tables 3 and 4, is the control input and the P and N fuzzy numbers shown in Table 4 are defined as

Positive: P,   Negative: N

Decision Lookup Table

Using Tables 3 and 4 and using the symmetrical property of fuzzy control rule, the linguistic rule table is constructed as shown in Table 5 [14]. By using the center-of-area (COA) method [5], a decision lookup table is then constructed in Table 6 [14]. Finally, through proper quantization, the actual control output is generated according to the contents of the decision lookup table. To enhance the error-driven adaptive capability, the output scaling factor G_o is selected to be error dependent as

$$G_o = \begin{cases} 0 & \text{for } e_{ro} \geq |e| \\ K_1 + K_2 \left(|e| - e_{ro} \right) & \text{for } e_{ro} < |e| \end{cases} \tag{11.9}$$

where e is the position model-following error, e_{ro} is the preset limit of the fuzzy controller (e_{ro} =0.001, the scaling is 1V = 5.23598 rad.), and K_1, K_2 are constant gains. Equation (11.9) is emphasized to improve the transient response speed and to allow the output from the fuzzy controller to be zero when the error becomes smaller than a preset limit.

Since the lumped dynamic model of the ultrasonic motor (USM) is currently unavailable, discussion of the stability of the proposed fuzzy controller is not included in this section.

Table 1: Quantized error and error change [14]

Error e(mV)	Error change Δe (mV)	Quantized level
$E \leq$ -800	$\Delta e \leq$ -800	-6
-800 <$e \leq$ -400	-800< $\Delta e \leq$ -400	-5
-400<$e \leq$ -200	-400< $\Delta e \leq$ -200	-4
-200<$e \leq$ -100	-200< $\Delta e \leq$ -100	-3
-100<$e \leq$ -50	-100< $\Delta e \leq$ -50	-2
-50<$e \leq$ -25	-50< $\Delta e \leq$ -25	-1
-25<$e \leq$ 25	-25< $\Delta e \leq$ 25	0
25<$e \leq$ 50	25< $\Delta e \leq$ 50	1
50<$e \leq$ 100	50< $\Delta e \leq$ 100	2
100<$e \leq$ 200	100< $\Delta e \leq$ 200	3
200<$e \leq$ 400	200 < $\Delta e \leq$ 400	4
400<$e \leq$ 800	400< $\Delta e \leq$ 800	5
800<e	800 < Δe	6

4 Neural Network Model-Following Control

4.1 Description of the Neural Network

A general three-layer neural network shown in Figure 5 [21] consists of the input (the i layer), hidden (the j layer) and output layer (the k layer). This network is adopted to implement the proposed neural network controller.

Input layer

$$net_i = x_i, \qquad O_i = f_i(net_i) = net_i \qquad (11.10)$$

where x_i represents the ith input to the node of input layer.

Hidden layer

$$net_j = \sum_i (W_{ji} O_i) + \theta_j, \qquad O_j = f_j(net_j) = \frac{1}{1 + e^{-net_j}} \qquad (11.11)$$

Output layer

$$net_k = \sum_j (W_{kj} O_j) + \theta_k, \qquad O_k = f_k(net_k) = \frac{1}{1 + e^{-net_k}} \qquad (11.12)$$

where W_{kj} are the connective weights between the hidden and the output layers, θ_k are the threshold values for the units in the output layer and f is the activation function, which is a sigmoidal function.

Table 2: The indices in the state plane [14]

$\begin{matrix}e\\\Delta e\end{matrix}$	NB	NM	NS	ZE	PS	PM	PB
NB				C_2			
NM	A_3	A_7		C_4		A_2	A_6
NS			A_{11}	C_6	A_{10}		
ZE	P_2	P_4	P_6	ZE	P_5	P_3	P_1
PS			A_{12}	C_5	A_9		
PM	A_4	A_8		C_3		A_1	A_5
PB				C_1			

Table 3: Fuzzy control rules for reference points [14]

Rule No.	e	Δe	CI	Reference Points
1	ZE	PB	PM	C_1
2	ZE	PM	PS	C_3
3	ZE	PS	PS	C_5
4	PB	ZE	PM	P_1

5	PM	ZE	PS	p_3
6	PS	ZE	PS	p_5
7	ZE	NB	NM	C_2
8	ZE	NM	NS	C_4
9	ZE	NS	NS	C_6
10	NB	ZE	NM	p_2
11	NM	ZE	NS	p_4
12	NS	ZE	NS	p_6
13	ZE	ZE	ZE	Set Point

Table 4: Fuzzy control rules for reference areas [14]

Rule No.	e	Δe	CI	Reference Aeras
14	P	P	P	$A_1 A_5$
15	P	N	ZE	$A_2 A_6$
16	N	N	N	$A_3 A_7$
17	N	P	ZE	$A_4 A_8$
18	PS	PS	PS	A_9
19	PS	NS	ZE	A_{10}
20	NS	NS	NS	A_{11}
21	NS	PS	ZE	A_{12}

Table 5: The linguistic rule table [14]

Δe CI \ e	NB	NM	NS	ZE	PS	PM	PB
NB	NB	NB	NM	NM	NS	ZE	ZE
NM	NB	NM	NM	NS	NS	ZE	ZE
NS	NM	NM	NS	NS	ZE	ZE	PS
ZE	NM	NS	NS	ZE	PS	PS	PM
PS	NS	ZE	ZE	PS	PS	PM	PM
PM	ZE	ZE	PS	PS	PM	PM	PB
PB	ZE	ZE	PS	PM	PM	PB	PB

Table 6: The decision lookup table [14]

Δe \ CI	e = -6	-5	-4	-3	-2	-1	0	1	2	3	4	5	6
-6	-6	-6	-6	-5	-4	-4	-4	-3	-2	-1	0	0	0
-5	-6	-6	-5	-4	-4	-4	-3	-2	-2	-1	0	0	0
-4	-6	-5	-4	-4	-4	-3	-2	-2	-2	-1	0	0	0
-3	-5	-4	-4	-4	-3	-2	-2	-2	-1	0	0	0	1
-2	-4	-4	-4	-3	-2	-2	-2	-1	0	0	0	1	2
-1	-4	-4	-3	-2	-2	-2	-1	0	1	1	1	2	3
0	-4	-3	-2	-2	-2	-1	0	1	2	2	2	3	4
1	-3	-2	-1	-1	-1	0	1	2	2	2	3	4	4
2	-2	-1	0	0	0	1	2	2	2	3	4	4	4
3	-1	0	0	0	1	2	2	2	3	4	4	4	5
4	0	0	0	1	2	2	2	3	4	4	4	5	6
5	0	0	0	1	2	2	3	4	4	4	5	6	6
6	0	0	0	1	2	3	4	4	4	5	6	6	6

4.2 On-Line Learning Algorithm

To describe the on-line learning algorithm of the proposed on-line trained neural network controller, the energy function E is defined as

$$E = \frac{1}{2}\sum_n \left(x_{mn,N} - x_{pn,N}\right)^2 = \frac{1}{2}\sum_n e_{mn,N}^2 \qquad (11.13)$$

where x_{mn} and x_{pn} represent the states of the reference model and the states of the actual plant; where n denotes the number of the states, e_{mn} is the state error vector and N indicates the number of training iterations.

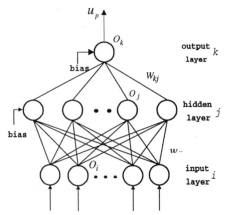

Figure 5: The general three-layer neural network [21].

Within each interval from N to $N+1$, the backpropagation algorithm [16-23] is used to update the weights of the output and hidden layers in the neural network controller according to the following equation:

$$W_{ji,N+1} = W_{ji,N} - \eta \frac{\partial E}{\partial W_{ji,N}} + \alpha \Delta W_{ji,N-1} \tag{11.14}$$

$$W_{kj,N+1} = W_{kj,N} - \eta \frac{\partial E}{\partial W_{kj,N}} + \alpha \Delta W_{kj,N-1} \tag{11.15}$$

where η is the learning rate, α is the momentum factor and $\Delta W_{ji,N-1} \left(\Delta W_{kj,N-1} \right)$ is the difference between $W_{ji,N} \left(W_{kj,N} \right)$ and $W_{ji,N-1} \left(W_{kj,N-1} \right)$.

The exact calculation of the Jacobian of the system, $\partial x_{pn} / \partial u_p$ in the derivation of $\partial E / \partial W_{ji}$ and $\partial E / \partial W_{kj}$, cannot be determined because of the unknown plant dynamic of the USM drive system. To increase the learning speed of the connective weights, according to [23], the derivative $\partial x_{pn} / \partial u_p$ can be approximated by its sign function, which is the ratio of x_{pn} and u_p. The bias of each neuron in hidden and output layers is also trained on-line using the same learning algorithm.

5 Fuzzy Neural Network Model-Following Control

A general four-layer fuzzy neural network (FNN) as shown in Figure 6 [31], which consists of the input (the i layer), membership (the j layer), rule (the k layer) and output layer (the o layer), is adopted to implement the proposed FNN controller. Nodes in the input layer represent input linguistic variables. Nodes in the membership layer act as the membership functions. Moreover, all the nodes in the rule layer form a fuzzy rule base.

5.1 Description of the Fuzzy Neural Network

Layer 1: input layer

For every node i in this layer, the net input and the net output are represented as

$$net_i^1 = x_i^1, \qquad y_i^1 = f_i^1 \left(net_i^1 \right) = net_i^1 \tag{11.16}$$

where x_i^1 represents the ith input to the node of layer 1.

Layer 2: membership layer

In this layer, each node performs a membership function. The Gaussian function is adopted as the membership function. For the *j*th node

$$net_j^2 = -\frac{\left(x_i^2 - m_{ij}\right)^2}{\left(\sigma_{ij}\right)^2}, \qquad y_j^2 = f_j^2\left(net_j^2\right) = \exp\left(net_j^2\right) \tag{11.17}$$

where m_{ij} and σ_{ij} are, respectively, the mean and the standard deviation of the Gaussian function in the *j*th term of the *i*th input linguistic variable x_i^2 to the node of layer 2.

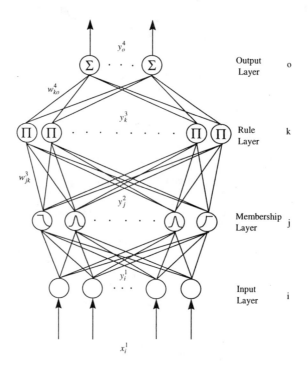

Figure 6: The general four-layer fuzzy neural network [31].

Layer 3: rule layer

Each node *k* in this layer is denoted by Π, which multiplies the incoming signal and outputs the result of product. For the *k*th rule node

$$net_k^3 = \prod_j w_{jk}^3 x_j^3, \qquad y_k^3 = f_k^3\left(net_k^3\right) = net_k^3 \tag{11.18}$$

where x_j^3 represents the jth input to the node of layer 3 and w_{jk}^3 is also assumed to be unity.

Layer 4: output layer

The single node o in this layer is a labeled Σ, which computes the overall output as the summation of all incoming signals

$$net_o^4 = \sum_k w_{ko}^4 x_k^4, \qquad y_o^4 = f_o^4(net_o^4) = net_o^4 \qquad (11.19)$$

where the link weight w_{ko}^4 is the output action strength of the oth output associated with the kth rule; x_k^4 represents the kth input to the node of layer 4; and y_o^4 is the output of the FNN.

5.2 On-Line Learning Algorithm

The central part of the learning algorithm for an FNN concerns how to recursively obtain a gradient vector in which each element in the learning algorithm is defined as the derivative of an energy function with respect to a parameter of the network. This is done by means of the chain rule and the method is generally referred to as the backpropagation learning rule, because the gradient vector is calculated in the direction opposite to the flow of the output of each node [24-31]. To describe the on-line learning algorithm of the FNN using the supervised gradient descent method, first the energy function E is defined as

$$E = \frac{1}{2}(x_{m1} - x_{p1})^2 \qquad (11.20)$$

where x_{m1} and x_{p1} represent the output of the reference model and rotor position of the USM. Then the learning algorithm based on backpropagation is described.

Layer 4

The error term to be propagated is given by

$$\delta_o^4 = -\frac{\partial E}{\partial net_o^4} = \left[-\frac{\partial E}{\partial y_o^4} \frac{\partial y_o^4}{\partial net_o^4} \right] \qquad (11.21)$$

and the weight is updated by the amount

$$\Delta w_{ko}^4 = -\frac{\partial E}{\partial w_{ko}^4} = \left[-\frac{\partial E}{\partial y_o^4} \frac{\partial y_o^4}{\partial net_o^4} \right] \left(\frac{\partial net_o^4}{\partial w_{ko}^4} \right) = \delta_o^4 x_k^4 \qquad (11.22)$$

The weights of the output layer are updated according to the following equation:

$$w_{ko}^4(N+1) = w_{ko}^4(N) + \eta_w \Delta w_{ko}^4 \tag{11.23}$$

The factor η_w is called the learning-rate parameter and N denotes the number of iteration.

Layer 3

Since the weights in this layer are unity, only the error term needs to be calculated and propagated.

$$\delta_k^3 = -\frac{\partial E}{\partial net_k^3} = \left[-\frac{\partial E}{\partial y_o^4} \frac{\partial y_o^4}{\partial net_o^4} \right] \left(\frac{\partial net_o^4}{\partial y_k^3} \frac{\partial y_k^3}{\partial net_k^3} \right) = \delta_o^4 w_{ko}^4 \tag{11.24}$$

Layer 2

The multiplication operation is performed in this layer. The error term is computed as follows:

$$\delta_j^2 = -\frac{\partial E}{\partial net_j^2} = \left[-\frac{\partial E}{\partial y_o^4} \frac{\partial y_o^4}{\partial net_o^4} \frac{\partial net_o^4}{\partial y_k^3} \frac{\partial y_k^3}{\partial net_k^3} \right] \left[\frac{\partial net_k^3}{\partial y_j^2} \frac{\partial y_j^2}{\partial net_j^2} \right] = \sum_k \delta_k^3 y_k^3 \tag{11.25}$$

and the update rule of m_{ij} is

$$\Delta m_{ij} = -\frac{\partial E}{\partial m_{ij}} = \left[-\frac{\partial E}{\partial y_j^2} \frac{\partial y_j^2}{\partial net_j^2} \frac{\partial net_j^2}{\partial m_{ij}} \right] = \delta_j^2 \frac{2(x_i^2 - m_{ij})}{(\sigma_{ij})^2} \tag{11.26}$$

and the update rule of σ_{ij} is

$$\Delta \sigma_{ij} = -\frac{\partial E}{\partial \sigma_{ij}} = \left[-\frac{\partial E}{\partial y_j^2} \frac{\partial y_j^2}{\partial net_j^2} \frac{\partial net_j^2}{\partial \sigma_{ij}} \right] = \delta_j^2 \frac{2(x_i^2 - m_{ij})^2}{(\sigma_{ij})^3} \tag{11.27}$$

The mean and standard deviation of the hidden layer are updated as follows:

$$m_{ij}(N+1) = m_{ij}(N) + \eta_m \Delta m_{ij} \tag{11.28}$$

$$\sigma_{ij}(N+1) = \sigma_{ij}(N) + \eta_\sigma \Delta \sigma_{ij} \tag{11.29}$$

The factor η_m and η_σ are the learning-rate parameters of the mean and the standard deviation of the Gaussian function.

The exact calculation of the Jacobian of the system, which is contained in $\partial E / \partial y_o^4$, cannot be determined due to the unknown of the plant dynamic of the USM drive system. Although the FNN identifier [27] can be implemented to calculate the Jacobian of the system, however, intensive computation effort is required. To overcome this problem and to increase the on-line learning rate of the connective weights, a delta adaptation law is proposed as follows:

$$\delta_o^4 \cong A\left(x_{m1} - x_{p1}\right) + \left(x_{m2} - x_{p2}\right) = A\,e_{m1} + e_{m2} \tag{11.30}$$

where x_{m2} and x_{p2} represent the derivatives of x_{m1} and x_{p1}; e_{m1} and e_{m2} denote the position and speed errors between the reference model and the actual plant; A is a positive constant. Selection of values for the learning-rate parameters has a significant effect on the network performance. If small values are given for the learning-rate parameters, convergence will be guaranteed with slow speed of convergence. On the other hand, if a large value is given for the learning-rate parameters, the system may become unstable. In order to train the FNN effectively, all the learning-rate parameters are also on-line training, using the algorithm mentioned in [27] based on the proposed delta adaptation law and all the parameters of the membership functions and connective weights are randomly initialized. Since the lumped dynamic model of the USM is unavailable, the closed-loop stability analysis and the simulation of the proposed FNN control system are impossible to carry out, however, the effectiveness of the on-line training FNN based on the proposed delta adaptation law are demonstrated using the experimental results.

6   The PC-Based Ultrasonic Motor Drive

Figure 7 [21,31] is the block diagram of the Pentium-based computer control system for the USM drive. A servo control card is installed in the control computer, which includes multi-channels of A/D, D/A, PIO and two encoder interface circuits. Digital filters and frequency multiplied by 4 circuits are built into the encoder interface circuits to increase the precision of position feedback. The execution interval for the position loop with three intelligent control algorithms is 2 msec. The USM is loaded with a single-axis moving table with 5 mm screw pitch and 550 mm travel. All the control algorithms are implemented in the PC using the "Turbo C" language.

7   Experimental Results

A 2nd-order transfer function of the following form with rising time set at 1.5 sec is chosen as the reference model [33]

$$\frac{\omega_n^2}{S^2 + 2\zeta\omega_n S + \omega_n^2} = \frac{6.7244}{S^2 + 5.1863S + 6.7244} \tag{11.31}$$

where

ζ : damping ratio (set at one for critical damping)

ω_n : undamped natural frequency

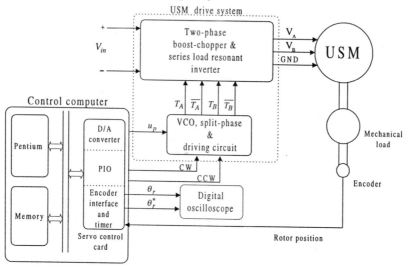

Figure 7: Computer-controlled USM drive system.

7.1 Fuzzy Model-Following Control

The configuration of the proposed fuzzy adaptive model-following position controller for the USM is shown in Figure 8 [14], in which the reference model is chosen according to the control specifications of the USM drive system.

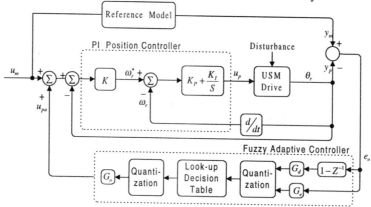

Figure 8: Configuration of the fuzzy adaptive model-following position controller for the USM [14].

At the nominal condition, since the position response y_p of the plant deviates from the response y_m of the reference model, the adaptation signal, u_{pa}, will be automatically generated by the fuzzy controller. Moreover, when parameter variations occur, the degradation of the tracking performance is greatly reduced by this error-driven adaptive mechanism.

First, a proportional-integral (PI) position controller is implemented to control the rotor position of the USM as shown in Figure 8. The gains of the PI controller must be obtained by trial and error to make the step-command tracking response of the rotor position of the USM close to the desired tracking performance. The gains of the PI controller are set at

$$K = 1, \ K_p = 1, \ K_I = 100 \tag{11.32}$$

The measured step responses of the reference model and rotor position due to a step command change of 2π are shown in Figure 9(a), with an obvious model-following error. A single-axis moving table, with 5 mm screw pitch and 550 mm travel, is directly coupled to the rotor of USM and the moving table is loaded with a 150 kg weight of iron disks. The load torque is approximate 100% of the rated holding-torque, which is 0.3 Nm, of the USM (USR-60). The measured step responses of the reference model and rotor position due to a step command change of 2π are shown in Figure 9(b), with a more obvious model-following error.

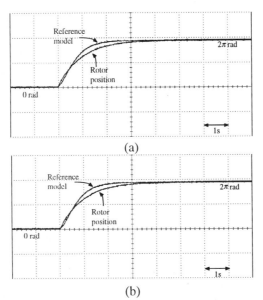

(a)

(b)

Figure 9: Measured rotor position and reference model responses without the connection of the fuzzy adaptive controller: (a) at no-load condition; (b) at heavy-load condition.

With the connection of the fuzzy adaptive model-following controller, the model-following error-driven adaptation signal will be generated to permit a favorable model-following control performance. The measured rotor position responses with the connection of the fuzzy adaptive controller are shown in Figure 10(a) to 10(c)

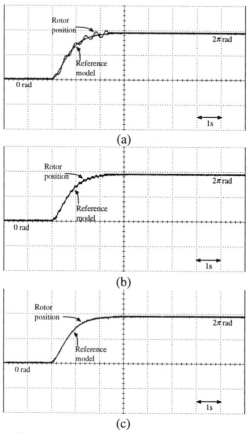

(a)

(b)

(c)

Figure 10: Measured rotor position and reference model responses due to step command change of 2π with the fuzzy adaptive controller connection at no-load condition: (a) $K_2 = 0.005$; (b) $K_2 = 0.01$; (c) $K_2 = 0.03$.

for K_2 in Equation (11.9) set at 0.005, 0.01 and 0.03, respectively and the K_1 is set at 0.01. Improvement of the control performance by augmenting the proposed fuzzy adaptive model-following controller can be observed from the results; and the model-following error is reduced much more by the increase in the value of K_2. The measured rotor position responses with the connection of the fuzzy adaptive controller at heavy-load condition are shown in Figure 11(a) to 11(c) for K_2 in Equation (11.9) set at 0.005, 0.01 and 0.03, respectively and the K_1 is also set at 0.01. The model-following error is also greatly reduced by the increase in the value of K_2. The robustness of the proposed fuzzy controller has been verified by the experimental results.

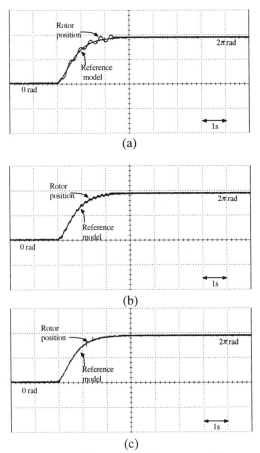

(a)

(b)

(c)

Figure 11: Measured rotor position and reference model responses due to step command change of 2π with the connection of the fuzzy adaptive controller at heavy-load condition: (a) $K_2 = 0.005$; (b) $K_2 = 0.01$; (c) $K_2 = 0.03$.

7.2 Neural Network Model-Following Control

The proposed on-line trained neural network model-following controller for the USM servo drive is shown in Figure 12 [21], where x_{m1} and x_{m2} are the states of the reference model; x_{p1} and x_{p2} are the rotor position and speed of the USM. In the proposed neural network controller, the error between the states of the plant and the reference model is used to train the connective weights of the neural network controller on line to permit a favorable model-following control performance.

The measured responses of the output of the reference model and rotor position with the PI controller due to a periodic step-command change of 2π are shown in Figure 13(b) and one cycle of the step responses is enlarged for detailed observation. From the experimental result, the desired step-command tracking response, which has been defined by the reference model, is difficult to satisfy using the PI controller.

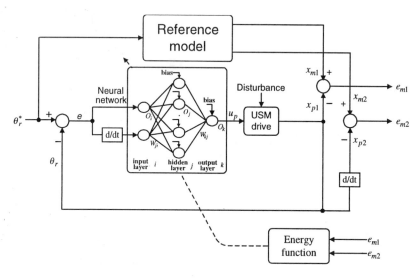

Figure 12: Configuration of the proposed neural network controller for USM servo drive [21].

Now the neural network model-following controller is implemented, the biases and connective weights of the neural network are initialized with random number. The values for the hidden layer nodes and learning rate parameters, K_2, have a significant effect on the network performance. The performance with different hidden layer nodes and learning rate parameters is tested. First, with the learning rate parameters, η and α, both setting at 0.01, the measured responses of the output of the reference model and rotor position due to a periodic step-command change of 2π for twenty and thirty hidden layer nodes are shown in Figure 14(a) through 14(b); one cycle of the step responses is enlarged for the observation of the on-line learning effect of the neural network. From the experimental results, the best control performance is obtained by using thirty hidden layer nodes. Moreover, accurate tracking response is obtained immediately and the accuracy in steady-state condition is within plus/minus one pulse for the precision of 4000 pulse/2π.

The experimental results of the proposed neural network controller are demonstrated. For comparison, a PI position controller is implemented to control the rotor position of the USM as shown in Figure 13(a) [21]. The gains of the PI controller are set in Equation (11.32).

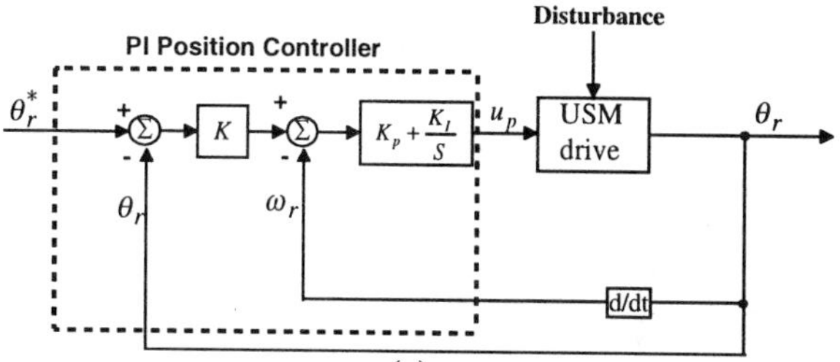

(a)

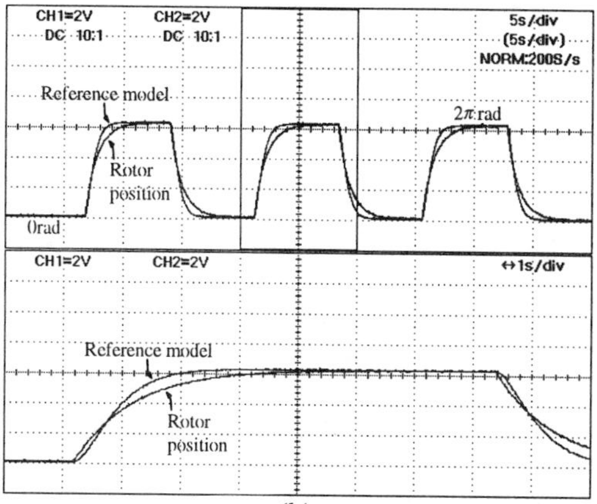

(b)

Figure 13: (a) Configuration of the PI position controller for USM servo drive. (b) Measured responses of reference model and rotor position using PI controller [21].

Next, the influence of different learning rate parameters are tested with thirty hidden layer nodes; usually, learning rate parameters must be small numbers to ensure that the network will settle to a solution. The measured response of the output of the reference model and rotor position due to a periodic step-command change of 2π for η and α, both setting at 0.02 and 0.03, is shown in Figure 15(a) and 15(b), respectively. Since divergence responses resulted for the value of 0.03, the learning rate parameters, η and α, both setting at 0.02 with thirty hidden layer nodes are adopted for the proposed neural network controller. Finally, the single-axis moving table is loaded with a 150 kg weight, which is approximately 100% of the rated holding-torque (0.3 Nm) of the USM. The measured responses of the output of the reference model and rotor position controller due to a periodic step-command change of 2π are shown in Figure 16. The effectiveness of the proposed neural network controller can be observed from the experimental result. Moreover, the experimental result with a load of 150 kg weight is almost identical to the experimental results without load as shown in Figure 15(a) due to the inherent high-torque driving capability of the USM. All the plots shown in Figures 13 through 16 start at time = 0 sec.

From the experimental results discussed above, accurate tracking response can be obtained by random initialization of the neural network weights due to the powerful on-line learning capability.

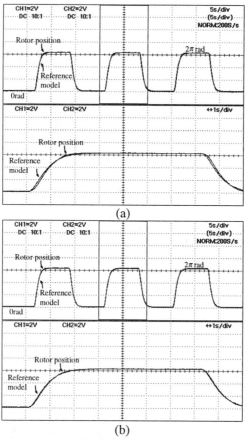

(a)

(b)

Figure 14: Measured responses of reference model and rotor position for different hidden layer nodes with $\alpha = \eta = 0.01$. (a) Twenty hidden layer nodes. (b) Thirty hidden layer nodes.

7.3 Fuzzy Neural Network Model-Following Control

In order to control the rotor position of the USM effectively, the FNN model-following controller is proposed in this section. In the FNN, the units in the input, membership, rule and output layer are two, six, nine and one, respectively. The configuration of the proposed FNN model-following controller for the USM is shown in Figure 17 [31], in which the reference model is chosen according to the control specifications and the position controller of the USM drive system is a PI type. The inputs of the FNN shown in Figure 17 are the state error vector.

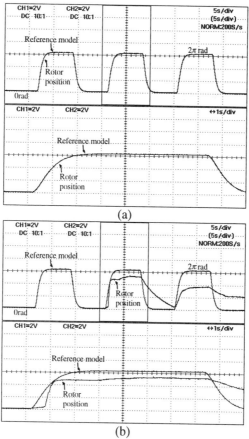

Figure 15: Measured responses of reference model and rotor position for different learning parameters with thirty hidden layer nodes. (a) $\alpha = \eta = 0.02$, (b) $\alpha = \eta = 0.03$.

Though good rotor responses can be obtained by a PI position controller, the gains of the controller have to be tuned by trial and error according to different operating conditions. Moreover, the control specifications of time-domain are difficult to satisfy. To obtain a favorable model-following characteristic, the FNN model-following controller is augmented. At the nominal condition, since the states of the USM drive deviates from the states of the reference model, the synthesis signal, $du_p\left(y_o^4\right)$, will be automatically generated by the FNN. Moreover, when parameter uncertainties or external disturbances occur, the degradation of the tracking performance is significantly reduced by this error-driven mechanism.

To test the control performance of the FNN model-following controller, another configuration of the FNN model-following controller is proposed in Figure 18 [31], in which the inner loop PI position controller is omitted and only the FNN model-following controller is used to control the position of the USM. Now the inputs of the FNN are position error and its derivative. The effectiveness of the proposed FNN model-following control structures will be verified by the following experiments.

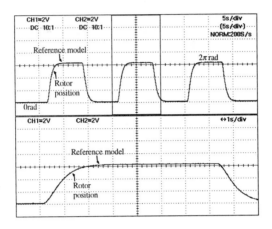

Figure 16: Measured responses of reference model and rotor position loaded with 150 *kg* weight.

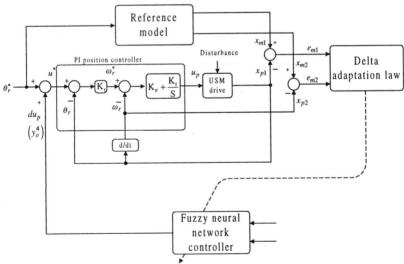

Figure 17: Configuration of the FNN model-following controller for the USM with PI position controller [31].

As shown in Figure 17, the gains of the PI controller are set as in Equation (11.32) and the connective weights and membership functions of the FNN are initialized with random number. Without the FNN model-following controller, the measured responses of the output of the reference model and rotor position due to a periodic step-command change of 2π radian are shown in Figure 19(a) [31], with an obvious model-following error. Now the FNN model-following controller is added and the model-following error driven synthesis signal will be generated to permit a favorable model-following control performance. The resulting responses of the output of the reference model and rotor position due to a periodic step-command change of 2π are shown in Figure 19(b) [31]. Since all the parameters of the FNN are randomly initialized, accurate tracking performance is obtained after one cycle on-line training of the FNN and the model-following error is then zero. To further test the performance of the FNN model-following controller, experimentation is carried out based on the control structure shown in Figure 18. The resulting responses of the output of the reference model and rotor position due to a periodic step-command change of 2π are shown in Figure 19(c) [31]. Perfect model-following response can also be obtained after one cycle on-line training of the FNN.

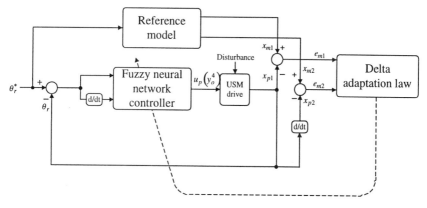

Figure 18: Configuration of the FNN model-following controller for the USM [31].

The same experiments are performed when the single-axis moving table is loaded with a 150 kg weight, which is approximate 100% of the rated holding-torque (0.3 Nm) of the USM. Based on the control structure shown in Figure 17 and without the addition of the FNN model-following controller, the measured responses of the output of the reference model and rotor position due to a periodic step-command change of 2π are shown in Figure 20(a) [31], with an obvious model-following error. Now that the FNN model-following controller is added, the resulting responses of the output of the reference model and rotor position due to a periodic step-command change of 2π are shown in Figure 20(b) [31]. Accurate tracking performance is also obtained after one cycle of training. Using only the FNN model-following controller to control the USM servo drive, the resulting responses of the output of the reference model and rotor position due to a periodic step-command change of 2π are shown in Figure 20(c) [31]. The model-following error is also reduced to zero after one cycle of training. The effectiveness of the proposed FNN model-following controller can be observed from the experimental results presented above. Since the rotor position response can accurately follow the output of the reference model using only the FNN model-following controller by observing the experimental results shown in Figures 19 and 20, the control structure shown in Figure 18 is better than the one shown in Figure 17. Moreover, the experimental results with a load of 150 kg weight are almost identical to the experimental results without load due to the inherent high-torque driving capability of the USM.

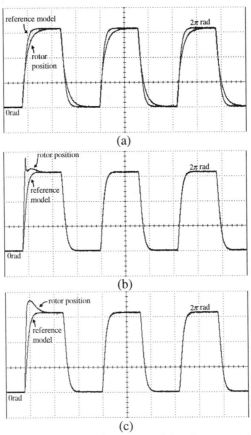

Figure 19: Measured responses of reference model and rotor position without load: (a) PI position controller without the FNN model-following controller; (b) PI position controller with the FNN model-following controller; (c) FNN model-following controller only [31].

8 Summary

Since the plant dynamic of the USM is complicated and contains high nonlinearity, the derivation of a lumped dynamic model is very difficult. Moreover, the motor parameters are dependent on operating temperature, load torque, rotor speed, rotating direction, applied voltage and static pressure force between the stator and the rotor. To drive the USM effectively, a new type of USM drive circuit was designed and implemented in this chapter. A two-phase chopper-inverter driving circuit was designed to provide a balanced two-phase voltage source for the USM. Each phase contains a boost-chopper DC-DC converter and a half-bridge series-resonant inverter with capacitor-parallel load. Moreover, to enhance the transient response and to increase the robustness of the USM drive system, three intelligent controllers, the fuzzy model-following controller, the neural network model-following controller and the FNN model-following controller, were proposed in the rotor position control loop of the USM drive. It is obvious that the rotor position tracking response can be controlled to closely follow the output of the reference model under a wide range of operating conditions.

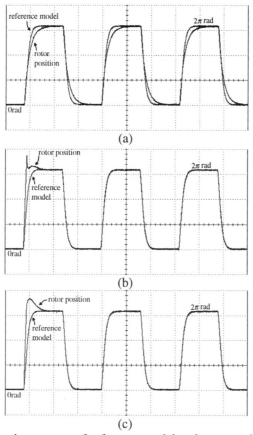

Figure 20: Measured responses of reference model and rotor position, loaded with 150 *kg* weight: (a) PI position controller without the FNN model-following controller; (b) PI position controller with the FNN model-following controller; (c) FNN model-following controller only [31].

References

[1] Sashida, T. and Kenjo, T. (1993), *An Introduction to Ultrasonic Motors*, Oxford Clarendon Press.

[2] Furuya, S., Maruhashi, T., Izuno, Y., and Nakaoka, M. (1990), Load Adaptive Frequency Tracking Control Implementation of Two-phase Resonant Inverter for Ultrasonic Motor, *Proc. IEEE-PESC*, pp. 17-24.

[3] Izuno, Y. and NaKaoKa, M. (1994), High Performance and High Precision Ultrasonic Motor-actuated Positioning Servo Drive System Using Improved Fuzzy Reasoning Controller, *Proc. IEEE-PESC*, pp. 1269-1274.

[4] Lin, F. J. and Kuo, L. C. (1997), A Novel Driving Circuit for the Ultrasonic Motor Servo Drive with Variable Structure Adaptive Model-Following Control, *IEE Proc. Electr. Power Appl.*, Vol. 144, No. 3, pp. 199-206.

[5] Lee, C. C. (1990), Fuzzy Logic in Control Systems: Fuzzy Logic Controller-Part I and Part II, *IEEE Trans. Syst., Man and Cybern.*, Vol. 20, No. 2, pp. 404-436.

[6] Yager, R. R. and Filev, D. P. (1994), *Essentials of Fuzzy Modeling and Control*, John Wiley & Sons, New York.

[7] Wang, L. X. (1994), *Adaptive Fuzzy Systems and Control: Design and Stability Analysis*, Prentice-Hall, New Jersey.

[8] Commuri, S. and Lewis, F. L. (1996), Adaptive Fuzzy Logic Control of Robot Manipulators, *Proc. IEEE Conf. Robotics and Automation*, pp. 2604-2609.

[9] Li, Y. F. and Lau, C. C. (1989), Development of Fuzzy Algorithm for Servo Systems, *IEEE Contr. Syst. Mag.*, Vol. 9, No. 3, pp. 65-72.

[10] Mir, S. A., Zinger, D. S., and Elbuluk, M. E. (1994), Fuzzy Controller for Inverter Fed Induction Machines, *IEEE Trans. Ind. Applicat.*, Vol. 30, No. 1, pp. 78-84.

[11] Narendra, K. S. and Annaswamy, A. M. (1989), *Stable Adaptive Systems*, Prentice Hall, New Jersey.

[12] Egami, T., Morita, H., and Tsuchiya, T. (1990), Efficiency Optimized Model Reference Adaptive Control System for a DC Motor, *IEEE Trans. Ind. Electron.*, Vol. 37, No. 1, pp. 28-33.

[13] Lin, F. J. and Liaw, C. M. (1993), Reference Model Selection and Adaptive Control for Induction Motor Servo Drive, *IEEE Trans. Automat. Contr.*, Vol. 38, No. 10, pp. 1594-1600.

[14] Lin, F. J. (1997), Fuzzy Adaptive Model-Following Position Control for Ultrasonic Motor, *IEEE Trans. Power Electron.*, Vol. 12, No. 2, pp. 261-268.

[15] Liaw, C. M. and Cheng, S. Y. (1995), Fuzzy Two-degrees-of-freedom Speed Controller for Motor Drives, *IEEE Trans. Ind. Electron.*, Vol. 42, No. 2, pp. 209-216.

[16] Narenda, K. S. and Parthasarathy, K. (1990), Identification and Control of Dynamical Systems Using Neural Networks, *IEEE Trans. Neural Networks*, Vol. 1, No. 1, pp. 4-27.

[17] Fukuda, T. and Shibata, T. (1992), Theory and Applications of Neural Networks for Industrial Control Systems, *IEEE Trans. Ind. Electron.*, Vol. 39, No. 6, pp. 472-491.

[18] Sastry, P. S., Santharam, G., and Unnikrishnan, K. P. (1994), Memory Neuron Networks for Identification and Control of Dynamical Systems, *IEEE Trans. Neural Networks*, Vol. 5, No. 2, pp. 306-319.

[19] Ku, C. C. and Lee, K. Y. (1995), Diagonal Recurrent Networks for Dynamic Systems Control, *IEEE Trans. Neural Networks*, Vol. 6, No. 1, pp. 144-156.

[20] Pao, Y. H. (1989), *Adaptive Pattern Recognition and Neural Networks*, Reading, MA. Addison-Wesley.

[21] Lin, F. J. and Wai, R. J. (1997), On-Line Trained Neural Network Controller for Ultrasonic Motor Servo Drive, *Proc. IEEE Conf. Power Electron. and Drive System*, pp. 363-388.

[22] Psaltis, D., Sideris, A., and Yamamura, A. A. (1988), A Multilayered Neural Network Controller, *IEEE Control Systems Magazine*, Vol. 8, No. 2, pp. 17-21.

[23] Zhang, Y., Sen, P., and Hearn, G. E. (1995), An On-line Trained Adaptive Neural Controller, *IEEE Control Systems Magazine*, Vol. 15, No. 5, pp. 67-75.

[24] Lin, C. T. and Lee, C. S. G. (1991), Neural-network-based Fuzzy Logic Control and Decision System, *IEEE Trans. Computers*, Vol. 40, No. 12, pp. 1320-1336.

[25] Horikawa, S., Furuhashi, T. and Uchikawa, Y. (1992), On Fuzzy Modeling Using Fuzzy Neural Networks with the Backpropagation Algorithm, *IEEE Trans. Neural Networks*, Vol. 3, No. 5, pp. 801-806.

[26] Ismael, A., Hussien, B. and McLaren, R. W. (1994), Fuzzy Neural Network Implementation of Self Tuning PID Control, *Proc. IEEE Int. Symp. Intelligent Control*, pp. 16-21.

[27] Chen, Y. C. and Teng, C. C. (1995), A Model Reference Control Structure Using a Fuzzy Neural Network, *Fuzzy Sets and Systems*, Vol. 73, pp. 291-312.

[28] Zhang, J. and Morris, A. J. (1995), Fuzzy Neural Networks for Nonlinear Systems Modelling, *IEE Proc.-Control Theory Appl.*, Vol. 142, No. 6, pp. 551-556.

[29] Jang, T. S. R. and Sun, C. T. (1995), Neural-fuzzy Modeling and Control, *Proc. IEEE*, Vol. 83, No. 3, pp. 378-405.

[30] Chao, C. T. and Teng, C. C. (1995), Implementation of Fuzzy Inference Systems Using a Normalized Fuzzy Neural Network, *Fuzzy Sets and Systems*, Vol. 75, pp. 17-31.

[31] Wai, R. J. and Lin, F. J. (1997), Fuzzy Neural Network Model-Following Controller for Ultrasonic Motor Servo Drive, *Proc. Automat. Contr. Conf.*, Taiwan, pp. 677-682.

[32] Steigerwald, R. L. (1988), A Comparison of Half-bridge Resonant Converter, *IEEE Trans. on Power Electronics.*, Vol. 3, No. 2, pp. 174-182.

[33] Liaw, C. M. and Lin, F. J. (1994), A Robust Speed Controller for Induction Motor Drives, *IEEE Trans. Ind. Electron.*, Vol. 241, No. 3, pp. 308-315.

12 | INTELLIGENT AUTOMATION OF HERRING ROE GRADING

S. Kurnianto
C. W. de Silva
E. A. Croft
R. G. Gosine
Industrial Automation Laboratory
Department of Mechanical Engineering
The University of British Columbia
Vancouver, Canada

In this chapter, the development of an automated system for herring roe grading is described in detail. The automation of the process has been considered as a desirable solution for improving the herring roe industry. The technology developed by us, in the Industrial Automation Laboratory, to define and quantify the quality of herring roe is presented. This includes sensing of shape, size, color, and firmness properties of herring roe, and representing them within the context of quality assessment. Decision making process associated with grading using each of these properties, and then obtaining an overall quality index through "intelligent fusion" of this information, is presented. This technology has been incorporated into an industrial prototype, which has been developed by us with the assistance of our industrial partner. Design and development of the prototype, including component selection and integration to realize a complete grading system, is presented. Experimental results obtained using the prototype grading system, which support the use of automation in the herring roe industry are presented and studied.

1 Introduction

Herring roe has become an important part of the fish industry in British Columbia (BC). In the past several years, the herring roe fishery has contributed revenue of more than 200 million dollar per year, and is continuing to make a significant impact on the BC economy. As in any other food industry, quality assessment and control play an important role in shaping the market conditions of herring roe. In particular, the commercial value of herring roe is heavily dependent on its assigned quality indicator. The standards that determine various "quality indices" or "grades" of herring roe are strongly linked to market requirements and customer satisfaction. These consumer-driven market conditions force the roe processing companies to provide high consistency and controllability in grading at a minimum cost. Moreover, in order to stay competitive within a wide spectrum of customer needs and constraints, a sufficiently large number of grading classifications must be provided by the roe processing companies. Unfortunately, herring roe industry is seasonal (about 3 months/year) and the grading tasks are still done manually. These workers have long shifts, and often work under difficult working conditions and time constraints. The need for trained manual labor, in seasonal industry, drives up the cost of processing. Thus, the two incompatible objectives, namely to attain a high consistency of grading and to keep the production cost at minimum, have generated a considerable strain on every roe processing company in BC to improve its product quality and cost effectiveness.

Over the years, herring roe has become a luxury product with a significant market in Japan and smaller markets in several other eastern countries. The traditional new year demand, along with up-scale restaurant needs, represents the bulk of the market. A small quantity goes for general household consumption. Due to market factors and limited supply, the price of high quality herring roe can become quite expensive. In BC alone, the total wholesale value of herring roe in the last four years is estimated to be over $800 million, which corresponds to a yearly average of $200 million [1]. The competition within the entire herring roe industry has also become stiffer over the years. The increased supply of herring roe from other countries, particularly from the growing American herring roe industry in Alaska and the recent development of herring roe imitations made from capelin roe, have affected the demand for the BC produced herring roe. The new competition has given rise to a situation where the traditional foreign buyers are demanding a high quality product at a considerably lower price. The roe buyers are scrutinizing the quality of herring roe more stringently during price negotiations. As a highly competitive industry, where there is constant pressure from both the customers and the competing roe processors, the herring roe industry requires an improvement in quality of the product and a grading process that is efficient and cost effective.

With these market conditions, the roe processors have begun to reconsider the process of grading, placing an emphasis in achieving the quality and efficiency objectives. Since grading is still carried out manually, the consistency of the graded product can be somewhat questionable. Grading is a repetitive and boring task, typically done under uncomfortable working conditions (noise, smell, dampness, extreme temperatures, etc.). Also, continuously high concentration is needed during grading, and the graders are required to remain standing on either side of the grading line. As a result, the worker's ability to grade roe will degrade toward the end of a shift, where the product reaches what is termed a mixed-grade pack.

All these observations point to the use of automation as a valid solution to the set of objectives described above. These considerations have motivated the research described in this chapter, for the development of an automated grading machine for herring roe [2]. The goal is to design an automated system that can grade products more consistently and, preferably, faster than workers.

2 Herring Roe Grading Process

A herring roe skein, normally referred to as "herring roe" or simply "roe," is a sac of tiny herring eggs. Each female herring produces two skeins, which are essentially the right and left ovaries of the fish, as shown in Figure 1. A roe has to be properly processed prior to human consumption.

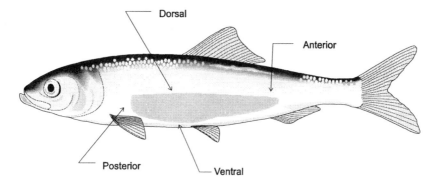

Figure 1: A roe skein inside a female herring.

The market value of herring roe depends on the assigned "quality" or "grade," which is determined based on several properties. The main properties that affect the grade of roe are shape, color, firmness, and size. A full shape, which corresponds to a fully-formed roe without deformity, is considered the most important grading property. The desirable color of herring roe is a yellowish golden hue. It is also important for the roe to be free from bloodstains. The typical consumer desires the texture of herring roe to be firm and crunchy. Soft and spongy roe have very little market value. Last, a large size roe, measured by its weight, commands a disproportionately higher price in the market. The actual price paid for each grade varies according to the existing conditions of the market. A representative sample of various grades of herring roe is shown in Figure 2.

The structure of the herring-roe market is well defined. However, this structure does not ensure market stability. Since the market is mainly consumer-driven, as mentioned before, the product quality is paramount. The manner in which the fish are handled prior to roe extraction and how the roe are processed affect most of the "quality" properties associated with herring roe. The three stages of processing and delivery to the market are the pre-extraction, the main grading, and the price negotiation. The large roe processing companies commonly have all these three stages in place at their plants. However, variations have been found to accommodate the production volume and market conditions.

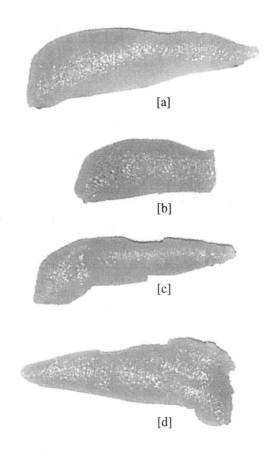

Figure 2: Typical grades of herring roe. [a] Grade 1. [b] Grade 2 (broken tail). [c] Grade 2-H (minor deformation). [d] Grade 2-C (cauliflower).

2.1 Pre-Extraction Stage

The pre-grading process refers to the handling and storage of fish from the point where they are caught to the time of roe extraction. In order to avoid spillage, it is important for the fish to be immediately cooled to 0°C after harvesting. Next, the fish must be frozen and stored at a temperature of -18°C or less to maintain good roe quality. Poor preprocessing can result in poor quality roe, mainly due to discoloration. For example, 'black tips' and bloodstains develop when the storage period prior to the freezing stage is excessive. Fragility, where the roe crumbles into tiny individual eggs, can also occur as a result of extended holding times prior to freezing. During the freezing process, a rapid rate of cooling is required to avoid the development of sponginess in the roe [3].

Although the storage and the freezing processes do not seem to cause an unsatisfactory deformation on the shape, they do alter some of the properties of the roe. These property changes can result in poorer grade.

Degrading of the processing quality is mainly due to inadequate equipment for storing and handling the roe. Most problems occur during the time when the fish are still in the fishing vessels or during holding prior to the roe being shipped to the processing plant. During the last two years, major improvements have been made in this pre-extraction stage. Most processing plants claim that the roe quality downgrading due to poor preprocessing is minimal.

2.2 Main Grading

The extraction and grading stages are central to the production of high quality roe. These processes include all the steps from the fish arriving at the plant up to the packaging of processed roe. All these activities are highly labor intensive. The overall process varies slightly from year to year and from processor to processor. The variations are made to accommodate the volume of production. The main grading stage is considered to include the roe extraction and the brining (firming and washing) processes as well.

In a well-designed processing plant, the overall grading line consists of several smaller grading lines. Each such pre-grading line reduces the volume to be processed at the subsequent line, by removing poor-quality roe. The steps involve visual inspection before the roe extraction, visual grading in between extraction and brining processes, and rigorous manual grading prior to packaging.

The roe processing plants employ a large number of workers. The tasks done by the workers in these plants are very repetitive. This is particularly true for the graders whose main task is to scan each incoming roe. Problems arise in a manual grading line because of the tendency of the workers to lose concentration when performing repetitive tasks for extended periods of time, particularly in a somewhat unpleasant environment. The situation becomes worse in a high volume operation where it is difficult for the graders to carefully examine all the roe. These conditions are known to reduce the consistency of grading. Degradation of the inspection quality due to human factors will lead to lower bids for herring roe in the market.

The fact that the herring roe industry is a seasonal one aggravates this situation. Roe processing companies have difficulties retaining their trained graders from season to season, and often have to train new graders in the beginning of each season. The introduction of new, inexperienced graders will reduce the speed and consistency of the grading operation. Thus, manual grading of herring roe is an expensive and somewhat inefficient process.

2.3 Price Negotiation

It is interesting to note that since the herring roe market is primarily driven by its wholesale buyers, it is not unusual to have sales taking place right at the processing plant. The wholesale buyers send their representatives to examine and analyze grading samples and make immediate purchasing negotiations. The samples from several containers of a single batch (usually from the same fish "catch") are analyzed by carefully re-grading each sample, piece by piece. This process is termed table grading. Every misgraded piece from the line is counted and the proportion of the table-graded samples is used by the customers to negotiate a price for the batch. Since table grading involves very careful scrutiny of the roe, pressure is put on the roe processing companies to lower their price if misgrading is found to be excessive. Often, expert graders from the roe processing company are seen to argue with the buyers over the grading criteria, and bargain for a fair price. Although this approach to wholesale marketing seems to be ineffective, it is the most common method practiced by the roe processors and the buyers. This situation has provided a challenge for development and implementation of automated machinery for grading roe. Machine grading is expected to be more consistent, repeatable, and uniform. This means the "quality" may be quantified and preserved without being affected by subjective human factors. Also, the cost of grading will be reduced through large-scale machine grading. But, the confidence of and support from expert graders of the roe processing companies is essential for successful implementation of automated grading.

3 Grading Technology

Since the herring roe grading process is primarily based on visual inspection, machine vision plays a dominant role in the development of technology for grading the features of herring roe. The use of machine vision in food industry is not a particularly new application. For example, machine vision techniques have been developed for apples [4], peaches [5], peanuts [6], raisins [7], and tomatoes [8]. In the seafood industry, several machine vision applications are used for grading fish and shrimp [9,10].

In herring roe grading, as described in this chapter, the use of common sensors and common methodology for detecting different features can greatly simplify the integration of the overall grading system. Accordingly, the approach of sensor generalization has been retained throughout the development of various technologies.   Initial development leading to a full-span automated grading technology for herring roe has been carried out in the Industrial Automation Laboratory of the University of British Columbia (IAL-UBC).

3.1 Shape Analysis

For shape grading, a representation that will determine the perceived "goodness" of herring roe shape is required. Two different steps are performed to analyze the shape. The first one identifies the shape as an overall figure of the roe, and the second one identifies a particular detail of breakage at the tip of a roe.

Knowledge-Based Shape Classification

This shape representation technique formulates a bi-classifier system that would reduce the bulk that is required if manually graded [11]. It should be noted that the system is initially not intended to grade herring roe into all of the numerous, and often changeable, existing grades.   However, it would pre-select the good roe (Grade 1) and leave the rest for manual grading.

The method represents herring roe shape by the area of equal segments along the major principal axis, termed "width measures."   In practice, the roe image is segmented by thresholding. To rid the binary image of unwanted noise and other objects, image processing algorithms, including smoothing and morphological techniques, are applied as a subsequent step. The image is then rotated about its centroid, so that the object's major principal axis coincides with the horizontal axis of the screen, where the axis is divided into 100 equal-length segments. For each segment, the pixel area is computed on both the dorsal and ventral sides of the principal axis to obtain the width measures. These width measures are weighted by the factor $\frac{100}{l}$ where l is the length of the object along the major principal axis; see Figure 3. Thus, every skein is represented by 200 width measures on both sides of the major principal axis. In terms of practicality, this method has the desired advantage toward robustness against any hair-like or thin extensions along the object's perimeter over a contour parameterization method.

In order to quantify the grade, the roe shape is classified by comparing this shape representation with the representations of roe skeins that had been previously classified as "good roe" by an expert grader.

The actual comparison is done by first smoothing the width measures using a Gaussian approximation filter and computing the tangent angles between each width measure to yield the final shape representation. The magnitude of the vector difference between the final representation of each inspected roe and those in a previously prepared database of "good roe" is measured. Finally, the least three difference values are averaged to yield a measure of the shape quality of the inspected roe. A classification threshold is then selected to discriminate between a good shape and a poor one.

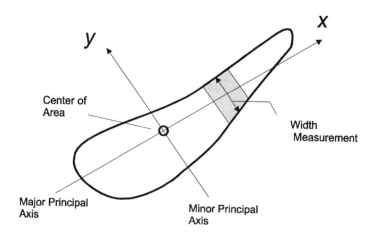

Figure 3: Shape representation for herring roe.

Since the classification measurement is based on a value that relates to the deviation of the inspected shape from a previously determined "good" one, the accuracy of the good quality shape selection can be altered by changing the classification threshold. However, the total classification accuracy (i.e., counting both good rejects and bad accepts) is known to be approximately 82%, which is too low to be reliable as a stand-alone system. Nevertheless, this system can considerably reduce the bulk for manual grading by picking out Grade 1 roe at a desired accuracy. Details of tradeoff between accuracy and the percentage of bulk removed are given in [11].

Broken Tail Detection

Shape grading of herring roe has been further refined by incorporating a method for detecting broken tails [12]. Although our initial shape grading method is able to detect the presence of major breaks, it was noticed that smaller breaks and sharp angular breaks, which mostly occurred at the anterior part (i.e., the tail section), could go undetected by the grading algorithm.

Based on interviews with herring roe grading experts, it was established that some of the breaks could be acceptable depending on their size, shape, and orientation. Our refinement to the original shape grader analyzes the anterior part of a roe skein for the occurrence of small breaks or angular breaks and determines whether or not they are acceptable based on the aforementioned criteria provided by grading experts. In the process, some of the features were found to be easily quantified, while others were very vaguely defined. For example, the size of a break must be at least 3/8 inches (1 cm) for it to be classified as broken, or the angle of a break must be between 35° and 145° from the dorsal side. However, the seriousness of the break depends on whether or not the part could be hidden by careful packaging for retail sale.

The first step in tail feature extraction is the detection of the tail itself. In good Grade 1 roe, the posterior end is larger in terms of its area than the anterior end. Thus, the detection of the roe orientation (or tail end) can be based on this fact.

Features based on geometrical properties are chosen to represent the shape of the tail of a skein. In order to calculate the geometrical properties, the tail boundary is approximated using piecewise linear segments, also termed polygonal approximation. Here, the initial line is obtained by joining two points on the boundary. The distance of each point on the boundary is computed. An admissible error of approximation that represents the maximum distance from the segment to the actual boundary is defined. If the maximum distance is greater than the admissible error, the curvature is split at that boundary point. Finally, the polygonal segments are obtained by joining all the splitting points. The method is described in detail in [13] and is illustrated below in Figure 4.

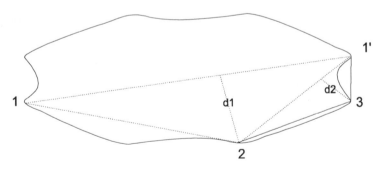

Figure 4: Polygonal approximation of boundary contour.

All the angles formed by the polygonal segments are calculated starting from the dorsal part. Any angles less than 35 degrees are eliminated, and the segments are joined together to form the dorsal facet. The first facet having an angle between 35 and 145 degrees with the subsequent facet would be considered as a tail facet. The last facet forming an angle greater than 35 degrees would be considered as the start of the ventral facet; see Figure 5.

After the extraction of the features, a decision-making system is employed to determine the occurrence of breakage on a roe tail. The system is based on the concepts of fuzzy logic. For each feature, three triangular-trapezoidal fuzzy states are developed corresponding to the fuzzy measures of small, medium, and large. These membership functions are shown in Figure 6.

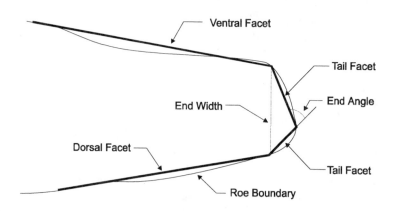

Figure 5: Features of herring roe tail.

All the tail facets form the tail end. From these facets four features are extracted:
- Total angle formed by facets (EndAngle)
- Number of facets (FacetNum)
- End width (EndSize)
- Maximum facet size (FacetSize)

The support values and the overlapping regions of the fuzzy states have been chosen according to expert knowledge and heuristics. The feature values are fuzzified by matching them with their respective membership functions. Seven fuzzy rules have been developed based on the opinion of grading experts to obtain a fuzzy decision. These fuzzy rules are shown in Figure 7. Using the compositional rule of inference, incorporating the min-max composition, the shape of the output solution can be obtained in fuzzy form.

With this method of inference, each fuzzy rule or proposition contributes to the shape of the output solution when it applies. The output of this system is the likelihood of the tail being broken (TailCut). The membership function of the TailCut value is shown in Figure 8. The centroidal defuzzification method is used to obtain a representative numerical value for the decision.

A preliminary test for this system was done on a sample of 60 skeins. Half of the sample consisted of roe with broken tails as determined by expert graders; the other half had unbroken tails. All these skeins had been classified as Grade 1 Roe by our initial shape classifier. The results are shown in Figure 9. It can be seen from the results that a significant improvement can be obtained using this refinement.

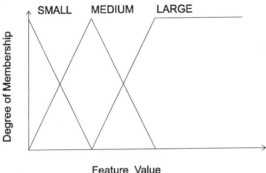

Figure 6: Fuzzy states of a tail feature.

1.	If EndSize=Small then TailCut=Small.
2.	If FacetNum=Small then TailCut=Small.
3.	If Endsize=Large and EndAngle=Small and FacetNum=Small then TailCut=Large.
4.	If Endsize=Large and EndAngle=Large and FacetNum=Medium and FacetSize=Small then TailCut=Large.
5.	If Endsize=Large and EndAngle=Large and FacetNum=Large and FacetSize=Small then TailCut=Large.
6.	If Endsize=Large and EndAngle=Large and FacetNum=Medium and FacetSize=Large then TailCut=Large.
7.	If Endsize=Large and EndAngle=Large and FacetNum=Large and FacetSize=Large then TailCut=Large.

Figure 7: Fuzzy rule set for broken (cut) tail detection.

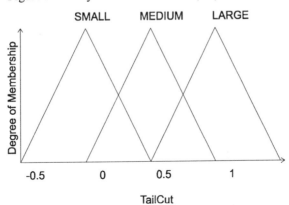

Figure 8: TailCut membership function.

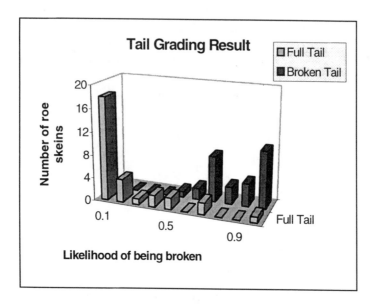

Figure 9: Preliminary results of the tail grading algorithm.

This improvement is intended to complement the shape classifiers, a two-step process. Therefore, a method of combining the measured values to arrive at a grading decision is required. Two different thresholds for the likelihood of tail being broken are used. These thresholds separate the output of the tail grading system into three categories: broken, unbroken, and undecided. A truth table, shown in Table 1, is used to incorporate these results into the existing shape classifier. For the undecided result, the shape classifier value is adjusted depending on the value provided by the tail grader. Consequently, a roe skein classified as Grade 1 shape with a borderline shape value and probably an unbroken tail will still be considered Grade 1, while the same skein with an 'undecided' tail grade will be downgraded. This method was tested on a sample of 160 skeins of roe, consisting of 60 skeins of roe with broken tail and 100 skeins of mixed Grade 1 and borderline shapes. The tail grading decision thresholds were selected as 0.35 and 0.5, respectively. The result of the classifying accuracy was 96%.

Table 1: Combined grading scheme

Tail Grader \ Shape Grader	Grade 1	Grade 2
Full Tail	Grade 1	Grade 2
Tough Decision	Shape Grade Value Increased	Grade 2
Broken	Grade 2	Grade 2

3.2 Ultrasonic Echo Imaging for Firmness Measurement

In order to analyze the firmness property of herring roe, a form of sensory data that relates to the physical condition within a roe skein is needed. Ultrasonic imaging provides this means. As with machine vision, the industrial use of ultrasound in measuring the texture property of food products is not new. For example, measurement of surface bruises has been done for apples [14]. The basis of the use of ultrasound in the firmness measurement of a food product is that the acoustic impedance of a material is related to its texture properties. The use of ultrasound has a considerable advantage over a mechanical sensor, since it allows for faster data acquisition. Furthermore, the availability of commercial ultrasonic scanning systems and the possibility of their smooth integration with other processing hardware for roe feature analysis (in particular, use of the same image processing hardware) heavily influences the decision to use the technology.

For firmness measurement of herring roe, features related to firmness investigation are extracted from the ultrasonic image produced by a medical ultrasound system with a mechanical sector probe at 10 MHz scanning frequency. This mechanical probe scans the object in a circular path, thereby producing a nonlinear image with the sections near the edges scanned at an angle to the vertical axis. In order to get a clear image, the probe has to be firmly in contact with the roe.

A technique based on statistical values, developed by researchers in IAL, provides a means of feature extraction and firmness classification [15]. The technique uses statistical features related to the image intensity. This is based on initial experimentation which shows that a soft roe gives rise to a brighter ultrasonic image than a firm skein.

The firmness analysis technique creates rectangular regions (patches) covering the almost linear section of the image, depicted in Figure 10. The width of these patches is kept constant, but the height is scaled according to the thickness of the roe image. The thickness is found by measuring the difference of the two bright (i.e., pixel intensity above a specified threshold value) rows in the image which correspond to the top surface of the roe skein and the bottom surface touching the support platform.

Two values are extracted from these patches. The first one is the overall mean brightness, computed using

$$M_{overall} = \frac{1}{N_p} \sum_{i=1}^{N_p} M_i \qquad (1)$$

with

$$M_i = \frac{1}{N} \sum_{x=c1}^{c2} \sum_{y=r1}^{r2} I(x, y) \tag{2}$$

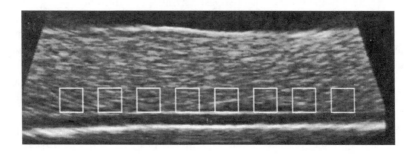

Figure 10: Ultrasonic depth image of a herring roe.

where M_i is the mean intensity of each patch, $I(x, y)$ is the intensity of a pixel at (x, y), $(c1, r1)$, and $(c2, r2)$ are the coordinates of the primary diagonals of the patches, N is the number of pixels in each patch, and N_p is the number of patches. The second value extracted is the average standard deviation of those of the patches, given by

$$\overline{SD} = \frac{1}{N_p} \sum_{i=1}^{N_p} SD_i \tag{3}$$

with

$$SD_i = \sqrt{\frac{1}{N-1} \sum_{x=c1}^{c2} \sum_{y=r1}^{r2} (I^2(x, y) - M_i^2)} \tag{4}$$

Two other features are derived using image-processing techniques. The image is first convoluted with the Sobel operator. The resulting edge-enhanced image is thresholded to separate the areas, called "structures," of high brightness value. The number of structures appearing in the image is computed to yield this third feature.

The fourth feature is computed using the same technique as the third feature, except that the original image is eroded twice and subsequently dilated two times using a 3 by 3 mask matrix.

Two classification methods are developed, combining these four features, to yield a decision describing the firmness of a roe skein. The first method uses a regression function to construct a continuous model for the stiffness value. The method is based on the correlation between the four features extracted from the ultrasound images with the stiffness value obtained from manual force-deflection measurements. Various regression models have been analyzed. The linear model, which still yields an error of the order of 10% of the estimated value, is found to provide the closest approximation of stiffness.

The second method employs the use of a fuzzy-logic rule base and an associated inference engine. The membership functions are constructed by using probability density functions, since there is a direct correlation here to statistical estimation of membership functions [16], and since there is a lack of available expertise in this type of analysis of herring roe. The logical interpretation of the membership functions is: the higher the probability density, the larger the possibility that the value is representative of the particular class. The probability density functions for each feature are obtained off-line for the two classes, firm and soft, from a batch of 160 roe skeins. The classifications are confirmed by an expert grader. Index values between 0 to 1 are obtained for each feature by normalizing the measurements to form the support set. The membership function curves are also normalized such that the largest value of each function is unity. A fuzzy rule base is constructed to generate a decision value for the roe firmness. The classification accuracy of the system is observed to be around 84%.

3.3 Vision-Based Weight Estimation

As an alternative to weight classification by direct weighing, an estimation of weight by the size of skein has been developed by the researchers of IAL [17]. This is justified by the fact that an on-line direct weighing system for herring roe skeins is not commercially available and would be very costly to manufacture. One of several reasons for this is that roe skeins are very light, with a weight of more than 40 grams being very rare. Moreover, the irregular shape of herring roe skeins requires a complex handling mechanism to allow an on-line direct weighing process. Furthermore, due to vibrations and other transient conditions, it is difficult to obtain a steady reading at the required speeds.

An alternative approach employs the use of geometric features of the object to approximate the weight of a skein. These geometric features are extracted from images obtained by both the camera and the ultrasonic imaging device.

Three main features, the area (A), the peripheral length (P), and the thickness (T) of a skein are extracted from the skein images. Using the multiple linear regression technique, five different combinations are tested. These combinations are defined as:

1. Area (A);
2. Area and Peripheral length ($A+P$);

3. Area and Thickness $(A+T)$;
4. Area, Peripheral length, and Thickness $(A+T+P)$;
5. Area, Ratio of Area to Peripheral length, and Thickness $(A+A/P+T)$.

The method is selected as it has the advantage of direct software implementation. Through experimentation, it was found that the combination of area, peripheral length, and thickness gives the best approximation model. The average error recorded is 1.26 grams for roe, with an average weight of 20 grams.

The method of linear regression analysis for estimating weight works best when a measurement of the skein thickness is available. Originally, this measurement is provided by the ultrasonic imaging sensor itself, which gives a depth scan. However, the ultrasonic imaging system has not been incorporated into the original prototype due to the mechanical complexity involved in implementing the contact probe. Moreover, recent improvements in herring roe processing, which has reduced the amount of spongy roe (detectable by ultrasound), have relaxed the need for ultrasonic imaging in grading to a certain extent. Accordingly, the linear model that is implemented depends only on the area of a roe. An area-peripheral length combination does not really give any significant effect, while increasing the computation time. A refined technique based on a statistical method has been developed as well, to improve the accuracy of the weight estimation.

This refinement uses a piecewise linear model to further improve the accuracy of weight estimation. The piecewise linear model divides the weight range into different segments. Each segment is analyzed independently to yield a linear regression model. A piecewise linear model has an advantage over a polynomial model, where discontinuities appear on the correlation model within the overall range of the correlated parameter.

Three weight ranges are selected corresponding to the small, medium, and large weight classifications, respectively, for herring roe. The linear models are obtained by performing a least-squares analysis to best fit the data, as follows:

$$Y = mx + b \qquad (6)$$

where,

$$m = \frac{n - (\sum xy)(\sum x)(\sum y)}{n(\sum (x^2)) - (\sum x)^2} \qquad (7)$$

$$b = \frac{(\sum y)(\sum (x^2)) - (\sum x)(\sum xy)}{n(\sum (x^2)) - (\sum x)^2} \qquad (8)$$

and n is the number of observations. The standard error introduced for each model with this method is given in Table 2.

Table 2: Standard error of the linear regression models.

Regression Model	Standard Error (%)
Small	8.8
Medium	8.5
Large	7.8

The average percentage error obtained by this method is worse than the industry requirement of ±5% for weighing systems. However, the weight estimation system is not intended for the purpose of exactly determining the weight of each roe. Instead, it is developed as an automated system for classifying roe into a few weight groups (i.e., a weight sizing system). Thus, misclassifications occur only where the weight ranges overlap; i.e., at the limiting value that separates small and medium roe, and, similarly, the weight limit that separates medium and large roe. In this situation, where a crisp separation between weight ranges is least desired, a fuzzy logic-based solution seems quite appropriate and would provide a subsequent step of enhancement [16].

Compared to weight classification by human visual inspection, the weight estimation presented herein greatly exceeds the size classification accuracy of a human grader. Table 3 gives the percentages of classification error by the automated weight estimation system and the corresponding percentages of classification error by human graders, for three sample batches of roe.

For the existing system, another method for implementing the regression estimation can be performed due to the fact that some roe processing companies employ statistical samples to determine the proportion of large, medium, and small skeins in their graded roe. A statistical enhancement method based on the probability of misclassification has been developed for this purpose. This method uses a probability of misclassification determined by statistical analysis of classification results in the range where misclassification occurs. A normal distribution function is used to represent the probability of misclassification over these ranges. This misclassification probability is then applied for each roe graded inside the misclassification range. The number of misclassified roe is estimated based on the total probability of misclassification for all Grade 1 roe. Table 4 shows the estimation figures compared to the actual number of misclassified roe based on a sample of mixed large, medium, and small size skeins. This method works well in giving a size-proportion estimation of classified roe based on the number of graded skeins. The average error is approximately 1.3%.

3.4 Color Grading

A premium Grade 1 skein of roe typically has a light golden yellowish color. Although of late, a post-processing technique has been developed to "bleach" the bloodstains in roe, a skein with a good natural color is always preferable and normally commands a better price.

Table 3: Comparison of machine and human weight classification error.

Grade 1 Size	Sample Size	Machine Classification Error	Human Classification Error
Small	100	5 %	30 %
Medium	150	5 %	20 %
Large	200	1 %	15 %

Table 4. Computer estimated distribution of weight estimation.

Bucket	Actual Distribution				Computer Estimated Distribution			Error (%)
	Small	Medium	Large		Small	Medium	Large	
1(Small)	79	11	0		78	12	0	1
2(Med)	11	183	17		11	187	13	2
3(Large)	0	22	221		0	20	223	0.8

Several color classification methods have been developed, using two images obtained from the red and the blue channels of RGB video signals [17]. The color analysis is performed only on the red-channel image, and the other image is used for segmentation of the roe from the background.

As roe generally has a light yellowish color, it absorbs most of the blue spectrum of the visible light. Thus, a roe will appear dark on the blue-channel image. A white background is used to give a good contrast to the roe image. Image segmentation by thresholding has been applied to find the location of the roe in the image.

Once the location of the skein on the screen is known, the analysis of its intensity in the red channel is performed. The red channel is chosen since roe reflects most of the visible red color spectrum, while the bloodstains absorb all of it. Three different methods are tested to classify roe color, described here in the order of complexity.

In the first method, the average intensity of a roe skein from the red-channel image is computed. As a darker color has lower intensity values, a good-colored roe will give a higher average intensity than a poor-colored one. A threshold value is used to separate a good-colored roe from a poor-colored one, by this method.

The second method is somewhat similar. However, instead of computing the average intensity for the whole section of a skein, the roe is sliced into several equivalent sections. The average intensity of each section is then calculated. In this way, the method could pinpoint where the bloodstains occurred on the roe skein.

The third method is performed by first constructing a density function of intensity for the roe segment. The area (A) and the average intensity $(\bar{I})$ are then computed. A 20% range of average intensity (p) is used to compute a threshold value using

$$th = \bar{I} - p \times \bar{I} \qquad (9)$$

Every pixel with an intensity value less than th is accumulated to give N. The ratio of the number of these pixels to the total area of the skein is used as a measurement of color.

$$Ratio = \frac{N}{A} \times 100 \qquad (10)$$

All these methods perform well in classifying roe on the basis of color. However, grading requirement for color is relatively loose and, as a result, even the first method would suffice to provide an acceptable level of color classification for roe skeins.

3.5 Fuzzy Decision-Making System

In order to combine measurements of various properties of roe to yield an overall grading decision for the roe, a fuzzy logic method has been developed [17]. For this purpose, the outputs from various property measurements are matched to their respective membership functions. Index values for the grading properties are defined, and the membership functions are designed accordingly. Five fuzzy states having triangular-trapezoidal membership shapes are used to represent the roe quality fuzzy values of very low, low, medium, high, and very high, respectively. The required fuzzy rules are developed with the assistance of expert graders to yield a fuzzy decision on the "goodness" of a roe skein. The compositional rule of inference is used to make decisions from the rule-base. Finally, a defuzzified output value is obtained by using the centroidal method [16]. The classification between good and poor grade roe is done by applying a threshold to this output.

4 Prototype Development

The industrial prototype development is initially intended only to implement the machine vision system. The main objective is to develop and integrate an automated grading machine for industrial use. Since an ultrasonic imaging system would require a complex mechanical system and the cost of the necessary components has been found to be rather high, it was decided to postpone the integration of the ultrasound probe for firmness measurement.

A general view of the automated herring roe grading system (HRGS) prototype is shown in Figure 11.

Figure 11: The automated herring roe grading system (HRGS) prototype.

The main physical components of the initial prototype are a conveyor system, an ejection mechanism for roe skeins, and a computer system for system control. All the control functions in the prototype are monitored and adjusted through the computer, with the exception of the conveyor belt speed, which is maintained constant and adjusted manually, if necessary. It is a straightforward task to use the computer to monitor and set the conveyor speed, when the machine is ready for industrial use. Figure 12 shows a schematic plan view of the HRGS prototype (excluding the computer).

Currently, roe is manually fed to the prototype. The maximum operating speed with all the grading functions operating, excluding the color grader, is 1 roe skein/s.

4.1 Conveyor System

A 3-inch wide belt is used to convey the roe past the sensory module and the ejection system. A white belt has been selected to provide the necessary contrast for the vision system. Two rollers, 2-1/2 inches in diameter, are used to support the belt at both ends. The top part of the belt travels over the flat surface of a 3×3 inch square aluminum section (hollow) that is 8½ feet long. Two guide-rails are used to ensure that the top of the belt stays flat. The topside of the conveying surface is 9 ft long. At the feeding end, two guide-alignment fingers are installed to center and align the skein toward the middle part of the belt.

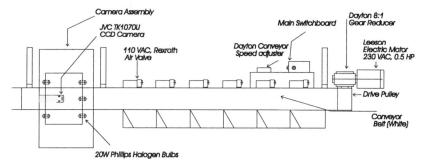

Figure 12: A schematic view of the herring roe grading system (HRGS).

The conveyor belt is driven by a 230 VAC, 1725 rpm electric motor, manufactured by Leeson Inc. The motor is connected to the drive pulley (and gear reducer) at one end of the conveyor, with an adjustable rail for the tension of the belt. A Dayton gearbox (speed reducer) with 8:1 speed ratio is used to reduce the rotation speed of the motor and thereby increase the available motor torque. The electric motor delivers a maximum output power of 0.5 hp.

To adjust the speed of the belt, a Dayton AC inverter is connected to the motor. The inverter can adjust the input current frequency with a resolution of 1 Hz. This drive mechanism, at the maximum rating of 60 Hz, provides a maximum belt speed of 24.4 in./s. Accordingly, the speed of the belt can be modified in 0.4 in./s increments.

4.2 Ejection Mechanism

Six ejection gates, activated by pneumatic valves, are used to separate the roe skeins into designated buckets. The valves are manufactured by Rexroth Inc. Each valve is connected to a single air compressor source. The compressor is equipped with a pressure regulator to maintain the required supply pressure to the valves.

The pneumatic valves are activated by applying an input voltage of 110V. This voltage is regulated by an OPTO I/O hardware module, which is connected to the computer through an ISA bus.

When actuated, the valve releases an air jet to push the roe off the belt. Buckets filled with brine are located in front of the ejection valves. The separation distance between the valves is 9 inches. The ejection valves are shown in Figure 13.

Figure 13: The HRGS ejection valves.

4.3 Sensory System

Two sensors are used in the present herring roe grading prototype system. A visual sensor serves as the shape/size grading sensor of the machine. A position sensor is used to locate the position of a particular skein along the belt; it is used to determine the timing for both the visual sensor and the ejection mechanism.

Visual Sensor

The visual sensor includes the digital camera setup and the illumination system. A 26 in. by 14 in. by 6 in. aluminum box is mounted above the conveyor belt, which serves as the camera enclosure. A camera-mounting frame, consisting of two panels to hold the camera, is bolted to the inside of the enclosure. The height of the face of the camera (not including the lens) is 4 ft above the conveyor belt. The camera-mounting unit is shown in Figure 14.

The camera used for the HRGS is a JVC color CCD camera, model number TK1070U, powered by a 12V DC power supply. This camera uses a CCD matrix with a resolution of 768 by 493 pixels as the sensing element. Although a higher resolution of the matrix provides a better image sensing capability, the actual image resolution is ultimately determined by the image acquisition hardware and the transfer signal between the camera and the acquisition hardware. The camera uses a 16 mm lens to give a viewing area of approximately 9 in.^{2} for 512 by 512 pixels of image resolution (as selected by the frame grabber).

The camera has an adjustable shutter speed. For the present HRGS, the shutter speed is set at 1/500 seconds (2 ms) to reduce the blurring effect due to the motion of the roe skeins during imaging. The shutter speed selection is based on the image resolution of 60 pixels/in. and a belt speed of 8.13 in./s. This translates into a moving image of 487.8 pixels/s. Hence, to reduce blurring, a shutter speed of

1/487.8 or less than 2.08 ms would be required. To provide a closely uniform background, a white 10-inch-square panel, the middle part being the conveyor belt, is positioned in the viewing area underneath the camera system.

Figure 14: Camera mounting unit.

A direct lighting technique is used in the HRGS, for illuminating the object within the viewing area. Six halogen light bulbs (20W/12V each) are used for illumination of roe. The power to the light bulbs is provided by a 12V DC, 192 watts power supply, capable of providing 16 amperes of current. A DC power supply was selected to ensure a constant light intensity in the viewing area. An aluminum cover with a white reflective lining is used to shield the viewing area from ambient lighting.

An important consideration in designing the illumination system for herring roe grading has been the elimination of shadows in the image. This is achieved by placing diffusers in front of the light sources. The orientation of the light bulbs has also been adjusted to minimize the shadows, by illuminating the roe skein along its length, i.e., parallel to the direction of the belt motion. Figure 15 shows the lighting arrangement in the HRGS.

The spatial constancy of illumination over the field of view is also achieved by using diffusers of different strengths for the light bulbs. Bright spots still remain at the four corners of the viewing area. However, since a roe skein appears only in the middle of the viewing area, this condition is considered to be acceptable.

To adjust for illumination changes over a period of time, due to a build up of salt and other foreign material on the lighting apparatus, changes in ambient lighting, etc., an automatic thresholding algorithm has been developed. The algorithm uses the gray level histogram of the image to separate the roe skein from the background. Figure 16 shows a typical histogram of the viewing area. The threshold is selected to lie between the group of low-intensity pixels (associated with the skein) and the set of high-intensity pixels (associated with the background). The most appropriate threshold has consistently been found to be the value halfway between the low intensity peak and the first local high-intensity peak. However, several experimental results have shown that under some circumstances the algorithm can fail. In view of this, a fixed intensity threshold is used with a calibration function, each time the HRGS is switched on, and the threshold is picked for achieving the best silhouette for a skein of good color roe.

Figure 15: Lighting orientation in the HRGS prototype.

Position Sensor

Optical sensors, manufactured by Omron, model E3X-A11, are used in the HRGS to detect the skein position on the belt. The optical sensors use the red spectrum of visible light and are equipped with amplifiers to adjust the light intensity. Three optical sensors are arranged in a column. Each sensor utilizes a transmitter and receiver pair, which is installed facing each other across the belt. The arrangement of the sensors in each column alternates between the transceiver and the receiver, to minimize the interference between each pair of sensors. The column arrangement is necessary to improve the accuracy of detecting skeins of various sizes and shapes. Irregular shape of the roe skein may make detection by a single sensor ineffective; for example, the forward section of the roe may be missed by a single sensor, resulting in the camera being triggered too late to capture the entire skein.

The digital output signals of the sensors are sent through logical OR operation using logic gates. Thus, any sensor detecting an incoming skein will send a high signal to the main processor. The output of the resulting signal is sent to the computer via a digital I/O acquisition board.

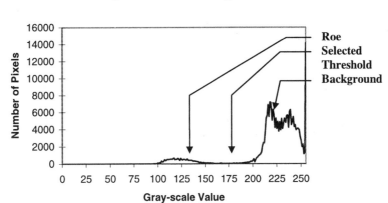

Figure 16: Gray level histogram of a roe image.

4.4 Prototype Control System

The control of the prototype, including the main processing hardware, is provided by a single system computer. Control functions have been developed to integrate each component within the working prototype.

Computer Hardware

A 486DX PC-100 MHz is set up as the control computer and the grading processor for the HRGS. The computer is equipped with a Sharp GPB-2 frame grabber/image processing board. The four camera inputs of the GPB are connected to the red, green, blue, and sync channels of the camera, respectively. Two monitors are provided for the computer. One monitor displays grading statistics and the current grading decisions, and the other shows the image from the GPB display bank for carrying out calibration procedures and result analysis. The GPB bank is capable of storing a 512 by 512 pixel image. However, since the video signal from the camera is used, the acquired image is 512 by 480 pixels for a full frame. The acquisition of a full frame takes 35 ms to complete.

For controlling the position sensor and the pneumatic valves, an OPTO 22 I/O board is used. The OPTO 22 is a digital input-output card capable of handling 24 digital input/output channels. Each channel is addressed by a single bit. Thus, 3 different bytes can be used for input/output requirements. In the HRGS only seven bits of a single byte are used. The first bit is used to obtain the input signal from the position sensor. A low signal means the optical position sensor is blocked. The other 6 bits are used to actuate the 6 pneumatic valves. The terminal block can accommodate I/O modules for converting the voltage output of the computer into the desired control voltage. Six output modules are used to generate the required voltage for the pneumatic valves.

Control Software

All the software functions including the control functions are written in C programming language. Since all of the grading functions were developed using Microsoft DOS on a PC platform, MS-DOS was used as the operating system for the HRGS. This selection eliminated the need for major code rewriting and recompilation of the grading software, as well as the hardware drivers and GPB image processing functions. The number of control tasks involved in the current HRGS is sufficiently limited so that the use of MS-DOS is adequate.

The control tasks required to complete the analysis of a roe skein in the HRGS are as follows:

1. Roe skein position detection.
2. Roe image acquisition.
3. Image analysis execution.
4. Gate identification according to grading decision.
5. Gate actuation.
6. Screen output.

However, in the physical system, the next skein must be detected and analyzed before the previous skein reaches its designated gate. Thus, the gate actuation function should be executed between the shape analysis processes. Moreover, the physical distance the roe skein travels, between the time of detection of the skein and the camera triggering, would require that the next position detection be executed before the previous grading process is completed. Thus, a multitasking, interrupt-driven environment is required.

Fortunately, the position detection and gate actuation functions require a very short processing time (less than 1 millisecond). Thus, they can be combined and executed multiple times during the main process by means of a periodic software interrupt. This periodic interrupt executes the position detection and gate actuation routines every 54 ms (18.2 Hz).

Since all the functions have to be executed in real-time, a timer of 1 ms resolution is required in DOS. The Intel 8254 programmable interval timer of the PC is used to fulfill this requirement. This same timer is also used for the software interrupt described above [2]. The main functions of the timer are to record the time needed to trigger the camera and actuate the ejection valves.

The grading functions, the weight estimation, and the broken tail detection algorithms use a full-frame resolution image, while the shape grading algorithm uses only half-resolution. Since GPB digitizes one full frame at a time, the weight estimation and broken tail detection are executed before executing the shape grading function. The color function is executed before the gate setup. However, for all tests on the HRGS, the color function was turned off, as it was not deemed important in plant operation at this time.

Since roe may travel past a maximum of 5 gates prior to ejection, it is possible that a later skein may arrive at its gate before the previous one is ejected. This problem is solved by using a queue line for gate activation. Thus, each time a grading decision is made, the gate number and the actuation time are put in the queue. The gate actuation function searches the queue to find an active gate number. A similar queue is used to close the opened ejection valves after a specified time interval. Figure 17 shows the flowchart of the HRGS control function.

With the control function and hardware configuration described above, the HRGS can grade roe skeins at a rate of 1 skein/second. The total processing time for one roe is approximately 950 ms, including the screen output. A breakdown of the average processing times for various grading functions is given in Table 5.

Table 5: Processing time of HRGS grading functions.

Function	Processing time (ms)
Weight estimation	130
Broken tail detection	330
Shape grader	440
Gate setup and screen output	50

Grading Decision Making

A very difficult challenge in developing the HRGS has been to keep up with the industry grading standards. Grading criteria change, depending heavily on the market situation. Thus, the HRGS software should allow the industrial user to easily adjust and redefine the roe skein grading categories.

Grading classification is done by applying an upper limit and a lower limit to each grading measurement corresponding to each gate. The skeins that meet the measurement criteria for a particular gate are assigned to that gate. Table 6 shows typical values for Grade 1 Large roe. Skeins that do not belong to any of the designated six gates are collected at the end of the belt. Care must be taken in assigning the gates, so that no conflicts arise. If a skein meets the criteria of more than one gate, the gate with the higher designated number (i.e., the later gate) is actuated.

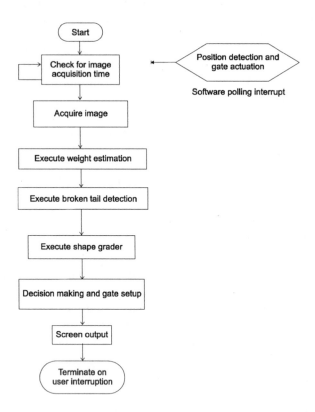

Figure 17: Flowchart of the HRGS control function.

Table 6: Gate setup parameters for grade 1 large roe.

Shape upper limit value	8.35
Shape lower limit value	0.0
Weight upper limit value	60.0
Weight lower limit value	20.0
Tail upper limit value	0.35
Tail lower limit value	0.0
Length upper limit value	10.0
Length lower limit value	3.0

For the grading parameters, the weight is calibrated in grams and the length in inches, as provided by the shape grader. Changing of the parameters corresponding to the gates is accessible using the user interface described next.

User Interface

The user interface provides a screen output, primarily for the benefit of the operator. The HRGS displays simple statistical data including the number of roe skeins sent to each gate. The measurement values for each function are shown on the screen for every skein. The total weight estimate of roe at each gate is also given. Figure 18 shows the statistical output screen.

For ease of modifications, all the parameters that are needed for the HRGS to perform the grading operation are saved in an ASCII file. A user-friendly interface has been developed to simplify modification of the grading categories. The user interface works independently from the system. At the start of operation, the user interface is called, and any modifications to the grading parameters made by the operator are saved to an ASCII file (the system is also provided with default values for these parameters). The main function is then called by the user interface, and CPU executes the HRGS main function. This function checks all the parameter data in the ASCII file and runs the machine accordingly.

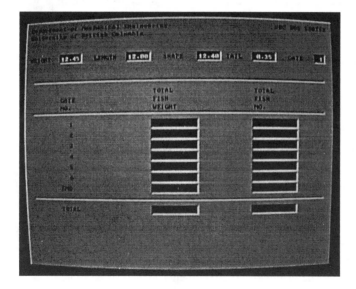

Figure 18: HRGS on-line output display.

5 Prototype Testing

The performance of the industrial prototype has been tested both in the laboratory environment and under real conditions of an industrial roe processing plant. For all experiments, ejection gates of the prototype machine are designated as follows:

Gate 1: Grade 1 Large roe classified according to good shape and large size.
Gate 2: Grade 1 Medium roe classified according to good shape and medium size.
Gate 3: Grade 1 Small roe classified according to good shape and small size.
Gate 4: Grade 2 roe classified according to broken tail condition and length.
Gate 5: Grade 2-H roe classified according to slight distortion in shape.
Gate 6: Grade 2-C roe classified according to heavy distortion in shape.
Gate 7: Any roe that does not fall into the above six grades.

"Gate 7" of the machine actually corresponds to the end of the belt where the skeins go off, since they are not ejected by the air jets. Bits and pieces of roe (Grade 3 roe) that do not meet any of the classification parameters described above would fall into this category.

The classification of Grade 2-H and 2-C is based on the shape grading values computed by comparing the shape of the roe to a database of 'ideal' roe. Thus, the Grade 2-H and 2-C categories do not really meet the actual industrial grading criteria. Since the initial system was not designed to differentiate between various types of grade 2 roe, two performance indices are computed for gates 4, 5, and 6 in this analysis. The first value shows the accuracy of the machine when trying to match the actual grade, and the second shows the accuracy of the machine for picking out the poor grades (i.e., separating Grade 1 roe from Grade 2 roe).

The shape classification thresholds have been adjusted according to the size of the skeins. The decision for broken tail is obtained using the truth table given in Table 1. Table 7 gives the assigned parameters for the prototype experiments.

The performance of the prototype machine is evaluated based on its ability to accurately classify roe by their shape and size.

5.1 Laboratory Experiments

To evaluate the system capabilities, three thorough tests were performed in the laboratory after tuning the system for optimum performance. The first experiment was designed to evaluate the machine performance in selecting roe, in each of the seven different grades. Each run was conducted on a sample of 50 roe skeins of the same grade. The results of the experiments are shown in Table 8. The borderline (BL) case is defined to be skeins with minor defects that could be classified as either Grade 1 or Grade 2 roe. In such cases, the shape value is usually close to the classification threshold.

The next experiment was done to simulate a real grading operation. A mixed batch of different grades, with a similar proportion of grades as found in a commercial grading batch of roe, was fed into the machine. The total number of skeins in the sample was 200. The proportion of the grades is shown in Table 9(a). No Grade 3 skeins were present in this experiment. The experimental results are given in Table 9(b).

Table 7: Machine parameters for experiments.

Belt Speed	8.13 inches/s
Intensity Threshold	155
Grade 1 Large shape values	Less than 8.35
Grade 1 Medium shape values	Less than 8.9
Grade 1 Small shape values	Less than 9.1
Grade 1 Large weight values	Greater than 20 grams
Grade 1 Medium weight values	Between 15 grams to 20 grams
Grade 1 Small weight values	Less than 15 grams
Grade 2	Broken tail condition (see Table 1)
Grade 2-H shape values	Between 8.35 to 11.5
Grade 2-C shape values	Greater than 11.5
Feeding Orientation	Head first, convex side up

Table 8: Prototype grading performance on individual grades.

Grade	Accuracy (%)	Accuracy with BL-case (%)
Grade 1 Large	100	100
Grade 1 Medium	70	70
Grade 1 Small	78	78
Grade 2	62	62
Grade 2-H	94	100
Grade 2-C	96	100
Grade 3	100	100
Total Bi-classifier accuracy	82	84

Experiments to determine the grading repeatability of the machine have been performed as well. However, since the system is a classifier, the repeatability of classifying the skeins would depend on the variation of the shape value and the classification threshold. Thus, a skein with a shape value near the classification threshold has a higher percentage of non-repeatability, while a skein with perfectly good shape and a shape value far from the classification threshold has 100% repeatability.

The variations in the shape value may be caused by such factors as lighting variations, roe orientation and position, and system noise. Experiments have shown that the average variation of the shape value is 0.4 with a standard deviation of 0.2. Since the shape measurement values range between 5 and 25, with the Grade 1 roe falling within 5 to 9, 10% of values within the Grade 1 classification range have a repeatability of less than 100%.

Table 9(a): Proportion of grades in the mixed sample.

Grade	Proportion (%)
Grade 1 Large	60
Grade 1 Medium	15
Grade 1 Small	8
Grade 2	4
Grade 2-H	10
Grade 2-C	3

Table 9(b): Prototype grading performance for the mixed sample.

Grade	Accuracy (%)	Accuracy with BL-case (%)
Grade 1 Large	86	91
Grade 1 Medium	100	100
Grade 1 Small	100	100
Grade 2	52	88
Grade 2-H	50	100
Grade 2-C	75	100

Another experiment was performed to evaluate the accuracy of the system as a size classifier. Since only one function is needed for this purpose, the speed of the system can be stepped up to 24.4 inches/second. A total of 465 Grade 1 skeins were tested. The proportion of different sizes in the test sample is given in Table 10(a). This includes the "sho-sho" size roe, which is a class of Grade 1 skeins weighing less than 10 grams. Table 10(b) summarizes the experimental results. The borderline (BL) case corresponds to skeins with a weight within 0.6 grams of the weight limit. Such roe would be considered borderline in an industrial setting since the weighing scales used are accurate only to ± 0.5 grams.

5.2 On-Site Production Test

On site experiments were carried out recently at a local plant of our industrial collaborator. The experiments were done on roe that previously had been washed and processed. Most of the 2-C grade skeins had been already selected from the line. However, because manual belt grading performance is relatively low, some 2-C grade skeins still appeared on the line. The classification criteria and the gate assignments used in these experiments were the same as described before, in the beginning of Section 5. The sizing operation was also used to test the overall capability of the grading prototype.

Table 10(a): Proportion of grade 1 sample.

Grade	Proportion (%)
Grade 1 Large	43
Grade 1 Medium	33
Grade 1 Small	18
Grade Sho-sho	6

Table 10(b): Prototype grading performance for sizing.

Grade	Accuracy (%)	Accuracy with BL-case (%)
Grade 1 Large	90	91
Grade 1 Medium	84	89
Grade 1 Small	97	99
Grade Sho-sho	96	96
Overall	89	92

Table 11 summarizes the results of these two experimental trials. The total weight of the samples used for the experiments was approximately 21 kilograms of roe skeins of various grades.

Table 11: prototype grading performance in a production line.

Grade	Accuracy (%)
Grade 1 Large	94
Grade 1 Medium	70
Grade 1 Small	81
Grade 2	35
Grade 2-H	41
Grade 2-C	62
Grade 3	38
Total Bi-classifier accuracy	88
Sizing Accuracy	92

5.3 Performance Evaluation and Possible Improvements

The experimental results clearly show that the machine is capable of grading Grade 1 large skeins at an accuracy level greater than 94%. The repeatability test of the machine shows that it is also capable of reliably grading non-borderline cases.

The laboratory results show a remarkable performance of the machine. However, the results were somewhat less impressive at the industrial site. There were some extraneous circumstances beyond our control that affected the results. The first problem involved the tail detection process. Most roe skeins are light yellow near the anterior part, which makes it difficult to set the right threshold for image segmentation. Moreover, during the calibration phase, it was found that part of the conveyor belt had turned yellowish in color. This affected the image thresholding process, particularly for the tail of the roe. One possible solution for achieving higher contrast would be to use a blue-colored conveyor belt.

In the industrial tests, many irregularities were found in the roe. For example, the algorithm for broken tail detection misclassified as broken roe, the occurrence of tail extensions that had curled around the tail of the roe. This misclassification was due to smoothing done by the feature extraction method used to handle the full-frame resolution analysis. The smoothing created a two-dimensional image that resembled one with a broken tail, as defined by the algorithm. Therefore, re synchronization of the camera fields is required before image segmentation.

The disproportionate number of small size skeins also lowered the performance of the machine. Although expert graders indicated that Grade 1 Large roe most often comprises about 70%-80% of the total production of Grade 1, the proportion can vary on a day-to-day basis. For example, on the second day of testing, the proportion of small size skeins was approximately 30%, the remainder divided between medium and large skeins. Most of the very small skeins (sho-sho roe) are considered as bits and pieces by the machine, resulting in the low accuracy of Grade 3 roe classification. To resolve this problem, the resolution of the shape grader should be increased using a full frame analysis. The processing time, however, will be increased as a result.

The weight estimation method gives an average accuracy of approximately 90–92% (sizing accuracy for Grade 1 Large roe was close to 98%, comprising 70% of a typical batch). The error in the weight estimation has significantly affected the accuracy of grading Grade 1 medium roe, where accuracy has dropped from 91% to 70% (90% of the error was due to thicker roe). The use of a thickness sensor has been proposed to improve the weight estimation.

The overall classification accuracy for separating out Grade 1 roe is around 82–88%. A higher accuracy may be difficult to realize using the current grading technology, since most of the defects are caused by three-dimensional properties of a skein, which cannot always be seen in a two-dimensional image.

The classification accuracy of Grade 2-H roe and 2-C roe is very low. This is as expected, since the grading method does not analyze the shape property of the grades in the same manner as the industry. The accuracy of Grade 2-C roe is higher than that of the 2-H roe since the grading criteria are more dependent on the two-dimensional skein shape. Unfortunately, the shape distortion characteristic of Grade 2-H roe is difficult to detect from two-dimensional images. Thus, the improvements necessary to reliably detect this type of roe will have to consider a three-dimensional analysis.

6 Summary

Automation of herring roe grading process has been chosen as a valid solution to improve the market condition and the quality of the products. In this chapter, the technology development, which has led to the conclusion and testing of an innovative herring roe grading system for industry, was described. The technologies are capable of quantizing the quality of a herring roe in terms of its shape, firmness, color, and size properties. The process to yield decision making, not only for each individual property but also for the overall process of grading, has been included in the discussion. A detailed description of each component in the initial industrial prototype for herring roe grading automation was given. This includes the operation of the control system and the development of a user-friendly interface screen.

Extensive experimentation was carried out both under laboratory environment and on-site in a production plant facility. This has shown the actual capability of the first industrial prototype and the promising prospects of automation in the herring roe grading industry. The current prototype achieved a grading accuracy of 94% in picking out Grade 1 roe from a production batch. Further development of sensor technology and its integration into the system should be undertaken. This will provide improved automation, and is likely to help improve the future of the herring roe industry in general.

Acknowledgment

Funding for the work described here has been provided by the Natural Sciences and Engineering Research Council (NSERC) of Canada Chair Professorship in Industrial Automation at the University of British Columbia. Further funding from the Garfield Weston Foundation and the Science Council of BC is gratefully acknowledged. Industrial collaboration with BC Packers, Ltd. and Ryco, Inc., in the development of the prototype machine, was very useful.

References

[1]   Ministry of Agriculture, Fisheries and Food, *The 1994 British Columbia Seafood Industry in Review*, British Columbia, Canada, 1994.

[2]   Kurnianto, S., *"Design, Development, and Integration of an Automated Herring Roe Grading System,"* M.A.Sc. Thesis, Department of Mechanical Engineering, The University of British Columbia, June 1997.

[3]   Huynh, M.D., Hildebrand, L., and Mackey J., "Preprocessing Factors Affecting Quality of Herring Roe," *Technical Report no. 11*, Fisheries Technology Division, BC Research, Vancouver, British Columbia, 1984.

[4]   Rehkugler, J.E. and Throop, J.A., "Apple Sorting With Machine Vision," *Transactions of the American Society of Agricultural Engineers*, Vol. 29, No. 5, pp. 1383-1397, 1986.

[5]   Miller, B.K. and Delwiche, M.J., "A Color Vision System for Peach Grading," *Transactions of the American Society of Agricultural Engineers*, Vol. 32, No. 4, pp. 1484-1490, 1989.

[6]   Dowell, F.E., "Automated Inspection of Peanut Grade Samples," *Proc. of the 1990 Conf. Food Processing Automation*, American Society of Agricultural Engineers, Lexington, Kentucky, pp. 275-280, 1990.

[7]   Okamura, N.K., Delwiche, M.J., and Thompson, J.F., "Raisin Grading by Machine Vision," *Transactions of the American Society of Agricultural Engineers*, Vol. 36, No. 2, pp. 485-492, 1993.

[8]   Sarkar, N. and Wolfe, R.R., "Computer Vision Based System for Quality Separation of Fresh Market Tomatoes," *Transactions of the American Society of Agricultural Engineers*, Vol. 28, No. 5, pp.1714-1718, 1985.

[9]   Strachan, N.J.C. and Murray, C.K., "Image Analysis in the Fish and Food Industry," *Fish Quality Control by Computer Vision*, Pau, L.F. and Olafsson, R., Eds., Marcel Dekker, New York, pp. 209-223, 1991.

[10] Ling, P.P., and Searcy, S.W., "Feature Extraction for a Machine Vision-Based Shrimp Deheader," *Transactions of the American Society of Agricultural Engineers*, Vol. 34, No. 6, pp. 2631-2636, 1991.

[11] Beatty A., *"2D Contour Shape Analysis for Automated Herring Roe Quality Grading by Computer Vision,"* M.Sc. Thesis, The University of British Columbia, December 1993.

[12] Croft, E.A., de Silva, C.W., and Kurnianto, S., "Sensor Technology Integration in an Intelligent Machine for Herring Roe Grading," *IEEE/ASME Transactions on Mechatronics*, Vol. 1, No. 3, pp. 204-215, 1996.

[13] Ramer, U., "An Iterative Procedure for Polygonal Approximation of Plane Curves," *Computer Graphics and Image Processing*, Vol. 1, No. 3, pp. 244-256, 1972.

[14] Upchurch, B.E., "Ultrasonic Measurement for Detecting Apple Bruises," *Transactions of the American Society of Agricultural Engineers*, Vol. 30, No. 3, pp. 803-809, 1987.

[15] de Silva, C.W., Gamage, L.B., and Gosine, R.G., "An Intelligent Firmness Sensor for an Automated Herring Roe Grader," *Int. Journal of Automation and Soft Computing*, Vol. 1, No. 1, pp. 99-114, 1995.

[16] de Silva, C.W., *Intelligent Control: Fuzzy Logic Applications*, CRC Press, Boca Raton, FL, 1995.

[17] Wu, Q.M., de Silva, C.W., Beatty, A., Cao, L., Gamage, L., Gosine, R., and Kurnianto, S., "Knowledge-Based System for On-Line Grading of Herring Roe," *Proc. IEEE Int. Conference on Systems, Man, and Cybernetics*, Vancouver, BC, Vol. 4, pp. 3742-3747, 1995.

13 | INTELLIGENT TECHNIQUES FOR VEHICLE DRIVING ASSISTANCE

N. Lefort-Piat, P. Morizet-Mahoudeaux, and V. Berge-Cherfaoui
CNRS-Heudiasyc Laboratory, Computer Science and Engineering Dept.
University of Compiègne, B.P. 20529, 60205 Compiègne
France

Cars of the next century will contain more electronic components than mechanical parts for improving safety, comfort, and efficiency on the road. Electronic systems are embedded in cars mainly to inform and help drivers. This chapter concerns the development of a driver warning assistant using Artificial Intelligence techniques to help the driver by giving advice during her/his driving or by generating alarms if a situation becomes dangerous. This driver support system integrates an autonomous environment perception device, a real-time decision system, and a dedicated man machine interface. The on-board perception system is used to give information on the dynamic and static environment of the vehicle. Other sensors are used to provide information on its internal state. From the fusion of the set of data, the expert system based copilot analyzes the current situation and generates the appropriate warnings to the driver.

This research program was developed under the Eureka Prometheus project. Our contribution is mainly in sensors data fusion, real-time diagnosis, and control of the vehicle using intelligent techniques.

1 Introduction

Cars of the next century will contain more electronic components. The role of electronics ranges from the low-level controlling of the functions of mechanical parts, such as ignition or braking, to the highest level of information integration, such as regulating the speed of the car or providing the driver with information for driving and manoeuvres. We will present some of the recent applications of intelligent techniques, which are currently under development, for providing drivers with ease of use and efficient driving aid tools.

Imagine a car fitted with a computer whose function is to diagnose the capacity of the vehicle to execute a manoeuvre in safe conditions. It is possible to define different levels of diagnosis corresponding to the description of the vehicle, its state, and its environment. For example, at the lowest level is the internal diagnosis, which is based on the collection and the analysis of the inner data received from sensors. The diagnostic function is then realized by fusion of the data coming from these sensors. At an upper level we can define the external diagnosis. It includes the processing of outer data related to the reception of external stimuli. This kind of diagnosis also includes the evaluation of the dynamic situation of the vehicle and of surrounding vehicles, thus permitting decisions to carry out specific manoeuvres. The uppermost level of diagnosis corresponds to the capacity of the vehicle to execute a manoeuvre which has been decided by other modules, based on the other levels of diagnosis.

Let us consider a simple example. Consider the function of diagnosing the ability of the vehicle to change lanes on a highway, in order to overtake another vehicle. What is required is the diagnosis of the conditions under which the manoeuvre, or a degraded one, may be executed safely. The expected behaviour of the system is as follows. If our car, X, is travelling faster than a car, Y, ahead, in the same lane, the system should 1) identify the obstacle, Y, 2) determine the behaviour of Y (speed, acceleration, deceleration, changing lanes), and 3) evaluate the conditions under which the overtaking manoeuvre (changing lanes, overtaking, and returning to the initial traffic lane) can be performed. It should prevent X from overtaking Y if a third car, Z, is coming faster from behind. In this case, it should send a message to the appropriate module for slowing X down. The system should also send a message to modify the planned manoeuvre if, during the overtaking of Y, Y modifies its behaviour. Finally, it should give an alert if the distance between X and Y falls short of X's minimum braking distance.

A prototype vehicle with its sensors is presented in Figure 1. The video cameras sense the vehicle's position on the road, detect obstacles, and recognize road signals; the laser telemeter measures speeds and distances; the infrared receiver obtains information from road signals, the on-board computer processes signals coming from the sensors and car actuators for performing the diagnostic function; displays messages concerning an intended manoeuvre; and controls the execution of the manoeuvre. The tasks of sensing, processing signals, and controlling are handled by five modules: the first perception module corresponds to the real-time functions of detecting the white lines and signals of the road and giving the vehicle's position in the lanes or the meaning of the signals; the second perception module corresponds to real-time functions for detecting and tracking obstacles; the third kinematics and dynamic module computes the behaviour of obstacles; the fourth module is a human-machine interface, embedded in the dashboard; the fifth module diagnoses the capacity of the vehicle to execute a manoeuvre according to security criteria.

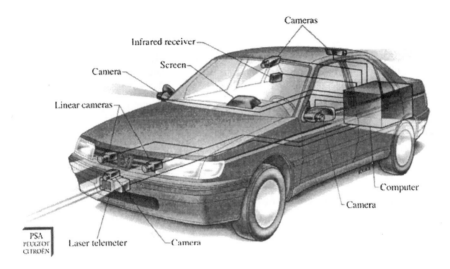

Figure 1: The prototype vehicle Prolab 2.

The required functions can be described in terms of perception, supervision, and information as shown in the functional architecture of Figure 2.

Development and use of these functions and the corresponding intelligent techniques will be presented in the following sections. Section 2 will present sensors and data fusion. Section 3 will present the modeling of the vehicle and its environment. Section 4 will present a real-time expert-system based architecture for the general controlling of the vehicle. Finally a discussion on future trends will be presented in Section 5

2 Multisensor Data Fusion

2.1 Introduction

The problems covered by multisensor data fusion concern diverse domains ranging from the fusion of satellite's images to obtain geographic information to the fusion of ultrasonic data to obtain a map of the environment in indoor robotics. Data fusion involves different techniques such as data association, data selection, and data interpretation. These techniques are largely described in [1]. Methods are chosen according to the application constraints. In intelligent vehicle applications, the objectives of the perception system are to obtain the best possible representation of the environment, as well as of the driver. The constraints in this domain are very high: embedded system, limited power and space, dynamic environment, high reliability, efficiency, and accuracy.

The perception system (sensor and processing) is embedded. The sensors are tied to the vehicle and, thus, they are subject to vibrations and to trajectories effects. It is necessary to know the state of vehicle, the driver intentions, and the static and dynamical environment to have an effective representation of the situation.

The proprioceptive sensors give information about the dynamic state of the vehicle including speed and acceleration. The driver sensors (comodo) allow observation of its intentions and actions. For example, left indicator normally advertises the intention of overtaking. The assistance system focuses on the danger of this manoeuvre.

The exteroceptive sensors give information about the static environment such as road type, lane number, and road signs. They also give information about the dynamic environment composed of the other vehicles, pedestrians, and bicycles.

All these sensors give raw data that must be processed in order to improve their semantic and to decrease their quantity. The data fusion can be defined as a set of techniques for integrating the complexity of the perception system: heterogeneous and/or redundant data, asynchronous data, inaccurate data, and incomplete data, into a common representational format. A distinction is made between low-level fusion (combination of low-level data coming from pre-processing sensor data), feature fusion (combination of features extracted from low-level data), and decision fusion (combination of high level information). Figure 3 presents these distinctions.

The low-level fusion is the combination of data coming from commensurate sensors. This process is the most user independent and it loses a minimum of information; sensor data are fused directly without approximation via feature extraction. Difficulties lie at the level of the association between the different raw-data (including matching, synchronization, and prediction).

The feature fusion is used to construct a representation of an object from different observations of its features. This is a heterogeneous data fusion. During this combination, the matching operation is simplified but, in comparison with the low-level data fusion, information is lost due to feature extraction. Estimation techniques, pattern recognition, and neural networks can be applied at this level.

High-level data fusion takes into account data coming from sensors but also *a priori* knowledge on applications. Human experts can give this knowledge or it is deduced from execution conditions of the applications. At this level, methods allowing combination of symbolic and numerical information are required. A number of methods including fuzzy logic, neural network, case-based reasoning can be used.

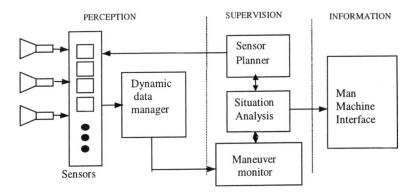

Figure 2: Functional architecture of the demonstrator.

In this part, we describe the perception system, sensor, and data fusion developed in the demonstrator Prolab 2 shown in Figure 1. In the first section, we present the sensors composing the perception system. Then, we describe the algorithms for data fusion at the features and decision level. Particular attention is given to the temporal aspects. The goal is to give the copilot the best image of the real situation. A dynamical image of the vehicle and its surrounding environment is also built.

2.2 Sensors

The copilot needs information to understand and foresee the situation. This information must be as complete and accurate as possible. This specification induces the use of a large number of sensors of various types and technologies.

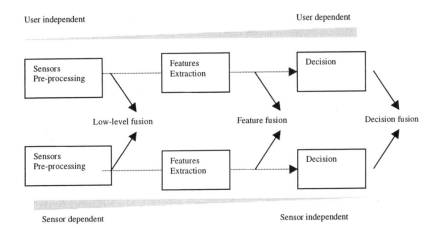

Figure 3: Fusion taxonomy.

Nevertheless, the quantity of devices involved in the perception system is limited because of a set of constraints like space, energy, computing power, and cost.

The environment of the vehicle is the same as a common car. There is no special equipment in the infrastructure. Thus, the system must be able to perceive the environment as well as the actual driver.

The first type of data perceived by the driver concerns the sensations of the movement of the vehicle such as acceleration, braking, turning, and the rate of engine rotation. Some proprioceptive sensors are set in the vehicle for lateral and longitudinal acceleration sensing or evaluating the state of the braking pedal. These sensors give information about vehicle's internal state.

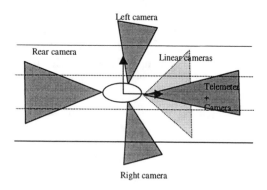

Figure 4: Sensors fields of view.

The driver drives the vehicle on the road according to driving rules. He/she must know the type of road he/she is driving on, what line mark is the border of the lane, and he/she should always be able to center the vehicle in the lane. A CCD camera has been placed beside the rear view mirror to obtain this information.

The driver must also take into account the surrounding environment composed of other vehicles, and the different obstacles like pedestrians, bicycles, and animals. Several sensors have been set in the vehicle to estimate the situation (position and velocity) of the obstacles:

- a 3D sensor, composed of a laser telemeter connected to a CCD camera, placed in the bumper to detect vehicles ahead,
- 2 stereo linear cameras placed beside the lights to detect front and lateral obstacles like pedestrians and bicycles,
- 2 lateral CCD cameras to detect lateral vehicles in a crossroad, and

- a rear CCD camera to detect following vehicles.

Sensor fields of view are presented in Figure 3. All the information is refered to in the vehicle reference.

2.2.1 Static Environment

The front and rear CCD cameras give information on the static environment. The software modules of the front one give the following information:

- The number and the location of the lanes and the type of lines.
- The vehicle's position and orientation according to the lane.
- The special horizontal signs like stop, turn, arrows, pedestrian crossing.

The method consists of an outline extraction and a linear approximation to detect the lines.

The software modules of the rear camera give the following information:

- the structure of the road,
- the lane positioning,
- the type of the lines.

The image is segmented into several interesting areas. Then special marks are searched. The software has been implemented on a special morphologic real-time computer based on an ASIC processor [2].

2.2.2 Dynamic Environment

The perception system gives information about moving obstacles independently from the vehicle motion. Several systems are used.

Telemeter/CCD camera sensor
This sensor is composed of a laser telemeter associated to a CCD camera. It works in two ways as follows:

- Detection mode: the front space is scanned and the relative position of the obstacles is given.
- Focused mode: the laser beam is focused on a specific obstacle and its relative position and velocity is given.

The method consists in partitioning an image in different depth plans corresponding to the different obstacles. The software algorithms are implemented in a transputer-based system named Transvision [3][4].

Stereo linear cameras

This perception system permits detection of all kinds of obstacles in front of our vehicle. It is composed of two linear cameras [5]. The algorithms give the relative position and velocity of the front obstacles. The method is based on edge detection on the two images in order to determine the points of interest, and to match them.

Lateral CCD cameras

The lateral cameras allow detecting the vehicles coming from the transversal roads in a crossroad. The algorithms are based on a spatio-temporal segmentation, then on an interpretation of the dynamical areas. The method is presented in [4].

Rear camera

This camera is used to detect rear obstacles. The algorithms used make it possible to compute the position of these vehicles.

The perception system gives information on the vehicle itself and its static and dynamic environments. However, these data are asynchronous, with a certain variable latency.

2.3 Temporal Data Fusion

It is necessary to know information about the vehicle itself and its static and dynamic environment to give reliable advice to the driver. The perception systems can give these data, but they are asynchronous with a variable latency. The image of the environment must also be easily usable by the supervision modules. The module that manages and fuses the data is named Dynamic Data Manager (DDM) [6]. It comprises several maps representing particular zones of the surrounding environment. Each map is composed of a set of objects that can be static, such as a stop line marked on a particular zone of the road, or dynamic, such as a car.

We describe, in the following, the structure and use of each type of different maps: the infrastructure map, the sensor maps (one for each sensor), the global map, and the copilot map. Figure 5 presents organization of the map.

2.3.1 The Static Environment Perception

The map of static environment allows obtaining information on the situation of the vehicle. The static environment is composed of the road elements, namely, the lane, the lines, the junctions, the stop lines, and the pedestrian crossing. As a matter of fact, the potential danger of an obstacle is different depending on the situation. Thus, according to the different contexts (turnpike, road, and crossroads), we have defined some zones of interest which correspond to the lane or a pedestrian crossing associated to a reference frame. On highway and road, the reference frames of the zones are situated on the lines and move with the vehicle. At junctions, they are situated on the line at the entrance of the junction. In this context, we also associate a zone of interest to the pedestrians crossing.

The infrastructure-map is composed of these objects. So, at any time, it is possible to know the situation of each zone. This means that the map can be updated at any time. The front camera gives information concerning the road markings which allows estimating the real position of the reference frames. It can be the distance and the orientation of the left line or the situation of a stop band or a pedestrian crossing referenced in the vehicle reference frame. In the situation given in Figure 5, the infrastructure map is composed of six zones of interest.

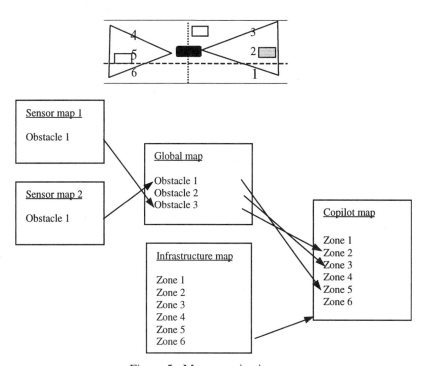

Figure 5 : Map organization.

2.3.2 The Dynamic Environment Perception

The dynamic environment is composed of the other vehicles like cars or trucks, but also bicycles and pedestrians. These dynamical objects, called obstacles, can move by themselves, independent of the movement of our vehicle.

The aim of this module is to survey the surrounding obstacles (seen or unseen). Despite numerous sensors, a blind zone may remain, or different sensor zones can overlap. Moreover, the sensors work asynchronously with time latency. In addition, the information can be required at any given time and must therefore represent the present time situation.

2.3.2.1 Definition of the sensor and global maps

The vehicle's surrounding environment is divided into several observation zones Z_i corresponding to the sensor i observation areas. For each sensor, we define a map called sensor-map (i), filled with the observed object-typed obstacles and updated with its own sampling period and delay time. The given data for each obstacle seen by the sensor are their position and speed. For each obstacle, a Kalman predictor/estimator filter [8] is used to update its state referenced in the reference frame of the vehicle. A possibility index gives information concerning the real existence of the obstacle. Then, in each map, there is a set of $obstacle_j$ seen by the $sensor_i$.

An obstacle k can be seen by two different sensors, and thus, belongs to two different sensor maps, or is unseen by any sensor, because it is in a blind zone. To have a global representation, a global map is composed of the same type of objects (called $obstacle_k$) as those of the sensor maps, but representing all obstacles. Whenever one or more sensors see the same obstacle, the data are predicted at the present time, and fused to update the state of the object of the global map by using another predictor/estimator filter. A logical link is created between the object $obstacle_j$ and the object $obstacle_k$. The inputs of the object $obstacle_k$ correspond to the outputs of the object $obstacle_j$. When the obstacle is unseen, the predictor filter updates the data. This double structure filtering is described in [7].

In Figure 5, there are three surrounding obstacles, thus three objects in the global map. Some links are created between objects of the sensor maps and the objects of the global map.

2.3.2.2 The different steps of the filtering operation

In the prediction step, the current state of the map and the proprioceptive reports are used to predict the future step of the map at the time of the next exteroceptive observation. In fact, the sensor map is always time-stamped with the date of the last exteroceptive observation t. The sampled period of the proprioceptive sensors is smaller than that of the exteroceptive sensors (50 ms and 200 ms). Hence, at each sample time of a proprioceptive sensor, the value of the report is stored. If the new exteroceptive observation is taken at the time $t + \bullet \bullet \bullet$, the prediction equation of the filter is applied to compute map information using proprioceptive data until the time $t + \bullet \bullet \bullet$.

In the *matching step*, the predictions are connected to the newly arrived observations. Matching is done between the list of observed obstacles L_{sensor} and the list of know obstacles $L_{sensor-map}$ projected in time $t + \bullet \bullet \bullet$.

The usual method consists of computing a Mahalanobis distance [1] between the objects of the two lists and to choose the set of couples of objects that minimize this distance. This distance takes into account the value of attributes concerning the relative position and the associated inaccuracy.

For each couple of objects (X_i, Y_j) of the list $L_{sensor-map}$ and the list L_{sensor}, we compute the distance $d_{i,j}$. We search the set of couples that minimize this distance. If the distance is higher than the limit d_{max}, the X_i and Y_i do not correspond. If they are matched, the estimation mechanism can be applied.

The *estimation step* of the fusion algorithm concerns the update of the sensor maps. The reports are combined with the state of the objects using the well-known equations of the Extended Kalman Filter [8].

2.3.2.3 Reliability definition

A reliable index is associated to each object of the global-map and infrastructure-map. This index quantifies the consistence of the object. It is computed by checking its attributes or its behaviour. If they are consistent and the object is regularly seen, the index increases until 1 (meaning being sure of the object existence). If the object is not seen, or its attributes are inconsistent, its reliability index decreases. If a new measurement occurs, a new object is created, and its reliability is initialized at a value depending on the estimate attributes. In ProLab2, a heuristic function is used to evaluate the reliability index according to the relative position of the scene or known object.

2.3.3 The Copilot Mapping

All the obstacles are referenced in the vehicle reference frame and are stored without a special order. This list of obstacles is difficult to use for the supervision modules; the copilot needs to know to which zone each obstacle belongs. Thus, a copilot map is used to place the obstacles in the different zones of interest.

This map is composed of the list of zones of interest. Each one contains the obstacles referenced in the reference frame associated with the zone. To realize this operation, information coming from proprioceptive sensors and from the infrastructure map are integrated (see Figure 5).

In the Kalman filtering theory, when an obstacle is not observed, its uncertainty increases with time. Meanwhile, when an obstacle is in the blind zone or behind another vehicle, it is not seen and it is sure that its position is not in the observable area of the sensors. On a highway, the position of the vehicle is limited by the side of the road and by the boundaries of the visible area. We use this *a priori* knowledge to limit the inaccuracy on the state of each obstacle.

2.4 Conclusion

The map fusion algorithms are implemented on the demonstrator Prolab2. The experimentation has shown the real-time performances of the perception system to produce a copilot map for the supervision modules. The fusion algorithms are adapted to real driving situations (highway and crossroad context). They allow managing the sea of information in space and in time, and are robust with the problem of false alarms.

Actually, we are studying the problems of data association in the process of data fusion. Indeed, the methods using Mahalanobis distance are based only on numerical features. We want to integrate symbolic attributes to describe an object, for example, the color of the vehicle. For this purpose, we are developing other techniques based on the Dempster-Shafer theory methods or based on the fuzzy set theory [9].

3 Vehicle Modeling for Supervision of Manoeuvres

3.1 Introduction

The previous functionality makes it possible to obtain a map of the environment as accurately as possible. Now, from the set of data contained in the copilot map, the decision part deals with the manoeuvres supervision. The manoeuvres supervision integrates an analysis and a diagnosis of the current situation to maintain a safe driving situation. The following sections present the different functionalities of the supervision part and the vehicle modeling developed for planning and monitoring the manoeuvres.

3.2 Supervision of Manoeuvres

The objective of the supervisor is to analyze the current situation in order to alarm or warn the driver when critical situations appear [10], [11]. The supervisor is based on the analysis of the situation and its probable evolution. This makes it possible to detect potential or real risks.

In case of emergency, the supervisor identifies the nature of the danger, its location in space, the danger level and proposes an assistance to the driver which is either an alarm indicating the presence of risk, either a suggestion concerning a possible manoeuvre (overtaking) or an action (brake, slow down) to avoid collisions. The prediction of the future situation allows computing the feasibility or the risk associated to this manoeuvre.

The supervision part of the copilot system can be hierarchically and functionally divided into three modules as presented in Figure 2.

3.3 Situation Analysis

This is to determine the state of the vehicle in its environment and to diagnose the current situation. From data provided by the different sensors set in the vehicle and merged by the dynamic data manager, a diagnosis of the situation is made. This diagnosis is based on knowledge concerning:

the static environment:

- kind of contexts: highway, cross-roads, or road,
- number of lanes,
- road markings;

the dynamic environment:

- presence of obstacles in zones of interest,
- relative position and speed of the obstacles,
- kind of obstacles: car, truck, bicycle, and pedestrian;

the intentions of the driver:

- provided by sensors set in the vehicle, like left or right indicator, steering angle.

This diagnosis integrates a *lateral controller* to monitor the lateral behaviour of the vehicle by checking its position in the lane, and a *longitudinal controller* to monitor the longitudinal velocity of the vehicle with respect to the highway code (speed limit), the current characteristics of the road (curvature), and the presence of obstacles.

The analysis is based on the dynamic control of the vehicle. The vehicle's dynamic model allows us to compute the effective potentialities of the vehicle in terms of acceleration capabilities and braking capacities. The model is a nonlinear model of three degrees of freedom [10].

3.3.1 The Dynamic Model of the Vehicle

In order to develop an effective driving assistance system, it is important to take into account the dynamic behaviour of the vehicle [12], [13]. The definition of this model allows evaluating the capabilities of the vehicle for undertaking a given manoeuvre (acceleration, braking potentialities). A 3-dof model is used to simulate the evolution of the vehicle in real time and to predict its future behaviour. The suspension system is omitted from the 3-dof model, leaving, therefore, the longitudinal speed (u), the lateral speed (v), and the yaw rate (r) relating to the reference vehicle to be considered.

The fundamental dynamic equation, the action and reaction forces, the small angle assumption, and the cancellation of the angle product are all required when establishing the 3-dof model. The small angle assumption restricts the manoeuvres that could be simulated by this model to those with small steering and sideslip angles such as lane change manoeuvre.

$$\dot{u} = \frac{1}{M}\left\{T_a - T_b + M.v.r - M.f.g + C_f.\delta.\frac{v}{u} + a.C_f.\delta.\frac{r}{u} + u^2.f.(k_1 - k_2)\right\}$$

$$\dot{v} = \frac{1}{M}\left\{(T_a - f_b.T_b).\delta - M.u.r + C_f.\delta - (C_f + C_r)\frac{v}{u} + (b.C_r - a.C_f)\frac{r}{u}\right\}$$

$$\dot{r} = \frac{1}{I_z}\left\{a.(T_a - f_b.T_b).\delta - M.f.h.u.r + a.C_f.\delta\right.$$

$$\left. -(a.C_f - b.C_r)\frac{v}{u} - (b^2.C_r + a^2.C_f)\frac{r}{u}\right\}$$

The several nonlinear forms which appear in this model result from the front wheel driving, braking and steering, the Coriolis acceleration, the drag and the lift. The lateral forces are nonlinear when a longitudinal speed variation is considered.

This model is used to compute, at any time, a set of parameters that details information concerning vehicle safety. A safety parameter is a datum that informs about the border of a safe behaviour of a situation. The situation analysis module considers these boundaries when evaluating the safety of a situation or its danger level. The vehicle's dynamic and the safety parameters estimation detailed in [13] must be accurate enough to foresee the safety area boundaries evolution.

3.3.2 Vehicle Following

An important part of the analysis concerns the verification of the safe distances between the vehicles. The safe distance is the distance that must be maintained by the driver to avoid a collision with an obstacle.

When vehicles are detected in the vicinity of the reference vehicle, safe distances are computed and compared with the current distances between these vehicles. If the safe distance is not respected, an alarm is sent to the driver to suggest her/him to brake.

Consider a particular situation such as given in Figure 6. In this situation, there is a vehicle Va in front of the reference vehicle Vr. The safety distance between the two vehicles is computed as follow:

$$D_s = D_r(V_r) + D(V_r, V_a, F) + D_c - D(V_a)$$

where

- D_s is the safety distance,
- $D_r(V_r)$ is the distance covered by the vehicle before the driver reacts to an event,
- $D(V_r, V_a, F)$ is the distance covered by the reference vehicle until it reaches the speed of the previous vehicle, with a braking force F. The distance covered is deduced by integrating speed. This latter is deduced by integrating the equation of the vehicle's longitudinal dynamic when applying one of the braking forces.
- D_c is a constant distance imposed between the two vehicles (if they are at the same speed),
- $D(V_a)$ is the distance covered by the front vehicle during the reaction and braking time.

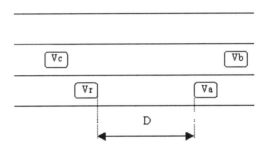

Figure 6 : Vehicle following.

3.3.3 Highway Access Manoeuvre

Several manoeuvres require knowledge about the time and distance covered by a vehicle in order to reach a given speed, like, for example, during highway access. These data deal with the engine power and reaction forces.

The available engine power and the traction force are first required. The available power is the difference between the maximal power (determined by the engine abacus) and the resistance power (determined from the reaction forces at the speed of the vehicle: the drag, the rolling resistance forces):

$$P_{available}(u) = P_{max\,imum} - P_{resis\,tan\,t}(u)$$

The tractive force available is then evaluated from the power available (the gearbox ratio is assumed to be constant).

$$T_{available} = \frac{P_{available}}{u}$$

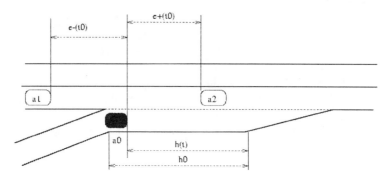

Figure 7: Highway access manoeuvre.

The vehicle's longitudinal dynamic is then considered with the available tractive force to evaluate the time and the distance until the vehicle reaches the reference speed.

The processing of insertion consists in determining a time law and the duration t_f of the insertion. The result is successful if

- the velocity of vehicle a_0 is such as $v_0(t_f) = v_1(t_f)$, where v_1 is the mean speed of the future lane,

- the access lane is long enough $h(t_f) > 0$,

- the chosen free space area is long enough such as between $[t_0, t_f]$:
  $e^-(t) + e^+(t) > d_{min}$ (d_{min} being the merging minimal distance).

3.3.4 A Lane-Changing Manoeuvre

The lane-changing or junction manoeuvres are sensitive as they appear to involve the most demanding sensor requirements.

The lane-changing safety is diagnosed by checking some conditions

- The left line is not a continuous line,
- If the manoeuvre is possible; the time and the distance needed to execute the manoeuvre are computed for this purpose.

The time and longitudinal distance to pass from one lane to another are estimated by the use of control methods, namely, the input/output linearization. They are considered only if the manoeuvre can be performed in a safe manner, i.e., if the steering angle and the sideslip angles respect the assumptions used to establish the 3-dof model.

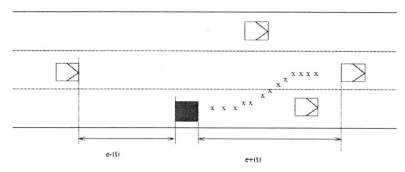

Figure 8: Lane-changing manoeuvre.

To estimate the two parameters, the behaviour of the vehicle is simulated using the dynamic model and assuming that obstacles have a uniform movement. Using these data, the *Situation Analysis* diagnoses if the manoeuvre is possible, by checking if the safety distances, $e^-(t)$ and $e^+(t)$, with the vehicles placed on the lane are respected at the beginning and end of the manoeuvre.

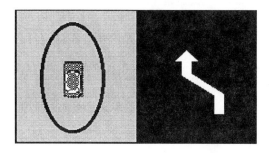

Figure 9: Overtaking is possible.

If this test is true, the *situation analysis* activates the *manoeuvre monitor* and the *danger controller*.

3.4 Manoeuvre Monitor

The *Manoeuvre Monitor* determines whether a desired manoeuvre is feasible with respect to the current situation of the vehicle and the surrounding world [14], [15]. If yes, the driver is informed (see Figure 9) the best way to perform the manoeuvre. Since the world's state can change during the course of the manoeuvre, the monitor has to control its execution by checking whether it remains feasible under current safety conditions. This is checked every period δt where δt is the response time of this module.

Figure 10 depicts the architecture of the monitor; it is composed of three parts: the manager, the manoeuvre planner, and the execution simulator.

The *manager* is the main module; it activates the two other modules and communicates information to the message manager. First, it activates the manoeuvre planner with the task of analyzing the possibility of performing the current manoeuvre, and it determines the most feasible way. The resulting messages are transmitted to the message manager. The planner generates a time-optimal trajectory for the vehicle to perform this analysis. This trajectory, called the nominal plan, is collision-free with the stationary and moving obstacles, and satisfies the dynamic constraints of the vehicle. It is made up of a geometrical path and a profile of velocity (the reader is referred to [16] for a presentation of the manoeuvre planner).

Since the nominal plan is generated using the current model of the world and a prediction of its future evolution, we have to check, at every instant, whether it is still possible to carry out the nominal plan. The execution simulator does this.

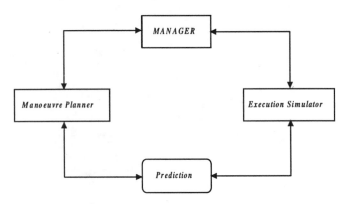

Figure 10: Manoeuvre monitor structure.

Basically, the simulator takes into account the current situation of the world at time t and a prediction of the future evolution of the world. This means that it takes into account the situation resulting from the driver's actions, which may be different from what the copilot has suggested. The result is a simulated plan representing the simulation of the execution of the nominal plan by the vehicle, regarding the current situation.

The manager, which decides if the manoeuvre is still possible, then processes the analysis of the resulting simulated plan by the driver. The diagnosis is then sent to the message manager. If it remains possible to carry out the manoeuvre, the message is "okay." Otherwise, two solutions are considered. First, the driver can continue the manoeuvre despite the new situation, but it has to do it in a certain way (by slowing down, stopping or accelerating). The driver is then informed. In the second solution, the driver cannot carry out the manoeuvre. In this case, the manager re-invokes the manoeuvre planner to find a new solution.

The techniques used by the manoeuvre monitor are based upon an architecture which is made up of behavioural rules combined with a potential field approach [14], [15].

The *Manoeuvre Monitor* checks the validity of the plan by analyzing continuously the current situation and adapting or invoking the trajectory planner if events occur during the execution of the plan [15].

3.5 Danger Controller

The role of the *danger controller* is to detect all dangerous events during the manoeuvre execution (reactive behaviour), whereas the role of the manoeuvre monitor is to give assistance to the driver (cognitive behaviour).

The *danger controller* is activated periodically every δt seconds (200 ms) to determine the level of danger during the manoeuvre and to warn the driver, if needed. The level of danger is computed by evaluating the safety distance between the reference vehicle and surrounding obstacles [11].

The *danger controller* module integrates the results given by the dynamic data manager and those of the dynamic characterization of the vehicle. It consists in realizing a classification of the obstacles according to their risk, verified at each moment, in the safety zone around the vehicle.

As soon as an obstacle is too close to the reference vehicle, this obstacle is considered as a dangerous obstacle and a message is sent to the display manager, which displays the corresponding icon. For example, in Figure 11, the rear obstacle intercepts the safety zone, so it is declared dangerous obstacle and a 'red' ellipse will be displayed on the screen with a deformation indicating the location of the risk for the vehicle (see Figure 12).

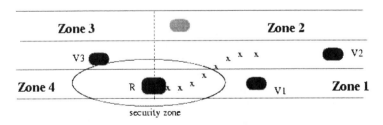

Figure 11: Security zone.

3.6 Requests Generation and Sensors Planning

The *situation analysis* module generates requests to the *sensor planner* to obtain more precise or relevant information on the environment. A virtual sensor is associated to each region of interest. The goal of a virtual sensor is to provide a specific kind of information about the content of a region like presence of obstacles, arrows, road works, etc. A virtual sensor can be associated to physical sensors and/or estimation algorithms (for blind zones).

Depending on the current situation, the *situation analysis* module decides which virtual sensors must be activated and their functioning mode, and it transmits its decision to the *sensors planner* as a request.

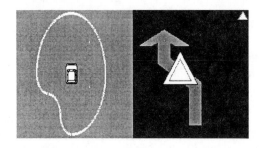

Figure 12: IHM Message: left rear vehicle dangerous.

Four kinds of virtual sensors have been defined:

1. *obstacles-DETECTION* to know if an obstacle is present or not in a specific zone of interest;

2. *obstacle-FOCUSING* to obtain, for a known obstacle, the accurate relative speed and position;

3. *obstacle-OBSERVATION* to obtain information on the obstacle concerning its shape, the state of its indicator, etc.

4. *road-OBSERVATION* to know if specific road markings exist like pedestrians crossing, stops lines, etc.

The *sensor planner* supervises the selection and execution of perception programs. From the requests sent by the decision level, it translates them into a series of low-level commands to accomplish the desired objective [17]. The selection of perception programs depends on the type of needed information, the current context (highway, two-way road, intersection), and the accuracy and response-time of vision programs.

In Figure 11, the reference vehicle R is in the right lane and wants to overtake the vehicle V1. In this situation, the Situation Analysis (SA) task

- generates requests to the Sensors Planner to ask information concerning the front vehicle,

  1. if the left indicator is on, *Request*: *obstacle-OBSERVATION, zone1, left indicator*

  2. what is its relative speed, *Request*: *obstacle-FOCUSING, zone1*

- makes a diagnosis concerning the lane-changing manoeuvre; it computes the time and the distance necessary to overtake and checks if the distances with the other obstacles allow engaging such manoeuvre.

- detects that the rear vehicle is close to the reference vehicle and thus sends the message "rear vehicle dangerous" and "overtaking is not possible."

The Human Machine Interface receives the messages and displays the corresponding icons such as shown in Figure 12.

3.7 The Driver Information Level

This information level is responsible for the interaction between the supervisor and the driver [18]. This module has to filter the messages generated by the supervisor to give the most appropriate information to the driver according to the selected mode. The driver can choose between three information modes: *alarm mode* is an emergency warning level; *advice mode* is a decision support interface intended to help the driver to undertake some manoeuvres such as overtaking (qualitative aid); *assistance mode* is the last step between copiloting and automated piloting. The system provides the driver with the best action to undertake regarding security parameters (quantitative aid, i.e., speed, timing).

3.8 Conclusion

The functionality of the decision part allows monitoring the behaviour of the driver-vehicle in order to maintain a safe driving situation. This functionality requires managing a large quantity of information (static and dynamic environment, driver intentions, and vehicle capacities). On the other hand, the monitoring of the driving situations implies respecting driving rules and reacting in real time.

In order to cope with these constraints, the expert system shell named SUPER developed in our laboratory is used. This expert system shell, SUPER, and its features are presented in the following sections.

4 On-Board Real-Time Expert System for Control of the Vehicle©

4.1 Introduction

The supervisor diagnostic system operates in real time. It should also be able to integrate a huge quantity of heterogeneous information. At the first level this information may not always correspond to precise, accurate, or reliable data. At the second level it is necessary to develop different kinds of mode developing models are developed to represent dynamic behaviour of the vehicle. These models are based on a cognitive representation of the driver's behaviour. All this information is integrated for decision at the third level. However, an exhaustive description of all situations and states, if available, would generate a combinational explosion. One of the main objectives of knowledge based systems was to find a means for managing the combinational aspect of the problem without the help of exhaustive tests algorithms. Non-exhaustive description relies on the capacity of a human expert in describing either the local behaviour of the physical system or the knowledge which is used for describing this local behaviour.

Conversely, it is not possible to have a deterministic algorithmic description of all modules describing the behaviour of the parts of the vehicle and its surroundings. Such a system involves interaction among its components or interaction with the environment. For example, if a laser radar were used to scan for obstacles, its setting should be changed from wide scanning to narrow focusing, in order to change its function from detecting an obstacle to tracking the obstacle. Based on the diagnosis deduced from the initial information coming from the radar, the system should then modify the setting of the radar according to the modification of the environment. It could then run a specific module to compute the dynamic data describing the behaviour of the obstacle.

The diagnostic module is a master module. It gives instructions to the other modules for setting their functioning modes and asks these modules for complementary information, if necessary. However, each module is autonomous. The role of the diagnostic module is to coordinate the other modules and to determine the vehicle's state, to evaluate the situation's danger level and to propose to the driver an adapted manoeuvre, via the user interface. The diagnostic module uses an expert system approach which is presented in the following sections. The architecture of our proposed system is presented in Figure 13.

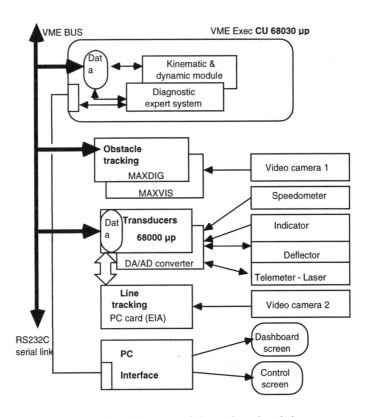

Figure 13: General architecture of the on-board real-time system.

4.2 Development of an Expert System for Control

The real-time expert system based on our expert system development environment SUPER has been developed. SUPER was developed for creating expert systems for maintaining and updating information about an evolving system as its state changes due to abnormal events such as faults, or as a consequence of external actions taken upon the system. It is equipped with a knowledge acquisition module to structure and compile the knowledge base as an AND/OR network. It runs with a looped inference engine which (1) selects an appropriate sequence of actions to be applied to the system in order to change its states, and (2) updates the database to reflect a change of state in response to an external action. It permits development of a multiple knowledge-based structure managing quantitative and qualitative variables, and propagating their domain constraints in the AND/OR networks of rules. The created expert system can finally be implemented in a real-time multi-tasking operating system and assigned the task of managing rule interruptions, asynchronous event or input handling, multi-tasking, and temporal reasoning.

This section presents the general characteristics of building the expert system for control. The real-time development is presented in the next section.

4.2.1 Building the Knowledge Base

The important requirement of our approach is the reliable rule base. This is possible only if our assumption holds that the controlled process follows a logic model. By "logic model" we mean that the same cause always generates the same effect. This is the case for most engineering systems, as long as we can cope with noisy signals. A solution to this problem is to develop knowledge sources which are specialized in invoking appropriate filtering techniques or degraded forms of reasoning such as incomplete or default reasoning. We'll discuss an example based on *algebraic constraints* later.

An example of a logic model is as follows. Whenever a car's engine does not start, the cause might be a lack of gasoline or an ignition fault. In the case of lack of gasoline (see Figure 14), this could be due to no fuel injection or an empty tank. The injection might not work because either the pump is broken or the battery is dead.

As this example shows, the elements with which the system must deal are logically linked to each other. The strategy, then, is to build these links at the time of knowledge acquisition. Once organized in the form of a network, the detection of loops, redundancies, contradictions, and inconsistencies between rules is almost straightforward. This permits the development of algorithms for keeping the knowledge base consistent as it incrementally increases.

As soon as the number of rules becomes large, building the network of rules is beyond the scope of a human operator. The knowledge acquisition module of SUPER helps in building the large rule base. For the sake of generality, we describe the components of the rules as pairs (**Item, Domain**), where **Item** is the descriptor of an entity such as temperature, pump, or battery charge, and **Domain** is the corresponding domain of variation such as [**-20°C., 45°C.**], {**dead, ok**}, or {**low, medium, high**. Rules are as follows:

IF <antecedents> THEN <conclusion>,

where **<antecedents>** is the conjunction of premises (**Item, Domain**) and the conclusion contains one such pair. SUPER then builds the AND/OR network of rules by linking two rules such that the conclusion of the first and one premise of the second have the same pair (**Item, Domain**) (see Figure 14). Whenever they contain a common item but have a different domain, the rules are not linked, but the algebraic domain constraints of the common item are computed, which we'll discuss later.

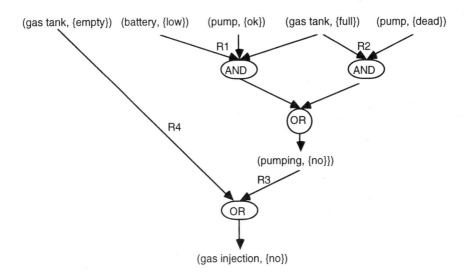

Figure 14: Example of a simple network. (The rules R₁ and R₃, and R₂ and R₃, are linked by the pair (pumping, {no}) which is a premise of R₃ and the conclusion of R₁ and R₂. The pair (gas tank, {full}) is a mandatory antecedent of (pumping, {no}), while (battery, {low}) is a possible antecedent).

We want to be sure that this network is logically consistent. In other words, linking rules do not involve contradictions, loops, or redundancy. To define this logic we must know the set of pairs of the network, which are related to any given pair. This set is called the antecedents set. We have defined two kinds of sets: the *mandatory antecedents* set and the *possible antecedents* set. We define a path from p_1 to p_n as any sequence of n linked rules R_i such that p_1 is a premise of R_1 and p_n is a conclusion of R_n. A pair's mandatory antecedents appear on any path of the network that leads to that pair, whereas the possible antecedents appear on at least one path. We've defined and implemented specific operators to compute these sets for each pair as the knowledge base incrementally increases during knowledge acquisition. We define the network's logical consistency as follows. Two pairs are *consistent* (inconsistent) if either of their **Items** are different or their **Domains** intersect. Two pairs are *inconsistent* if they have the same **Item** and their **Domains** do not intersect. This definition extends to sets of pairs, and the consistency of the network is evaluated by comparing the mandatory antecedents sets of pairs which are linked by rules.

The network structure allows a high level of integration for the internal representation of the information about the links between the rules and the propositions. It is close to a compilation of the network. For instance, it is easy to determine, for any rule, the number (i.e., the pointers) of the rules that are needed to prove its premises. It is also easy to know where each proposition appears as a premise. The knowledge base is thus a compiled network constituting a linked list that can be quickly scanned. This is the basis of the system's good response time.

4.2.2 Basic Functioning of the Expert System for Control

The control of an evolving process imposes three requirements. First, it is necessary to define a particular model to represent the changes of state of the controlled system in response to external actions. Second, the inference procedure of the expert system in charge of controlling the process must run continuously. Third, the system must update the database containing the parameters for modeling the successive changes of state of the process in response to external actions, and must keep that database consistent as data change.

Representing change of state. Let's consider the simple example of the function of controlling the use of the car's left-and right-turn indicators at an intersection, depending on the driving conditions. The system should check if the left (or right) indicator is turned on in accordance with a change of direction. (The image-processing module can inform the expert system of a change of direction, after the video camera detects a junction and signals marked on the road indicate a right or a left turn). It should then check if the used indicator corresponds to an authorized turn. If that is not the case (for example, if only a left turn is authorized, but the right indicator has been turned on), it should send the appropriate message to the user. If the car is going to move in the authorized direction, but the lateral sensors recognize a car coming across, then the system should also notify the user.

The primary aim of state-change rules is not to represent the changing behaviour of the process, but rather to represent the successive logical steps that the process will encounter if 1) a state, called initial, is encountered, and 2) an action allowing a state change occurs. The only things that must be represented are the initial state, the final state, and the trigger that makes the system change its state. Considering the example above, let the initial state be (**left-indicator, off**). If the system or the driver turns the indicator on (the state-change action), then the final state should be (**left-indicator, on**). It is important to note that the action does not (and should not) belong to the sequence used by the expert to infer that state. Effectively, given some state of the process, the action that makes the system evolve corresponds to the expert's reasoning. For example, the expert's reasoning is

> **IF junction AND turn-left**
> **THEN I must put the left-indicator on**

The corresponding rule is in the network representing the knowledge base. On the other hand, the state change rules represent the sequence that models the change of state given the initial state and the action taken. For example,

For example, let us consider a given KBS which contains a rule R_1;

if (a, true) and (b, true) then (c, true),

and the data base;

{(a, true), (b, true)}

after the execution of R_1, the database becomes

{(a, true), (b, true), (c, true)}

If the value of b changes from true to false after the inference of (c, true) by the IP according to R_1, the data base consistency maintenance process of SUPER will remove (c, true) from the data base since it is not consistent with the new value of b [13]. The database then becomes

{(a, true), (b, false)}.

On the other hand, if the value of b changes after the selection of R_1 by the IP, but before its execution, the consistency maintenance process will not find any contradiction. If the IP simply resumes at this point, it will infer R_1 and therefore insert (c, true). The database will finally be

{(a, true), (b, false), (c, true)},

which is not consistent.

The best solution would be to allow insertion of data only after the completion of the IP. However, this might take too much time and further degradation of real-time performance might not be warranted. Our solution breaks the IP into operations that are as atomic as possible (an atomic operation is, by definition, indivisible). Consequently, the reasoning can be interrupted safely between these operations. The insertion of the new data and its consequences are taken into account by the updating algorithm of SUPER. This algorithm tests the consistency of a new datum with all existing data in the database, and checks for its integrity with all previously used rules. The database is updated accordingly. For simplicity's sake, we'll now call this algorithm the *resume algorithm*.

A multitasking environment implies that several tasks might have to communicate information to the KBS at the same time. Because an interactive real-time system is primarily driven by external data inputs, the system should recognise the inputs even if they occur during an IP atomic operation. We use the mailing functions of the multi-tasking kernel to record such information. Between each atomic operation, the IP inspects its mailbox. If new data are present, they are inserted and the resume algorithm is set. The major advantage of using the mailing facilities is that we can define different strategies according to the workload. Such strategies include priority levels attributed to the mail, thresholds defining the maximum number of pending mail, and reduction of dataflows by filtering techniques.

Until now, we have assumed that the IP runs continuously. In fact, there may be intervals where nothing occurs. During these intervals, the IP need not be active. To account for this, we've implemented three control strategies for the IP: reflex activation (data-driven), periodic activation (temporally driven), and focus of attention.

4.3.3 Control Strategies for the IP

Reflex activation is an IP activation in response to an event, which is different from the interrupt mechanism. In the situation considered here, the IP is inactive when a given task needs to pass some information to the IP. We have developed several methods for reactivating the IP. The strategy described here can be considered as the most significant one. It is based on the multitasking kernel mail services.

Any task which is waiting for a mail is inactive. It can automatically be reactivated by the mail services as soon as a related mail occurs. An inference task (T_i) is scheduled when the IP is over. Its duty is to wait for new mail. Another task, (T_a), needs to send a related event, to wake up the IP by sending its information to T_i (see Figure 16). This strategy works whatever the status of the IP, because the task T_a does not have to test the IP activity before sending its mail. If the IP is active, the mail will be read at the end of the current atomic operation; in the other case, it will work as shown in Figure 16.

a) The inference task is active: between each cycle, the IP inspects its mailbox. When the IP ends, the inference task becomes inactive.

b) The inference task is inactive: it is waiting for a mail. When the mail is delivered, the task is reactivated.

For *Periodic activation*, the IP analyzes specific rules at a given sample rate and infers them whenever their conditions are satisfied. Such rules are commonly called *scanning*, *time-triggered*, or *temporally driven* rules. The multitasking functions allow the definition of cyclic tasks driven by a clock-interrupt mechanism. Thus, cyclic inference tasks are likely to cause consistency problems if they interrupt another inference task running on the same database. This can be managed either by using the mailing facilities or by interrupt handling, which we'll describe below.

Focus of attention must be set according to periodic activation. Effectively, because periodic activation concerns only specific rules, the IP must be able to focus on these rules. This control strategy differs from the previous ones since it is not used to activate the IP, but to temporarily drive its activity on a specific research space. For this purpose, the strategy splits the knowledge base into logical subsets. We define meta-rules such that the action of a given meta-rule of a subset of rules, S_1, drives the investigation on another subset of rules, S_2. The method considerably reduces the IP response time because it does not have to examine the entire rule set. Another benefit of partitioning the knowledge base is that, during the processing of a subset, the insertion of a datum not related to this subset will not make the system change its strategy. In this case, the resume algorithm will not be invoked, thus saving time. The drawback lies in the fact that the IP may miss an important event. It is possible to cope with this problem by linking a monitoring routine to the datum representing the event. This routine is called each time this datum changes. It is therefore being executed when the datum is inserted into the database. Its task is

i)   to set the current rule subset to a subset dealing with the datum,
ii)   to create and start a second IP on the new subset,
iii)   to go back to the first process.

The second IP is not memory consuming since the IP is reentrant. Furthermore, if the second IP inserts data relating to the rule subset of the first IP, the first IP resume algorithm will take into account this new information.

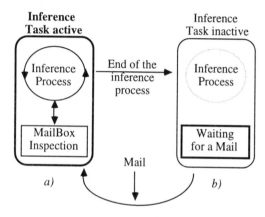

Figure 16: Description of two different states of an inference task.

4.3.4   Interrupt Handling

Asynchronous urgent external events constrain the KBS to temporarily suspend its activity and analyze the associated rules. Interruption of the IP might cause inconsistency problems during recovery. On the other hand, the IP can be safely reset to its beginning. The interrupt-handling procedure

(1) sets the inference process task as inactive,
(2) inserts the piece of information related to the interrupt into the data base,
(3) creates a second inference task and drives this task's IP on the rule subset that is known to be related to the interrupt,
(4) sets this task as the next task to run, and
(5) issues a call to the scheduler to resume the handling task.

The scheduler then selects the new inference task, which executes instead of the interrupted inference task. When the new task terminates, it resets the previous inference task. Thus, its state is set active and the scheduler can select it. When selected again, this previous task will resume its reasoning from the outset.

This method emphasizes the urgency of the event that has interrupted the KBS. By focusing the new IP on the rules associated with the interrupt, the KBS need not determine the event's level of significance according to the current reasoning state or look for the rules to analyze. On the other hand, after completion of the interrupt processing, the KBS can draw its consequence, using the resume algorithm.

4.3.5   Temporal Reasoning and Multiagent KBS

The use of a real-time multi-tasking environment allows multiple settings of the KBS. Access to the kernel timing facilities makes it possible to express temporal relations.

A real-time KBS must continually refer to what has happened, is happening, and might possibly happen. Moreover, it must cope with the problem of interacting with a constantly evolving world. For planning, it must be able to predict change, propose and trigger actions on the basis of these predictions, and notice when certain predictions are no longer warranted. All of this requires an enormous amount of complex independent information. Nevertheless, the particular application domain makes it possible to define a simpler representation of time that fulfills some of the primary requirements without losing the KBS response time efficiency.

Rather than deal with a general temporal logic, we can restrict the reasoning requirements to express temporal relations as the position of events on a time axis (before, after, etc.). To express these temporal relations and to realize a multiagent KBS, we use the multitasking real-time kernel. The kernel's monitoring routine mechanism manages temporal relations (see Table 1).

For example, for the duration analysis, we've developed the use of an agenda as follows:

(1) The agenda contains control facts which prevent the execution of rules with time parameters.
(2) The actions in the agenda are sorted chronologically and execute only if the clock corresponds to the time parameter.

With this strategy, the IP does not have to reconsider such rules at each clock step.

For example, consider the expression

"if the sensor is faulty for more than 30 seconds then ..."

it may be expressed within a rule as

IF (sensor, faulty) AND (time$_{\text{faulty-sensor}}$, out$_{>30}$) THEN ...

where **(time$_{\text{faulty-sensor}}$, out$_{>30}$)** corresponds to the control fact. Each time the sensor is faulty, the monitoring routine linked to the sensor puts instructions into the agenda for inserting **(time$_{\text{faulty-sensor}}$, out$_{>30}$)** in the database thirty seconds later. If, meanwhile, the sensor is not faulty any more, the monitoring routine will delete this action from the agenda. Otherwise, **(time$_{\text{faulty-sensor}}$, out$_{>30}$)** will be inserted into the database and the corresponding rule will be put into the work space.

Table 1: Temporal relations

	Relations	Examples
(1)	Comparison with current time	t. is less than 3 hours 10 minutes 25 seconds
(2)	Duration analysis	The sensor has been faulty for 30 seconds
(3)	Temporal correlation	Event e_j follows e_i within a time interval shorter than 10 seconds
(4)	Frequency analysis	e_i occurs x times during a given interval

The other temporal relations have been implemented by the monitoring routines, thus allowing selective dating functions as well as timing computations.

The multitasking environment makes it possible to run several KBS at the same time. This means that several domains of the whole research space can be explored concurrently. The advantage of this solution is that the knowledge base can be split into specific topics which may be analyzed only if needed. Moreover, when these topics are studied, they can be analyzed at the same time as the others, according to their priority level. The whole IP can then be considered as parallel.

The drawback of such a realization is the difficulty of checking the consistency of the multiple databases and knowledge bases. We overcame this difficulty in two different ways. The simplest is to logically split the knowledge base into several rule subsets and to link a separate IP to each subset. In this case, the consistency maintenance algorithm can work on the entire knowledge base and therefore does not need to be modified.

The second solution is to set up an architecture where communication between the different agents relies on information mailing instead of data sharing. In this configuration, the agents are totally independent, each using its own database. Therefore, the consistency maintenance algorithm has to be modified accordingly.

4.4 Conclusion

The diagnostic module is based on two expert systems. One system controls the vehicle's state and surroundings, and contains 120 rules. The second handles the safe execution of the manoeuvre, and contains 70 rules.

We tested the system in approximately 20 real situations on the road. Let us present, for example, the actual behaviour of the system for controlling the ability of the vehicle to change lanes or overtake another vehicle. Photos taken from the prototype vehicle X (see Figure 17) show the messages projected on the dashboard screen according to the state of the manoeuvre and the surroundings during an overtaking manoeuvre. The screen reproduces the vehicle's position with regard to the road. X cannot overtake the vehicle Y, ahead in the same lane, because the left lane is not free (not seen on the photo is a car coming from behind and detected by the video camera). X must keep sufficiently far behind Y and, if the driver does not brake, the system might brake automatically. The speed of the cars during the tests ranged from sixty to a hundred kilometres per hour.

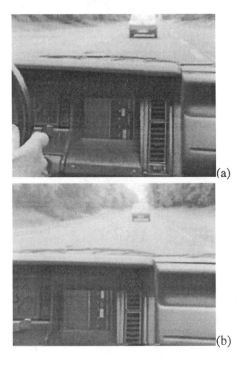

(a)

(b)

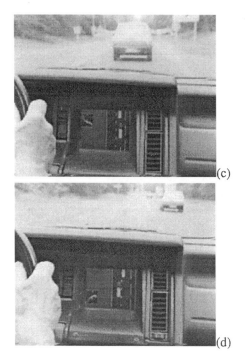

(c)

(d)

(e)

Figure 17: Photos taken from vehicle X during an overtaking manoeuvre: (a) the system detects a vehicle running more slowly; (b) the distance between the two vehicles is safe; (c) the overtaking manoeuvre is authorized; (d) the overtaking manoeuvre must be interrupted; (e) the distance between the two vehicles is too short.

5 Summary

Prolab 2 has been effectively demonstrated using real experiments. This allowed us to optimize the global behaviour of the copilot. Intelligent techniques, such as expert systems and data fusion, have proven their capabilities in such difficult situations as road driving.

The current work proposes to analyze driver-vehicle behaviours, or driver vigilance, according to driving situations. Our research concerns the development of new perception systems adapted to the automobile domain. It also concerns the design of new methodologies for developing real-time embedded applications and implementing them on heterogeneous hardware architectures.

Notation

u	Longitudinal speed
v	Lateral speed
r	Yaw rate
T_a, T_b	Braking and acceleration forces
δ	Wheel steering angle
a	Distance between the center of gravity and the center of the front wheel
b	Distance between the center of gravity and the center of the rear wheel
f	Rolling resistance ratio
g	Universal gravitational
$K_1 = 0.5.\rho.S_x C_z$	
$K_2 = 0.5.\rho.S_x C_x$	
M	Total mass
C_r	Rear spring roll stiffness
C_f	Front spring roll stiffness
I_z	Moment of inertia in relation to z axis

References

[1] Hall, D. L. (1992), Mathematical Techniques in Multisensor Data Fusion, *Artech. House.*

[2] Beucher, S. and Bilodeau, M. (1994), Road segmentation and obstacle detection using a fast watershed transformation, *Intelligent Vehicles'94*, pp. 296-300.

[3] Gallice, J., Collange, F., Alizon, J., and Trassoudaine, L. (1992), A camera/telemeter multisensory system for tracking and obstacle detection, *Proc. of 25th Int. Symposium on Automotive Technology and Automation*, pp. 147-154.

[4] Bellon, A., Derutin, J. P., Heitz, F., and Ricquebourg, Y. (1994), Real-time collision avoidance at road-crossings on board the Prometheus-Prolab2 vehicle, *Intelligent Vehicles'94*.

[5] Postaire, J. G. and Burie, J. C. (1993), A new edge matching procedure for obstacle detection by linear stereo vision, *IEEE Intelligent Vehicles Symposium'93*, pp. 414-419.

[6] Rombaut, M. and Eleter, B. (1993), On board real-time assistance driving system architecture, *Road Vehicle Automotion ROVA'93*, pp. 134-143.

[7]   Meizel., D. and Rombaut, M. (1994), Dynamic Temporal Multisensor Fusion in the Prometheus Prolab2 Demonstrator, *Proc. of the IEEE Int. Conf. on Robotics and Automation.*

[8]   Bar-Shalom, Y. and Fortmann, T. E. (1988), Tracking and data association, *Mathematics in Sciences and Engineering.* Vol. 179, Academic Press, 1988.

[9]   Rombaut, M. and Cherfaoui, V. (1997), Decision making in data fusion using Demster-Shafer's Theory, 3rd *IFAC Symposium on Intelligent Components and Instruments for Control Applications*, pp. 375-379.

[10] LeFort, N. and AbouKhaled, O. (1994), Real-time Diagnosis of Traffic Situations, *Proc. of the AIRTC Symposium on Artificial Intelligence on Real-time Control.*

[11] AbouKhaled. O., Hadi. R., and LeFort-Piat. N. (1995), A Real-time Supervision of a Vehicle in a Dynamic Environment, *Proc. of the IFAC Intelligent Autonomous Vehicles.*

[12] Rombaut. M. and Alloum. A. (1993), A Safety Indicator System for Driving Assistance, *In Road Vehicle Automation ROVA'93*, pp. 144-155.

[13] Alloum, A. (1997), Vehicle Modeling Aspects for an Embedded Real-time Application, *Vehicle System Dynamics*, pp. 187-219.

[14] Hassoun, M. and Laugier, C. (1992), Motion Control of a Car-like Robot: Potential Field and Multi-agent Approaches, *Proc. of the Int. Workshop of Intelligent Robots and Systems.*

[15] Hassoun, M. and Laugier, C. (1995), An Architecture for Planning and Control the Motion of a Car-like Robot, *Int. Journal of Computers and Mathematics with Applications.*

[16] Fraichard, T.H. and Laugier, C. (1994), Path-velocity Decomposition Revisited and Applied to Dynamic Trajectory Planning, *Proc. of the IEEE Int. Conf. on Robotics and Automation.*

[17] Shekhar, C., Gaudin, V., Moisan, S., and Thonnat, M. (1994), Real-time Supervision of Perception Programs for Prolab2, *IMACS Int. Symposium on Signal Processing and Neural Networks*, pp. 11-14.

[18] Pleczon, P. and Chalard, S. (1993), The Human-Machine Interface of Prolab2 Copilot, *Proc. of the IEEE Intelligent Vehicles Symposium.*

[19] Morizet-Mahoudeaux, P. (1996), On-board and Real-Time Expert Control, *IEEE Expert Intelligent Systems and their Applications*, Vol. 11, N. 4, pp. 71-81.

14 | INTELLIGENT TECHNIQUES IN AIR TRAFFIC MANAGEMENT

Yoshitaka Kuwata
Laboratory for Information Technology
NTT Data Corporation
Japan

Tsutomu Sugimoto
ATC Project
Public Administration System Sector
NTT Data Corporation
Japan

The Public Administration System Sector in NTT Data Corp. has been developing Air Traffic Management systems for the Japanese Ministry of Transport for more than two decades. About five years ago, the AI research group in NTT Data was involved in a joint project for the development of the next generation of an ATM system, as the current ATM system requires highly sophisticated human support mechanisms. The air traffic flow is expected to grow very rapidly in the near future in Japan, and the management of air traffic flow will become a very important issue. As it is very difficult for human air traffic controllers to make decisions in real time, the advanced ATM systems are expected to help human controllers. These systems, which rely on human expert knowledge to manage air traffic flow, will be the key component of the next generation of air traffic flow management. As the final decision must be made by a human Air Traffic Controller, ATC systems work as assistants for human controllers. In this chapter, we describe the latest technologies used in the Air Traffic Management systems. In the first Section, we briefly introduce the current Air Traffic Management systems and their potential problems. In order to solve these problems, the expectations for the next generations of ATC systems are described in Section two, as well as many examples of intelligent ATC systems. In Section three, we explain the model of Air Traffic Flow Management problem, as an example of ATM system. In the following section, we introduce our real-time scheduling algorithm used in the Air Traffic Flow Management system. In Section four, ATFM test system named 'AirTFM' is described followed by our real-time search algorithm for ATC scheduling. We conclude this chapter in Section six.

0-8493-9805-3/99/$0.00+$.50
©1999 by CRC Press LLC

1 Introduction

1.1 Future Air Navigation Systems (FANS)

In 1983, ICAO (International Civil Aviation Organization) established a special committee on the Future Air Navigation System (FANS). The committee was asked to develop a concept for a communication, navigation, surveillance, and air traffic management (CNS/ATM) system to support civil aviation into the 21st century. The FANS completed its work and submitted its final report (Doc 9524, FANS/4) which contains the global concept of new Communication, Navigation, and Surveillance/Air Traffic Management (CNS/ATM) systems in May, 1988.

After review of the FANS/4 report, the ICAO Council charged the existing FANS Committee, on an interim basis, with the new task of advising the Council on the over-all monitoring, coordination of development, and transition planning to ensure that the implementation of the new CNS/ATM systems would take place on a global basis in a cost-effective manner and in a balanced way between air navigation systems and geographical areas. In July 1989, the Council approved the establishment of a new Special Committee of the Council for the Monitoring and Coordination of Development and Transition Planning for the Future Air Navigation Systems. The new committee was commonly referred to as FANS Phase II. The FANS Phase II successfully completed its work with the various products such as the CNS/ATM cost benefit analysis guidance, the global coordinated plan for transition to the ICAO CNS/ATM systems, and the ICAO CNS/ATM systems "coping with air traffic demands" in October, 1993.

1.2 Air Traffic Management

The integration of the new Communications, Navigation, and Surveillance (CNS) systems under the FANS concept of ICAO will make it possible to realize a broad range of ATM benefits that will enhance safety, reduce delays, increase capacity, enhance system flexibility, and reduce operating costs. ATM consists of a ground part and an air part, where both parts are needed to ensure a safe and efficient movement of aircraft during all phases of operations. The execution of ATM calls for a close integration of the ground part and the air part through well-defined procedures and interfaces.

The airborne part of ATM consists of the functional capability interfacing with the ground part to attain the general objectives of ATM. The ground part of ATM comprises the functions of Air Traffic Services (ATS), Air Traffic Flow Management (ATFM), and Air Space management (ASM), where the ATS function was considered to be the primary component of ATM. ATS can be divided into the following subfunctions, i.e., Air Traffic Control (ATC), Flight Information Service (FIS), and Alerting Service(AL).

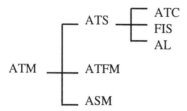

Figure 1: Classification of ATM.

In general, the new ATM system will enable aircraft operators to meet their planned times of departure and arrival and adhere to their preferred flight profiles with minimum constraints and no compromise of safety. The future ATM system will result in

(i)   improvement of air traffic safety :
-     establishment of communication link between pilot and ATS provider
-     surveillance service using new technology
-     reduction of communication error
(ii)   realization of efficient air traffic
-     selection of most effective route and altitude
-     formation of optimum air traffic flow
-     minimization of time and cost
(iii)   enhancement of air traffic capacity
-     aeronautical satellite system
-     area navigation
(iv)   efficient air navigation system
(v)   reduction of pilot and controller workload
-     automated position report
-     data communication of routine messages

1.3 ATM System Issues for FANS

The main objectives of the ATC service are to prevent collisions between aircraft and between aircraft and obstructions on the maneuvering area and to expedite and maintain an orderly flow of air traffic.

One of the reasons for introducing intelligent technology is automation. The concept of automation in aviation is not human-free, but controller-in-the-loop or Human Centered Automation. The ATM system acts as an efficient tool to support decision making of an air traffic controller. Several intelligent ATC systems calculate optimized time of arrival or of starting descent for approach and generate control instructions. In this area, the aspect of real time is strongly needed, so it is very important to consider human factors.

The main objective of ATFM is to ensure an optimum flow of air traffic to or through areas during times when demand exceeds or is expected to exceed the available capacity of the ATC system. ATFM assists ATC in meeting its objectives and achieving the most efficient utilization of available airspace and airport capacity while keeping delay cost to a minimum.

1.4 Applying AI Technology to ATC

There are two main requirements in which AI technology can be advantageously applied.

(1) Safety
Because controllers are humans, the ATC systems need to avoid human error. It is not a good idea to fully automate the ATC process, but it will be possible to help human controllers by giving advice.

(2) Efficiency
By optimizing the schedule, it will be possible to make good use of resources such as air traffic flow capacity, equipment at airports, fuel, and time.

In order to achieve these two requirements, the following technology is expected to be applied.

1.4.1 Scheduling and Planning

As many ATM problems are scheduling and planning problems., AI techniques for these problems can be applied. For example, ATFM problem can be formalized as a scheduling problem with various constraints, which will be described as AirTFM in Section 3 in detail. Similarly, ATS can also be modeled as scheduling.

Although there are common scheduling and planning problems, a predicted scheduling and planning are impractical in many ATM problems, because there are many factors, such as weather conditions, that need to be forecast for the planning and they are difficult to predict. In such cases, a balance of reactive planning and predicted planning is needed. We need to build initial predicted plans and change them according to actual situations. These problems are also *real-time* problems, with a deadline.

1.4.2 Agent Technology

For complex systems, it is useful to model them as agent systems. For example, in OASIS, aircraft are modeled as individual agents. Each agent monitors the status of aircraft. It also acts as a sub-system which optimizes its own benefits. The agent system model is used to decide the next action, by balancing the benefits of each agent. The modeling of a multi-agent system is also very flexible, as we can change an individual agents without changing the other.

2 Intelligent ATC Systems

Three intelligent ATC systems are shown in this chapter. The first is the OASIS system [10][11][12] in Australia which was developed by the Australian Artificial Intelligence Institute. OASIS is based on an agent-oriented paradigm. The second is COMPAS [14] in Germany. COMPAS is a flow control and metering system developed for the German Civil Aviation Authority by the German Aerospace Research Center, DLR. The third is CTAS [15] in U.S.A. CTAS is under development at NASA-Ames Research Center for the U.S.A. Federal Aviation Administration.

2.1 OASIS

OASIS was developed to help ease air traffic congestion. The system achieves this by maximizing runway utilization, achieved through arranging landing aircraft into an optimal order, assigning the landing time, and then monitoring the progress of each individual aircraft in real time.

OASIS is designed by sub-dividing the air traffic management task into its major parts and designing separate agents to solve each of those sub-problems. Each agent solves its part of the task independently, and cooperates with the others to produce the overall system behavior.

Agents communicate with each other and with the environment using messages. Messages are sent and received asynchronously, and are assumed to have assured delivery. However, no guarantees are given or assumed about the processing of the message once it has reached its recipient.

Agents have integrity. Facts, goals, and intentions that are part of the internal state of the agent cannot be manipulated from the outside. For example, if an agent needs the cooperation of another agent, it must request that cooperation using messages. Moreover, the internal state of an agent is private. The only way for an agent to find out the belief, goal, or intention of another agent is to send a message to that agent asking for the information.

As the environment changes, agents must decide how to act. If that deliberation continues for too long, the agent may find that the facts, goals, and state of the world on which the deliberation is based may no longer reflect the current situation.

Hence, each agent must be able to reflect the rate of change in the world and the tasks to be done, in order to effectively use its limited resources.

This design enables us to tailor each individual agent to the sub-problem it is solving. It allows for simplicity of design, high robustness, and dynamically variable reactivity to external events. This agent-oriented design goes beyond the traditional subroutine concept or object-oriented design, as it crucially depends on each agent being an autonomous reasoner.

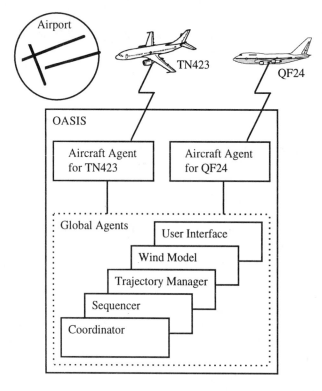

Figure 2: The structure of OASIS.
(©1992 by the Australian Artificial Intelligence Institute. Reprinted by permission.)

OASIS is designed using two classes of agents. First, those that handle inter-aircraft coordination and reasoning, called global agents; and second, those that perform computation or reasoning relevant to each aircraft individually, called aircraft agents. Figure 2 shows how each aircraft has an agent associated with it, including those aircraft anticipated to arrive from flight plans and departure time messages, but not yet detected on radar.

There are five global agents. The COORDINATOR agent serves as the task manager, coordinating the activities of the other global and aircraft agents. The SEQUENCER agent uses search techniques to arrange the aircraft in a least delay/cost sequence. The TRAJECTORY CHECKER agent verifies that instructions proposed by the system do not cause aircraft to violate statutory separation requirements. The WIND MODEL agent uses wind observations made by individual aircraft agents for predicting the wind field that aircraft are likely to encounter. Finally, the USE INTERFACE agent serves as the single point of communication with the Flow Director, managing all user interactions.

The system assigns an agent to each approaching aircraft intending to land at the airport. The AIRCRAFT agents contain all aircraft-specific data required by the system. The AIRCRAFT agent also assimilates position, speed, and altitude reports from real-time radar data. The tasks performed by an AIRCRAFT agent include estimates of when the aircraft will land, monitoring that the progress of the aircraft is as planned, and planning the trajectory of the aircraft.

2.2 COMPAS

COMPAS (Computer Oriented Metering Planning and Advisory System) is an approach sequencing advisory system in Germany and has been in operation at Frankfurt airport since 1988. COMPAS achieves five tasks as follows:

- Support of the air traffic controller: COMPAS displays advice that can be considered by the air traffic controller.
- Avoidance of arrival congestion: safe and quick guidance of aircraft at the arrival routes and early discovery of traffic peaks, metering hints.
- Influence of arrival sequence: calculation of aircraft characteristics for optimal sequence at a certain flow value.
- Reducing workload of the air traffic controller: planning and guidance of the arriving traffic.

COMPAS has many functions as follows:

- Acquisition of dynamic data from existing air traffic control systems – flight plan, weather data, radar data.
- Taking into consideration static data – airspace structure, runway configuration MET zones, aircraft performance.
- Calculation of flight progress / flight profile based on dynamic and static data.
- Generation of optimum landing sequence and time schedule.
- Generation of en-route / terminal control area control advice.
- Display of planned arrival sequence.
- Processing of controller inputs.

2.3 CTAS

CTAS (Center-TRACON Automation System) is an approach sequencing advisory system in the U.S.A. "Center" stands for Air Route Traffic Control Center (ARTCC) which is an air traffic control facility whose primary function is to provide separation of aircraft en route between airports. TRACON stands for terminal radar approach control, which is a terminal air traffic control facility associated with an air traffic control tower that uses radar to provide approach control services to aircraft.

CTAS consists of three separate but integrated automation tools that provide computer-generated intelligence to perform the task of assisting Air Traffic Manager and Controllers in the management and control of arrival traffic at high density airports. One tool, the Traffic Management Advisor (TMA), establishes optimized arrival sequences and scheduled landing times for arriving aircraft as they enter Center airspace. This is installed in the Traffic Management Units at both en route traffic control facilities and TRACONs. The second tool, Descend Advisor (DA), generates control advisories in Center airspace to allow aircraft to meet the TMA scheduled times. En route controllers use the tool. The third tool, Final Approach Spacing Tool (FAST), generates control advisories within the TRACON as a continuation of the process. A field test of DA was conducted at the Denver ARTCC in 1994 [5]. FAST was tested at Dallas Fort Worth International Airport TRACON from February to May 1996.

3 Intelligent Air Traffic Flow Management

3.1 A Model of Air Traffic Flow Management

A simplified model of the ATFM problem is shown in Figure 3. The air traffic control area is divided into small parts named "sectors." One or two human controllers are assigned to each sector and are responsible for controlling planes within that sector. The controllers' job is to organize air traffic flow within the sector, such as by directing planes to specific altitudes, in order to avoid collisions and keep flights safe.

As controllers must communicate with each plane's pilot, the capacity of each sector is defined by the human controller's capability rather than the geometric constraints of the sector. In one model of sector capacity, described in [16], the capacity of each sector is modeled by the capability of human communication through a radio link.

Because capacity depends on controllers, and, thus, on human factors, it is easily influenced by various factors like weather conditions. In bad weather, because the controllers must monitor locations of planes and direct the pilots more frequently, the capacity of a sector is reduced. Emergency situations and handling prioritized planes can similarly reduce a sector's capacity.

Central controllers are responsible for all flights, not just for a single sector. Their job is to authorize flight plans for each plane and suggest changes to avoid overflow in sectors. They are considered mediators rather than central schedulers. The original flight plans are made by carrier companies—controllers provide advice only when traffic flow becomes critical. This is because these flight plans are already optimized by the carrier company from their local point of view, and they prefer that they be executed without modification, if possible.

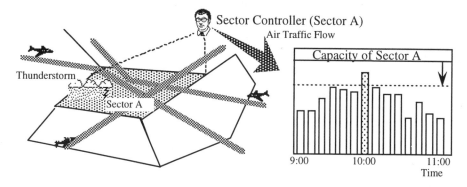

Figure 3: The model of ATFM problem.
Example Scenario: Three planes are scheduled to pass through Sector A as shown on the left. Expected air traffic flow over time is plotted as a histogram on the right. Because of a thunderstorm in sector A, capacity is reduced to the dotted line in the histogram, and capacity overflow is expected at time 10:00. Thus, we need to re-direct planes which pass sector A at time 10:00.
(©1997 by Elsevier Science. Reprinted by permission.)

3.2 Scheduling for ATFM

The goal of the ATFM system is to help central controllers make air traffic flow as efficient as possible. It calculates the expected positions of planes and counts them over time in order to predict the total traffic of each sector. If overflow is predicted, the central controller has three possible plans of action: (a) delay departure time for each plane, (b) suggest alternative routes to pilots, and (c) hold planes already in-flight before they enter the sector where overflow is expected. Plan (c) is inefficient because it has almost the same result as setting departure delays, plan (a), as well as increasing traffic flow in other sectors and wasting fuel. Therefore, holding planes is the plan of last resort, and is done only for planes already in-flight, not for planes that have only been scheduled.

As the first step, our ATFM system model controls only scheduled planes using plans (a) and (b). We leave the problem of handling in-flight planes as future work. As the scheduling of ATFM is considered an optimization problem, in which we need to choose optimal combination of plan change, we can solve the problem by using search technique.

Figure 4 shows an example scenario of sector overflow. Three airplanes, JAL001, ANA002, and SX003 are scheduled to pass throught sector A, B, and C. The maximum flow capacity of these sectors are set at 2, for simplicity. The expected flow of sector B is plotted in the lower part of Figure 4. Because all three planes are planned to pass the sector B, the sector is overflowed from time 15 to 20. When the overflow is detected at sector B, these three planes become the target plane.

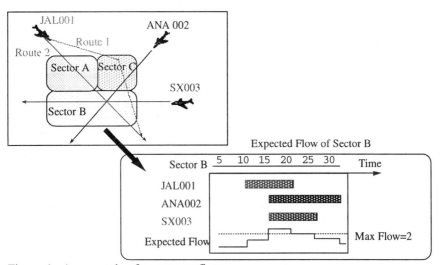

Figure 4: An example of sector overflow.
(Upper) Three airplanes, JAL001, ANA002, and SX003 are planned to pass the sector B. (Lower) The expected flow of Sector B is plotted.
(©1997 by Elsevier Science. Reprinted by permission.)

To resolve the overflow problem at sector B, we can explore an alternative plan using search methods. Figure 5 shows a search tree, in which the root node represents the original plan. From the original plan, three plan changes are possible; changing ANA002's departure time, changing route of JAL001, and changing the departure time of SX003. These three candidates are presented as child nodes of the root node in the search tree. We need to evaluate each plan in order to determine which changes are better. Because the total delay time is considered as one of the most important evaluation factor for plans, the delay time of each plane is plotted in Figure 5. When the new plans don't resolve overflow, we need to consider more changes over these first changes. Two new plans are generated over the third plan. In Figure 5, these plans are represented as new child nodes below the third node. We need to iterate these steps until at least one acceptable plan is found. Although these steps are very simple, it is impossible, in practice, to check all possible plans. This is due to the typical number of planes, needed to handled at one sector, is about 50 to 100. Assuming 100 target planes with three possible changes, for example, the total search space becomes the factorial of 300.

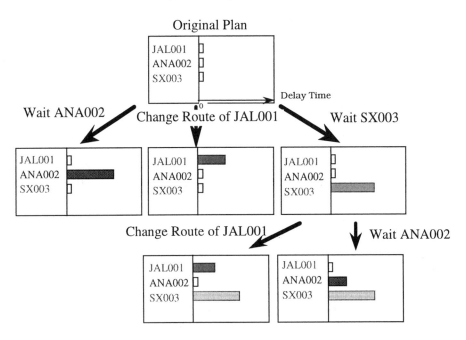

Figure 5: Search tree.
(©1997 by Elsevier Science. Reprinted by permission.)

The actual search space is huge and we have limited time for the search. Thus, we need to balance calculation time and the quality of search results; i.e., real-time search algorithm is important We also introduce the heuristics as described in the following section.

3.3 Heuristics

Two kinds of heuristics are relevant to the problem of managing air traffic flow: those that evaluate the result of re scheduling, and those for quickly finding the best solution. As the ATFM is a newly designed system still in the evaluation stage, there are no human controller heuristics for these tasks. The following evaluation heuristics are generally thought to be good, however.

Evaluation Heuristics

(1) *Minimize delays in departure.*
It is desirable to re-build plans so as to minimize total delays in departure.
(2) *Minimize variance in departure delay times.*
In some cases, setting a large delay for just a few planes can solve the overflow problem. This solution is not appropriate, however, because it is unfair. It is preferable to have well-balanced delays among planes. Therefore, we introduce this evaluation heuristic: the smaller the variance in delay time, the better the modification plan.

We introduce the following heuristics to minimize delays in departure and to prune search space.

Search Order Heuristics

(1) *First-come, first-serve (FCFS) order.*
This is a rule, rather than heuristic for finding the best solution. It is considered fair to assign a higher priority to those planes that enter the sector earlier. Alternately, one can assign FCFS priority according to the time a flight plan is submitted to FDP.
(2) *Assign long distance flights first.*
Delaying long distance flights in the last stage of scheduling is likely to affect many other sectors, thus it is better to consider them in the early scheduling stages.

4 The Air Traffic Simulation Test

4.1 Interactive Plan Steering Architecture

We designed the Air Traffic Flow Management System, named "AirTFM," as a flexible test rather than a simple application. It is intended for use in experiments on the interactions of expert systems with humans. We are also using it for evaluating the performance of real-time search algorithms under realistic conditions in the ATC field. The architecture of AirTFM is shown in Figure 6. This architecture is based on Interactive Plan Steering (IPS), developed by the Experimental Knowledge Systems Laboratory at the University of Massachusetts, Amherst [8,9]. The basic idea of IPS is to provide a mechanism for assisting humans with plan execution. It can be very difficult for humans to predict problems in the execution of a large plan such as those used by ATFM. In IPS, pathology detection components predict problems in plan execution. This is followed by a visualization phase that illustrates the problem to humans. For example, in ATFM, it might point out the overflow of sector T23 at time 10:00 by highlighting the problem on a simulation screen.

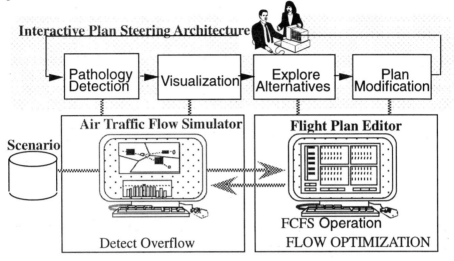

Figure 6: 'AirTFM' test.
       The Air Traffic Flow Simulator (ATFS) and Flight Plan Editor (FPE) have IPS functionality built in to help humans steer plans at runtime.
       (©1997 by Elsevier Science. Reprinted by permission.)

After humans notice the problem, they can explore alternative plans supported by the simulator in the IPS architecture. This step invokes "what-if" style questions. For example, they might ask what will happen if we redirect JAL-001 to a northern route rather than a Pacific route. After verifying the proposed changes, the original plan is actually modified, and continues executing.

4.2 AirTFM – the ATFM Test

We integrated this functionality into two components of AirTFM. One is an Air Traffic Flow Simulator (ATFS), which calculates the position of each plane based on flight plans in the scenario. ATFS also predicts the amount of flow in each sector, which corresponds to pathology detection. ATFS's function is to control time in the simulated world. The traffic situation in the scenario can be displayed at any time. Simulation speed can easily be specified by using the "real-time knob" on the simulator's panel. For example, setting the real-time knob to "60" produces a ratio of 1 minute of simulation time to 1 second of processor time. By speeding up simulation, we can evaluate the result of the controllers' modifications before they are actually executed. Backward simulation is also implemented to help uncover the causes of problems.

The other component of AirTFM is a Flight Plan Editor (FPE). The FPE allows a human controller to view and edit flight plans under consideration. The FPE contains two functions that may be used to advise the human controller. The first, FCFS, or first-come-first-served operation, is used to calculate the minimum delay for each plane, assuming priority is based on the time the plane entered the sector. The second, Flow Optimization, is intended to calculate candidates for modification that minimize total departure delay. It uses the adaptive heuristic search algorithm discussed in the previous section.

ATFS uses a scenario that contains sector information and flight plans associated with a time to be submitted to the simulator. In the current scenario, random factors, like unexpected changes in plane speed, are not included. Therefore, the prediction is 100 percent accurate; given the same scenario, the simulator produces exactly the same situations every time, with the exception of any changes made by human operators via ATFS's control panel and FPE's. Figures 7 and 8 show a snapshot of the ATFS and FPE screens.

ATFS is written in C using the SL-GMS[1] graphic modeling system for rapid prototyping. FPE is written in C++, using our original search class library (SCL) and SL-GMS. Both components of AirTFM are implemented on a Sparcstation 20 with the X11 window system running on SunOS 4.1.3. Both components run with reasonable speed under this configuration.

The AirTFM components are designed to communicate with each other by socket mechanisms via Ethernet, thus they can also be run on separate machines. More than one ATFS or FPE can run at the same time. For example, we can run two FPEs in order to compare the quality of solutions given different calculation times. This feature is also useful for executing multiple simulations under different assumptions before the final decision is made.

5 Real-Time Search Algorithm for Air Traffic Flow Management

5.1 Real-Time Search Algorithms

5.1.1 Real-Time Planning and Scheduling Algorithms

Several scheduling and planning algorithms have been proposed for real-time problem solving in which an explicit deadline exists. The task of these algorithms

---

[1]SL-GMS, an object-oriented graphical modeling system is a product of the Sherrill-Lubinski Corporation in U.S.

is to balance the quality of the solution against thinking time.

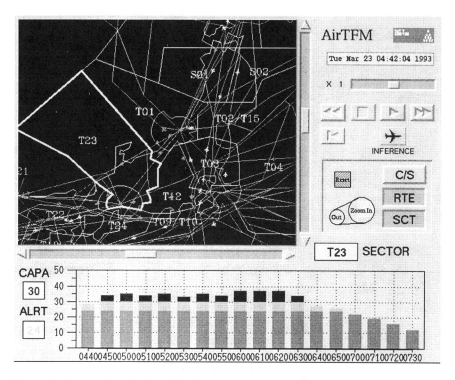

Figure 7: A snapshot of air traffic flow simulator on AirTFM.

The top left area shows the locations of planes on a map. A human operator can scroll the map to view any area in Japan and zoom in on sectors for a closer look. Buttons in the top right corner allow them to control simulator time. The histogram in the bottom of the panel shows predicted traffic flow in specified sectors for the next three hours, in 10-minute intervals. The operator can directly indicate sectors of interest by clicking on a sector in the map display area. Highlighted areas indicate sectors where the predicted traffic flow is expected to exceed the area's capacity.

(©1997 by Elsevier Science. Reprinted by permission.)

Incremental algorithms solve problems gradually and can produce the best known solution at whatever time they are told to stop thinking. Assuming the existence of incremental algorithms, Boddy and Dean [1,2] proposed 'Anytime Algorithms,' which schedule planning and execution time before the plan's execution in order to maximize utility of action. Anytime algorithms are based on the trade-off between the expected increase in plan quality when given more time to think versus the potential decrease in benefits caused by delaying action.

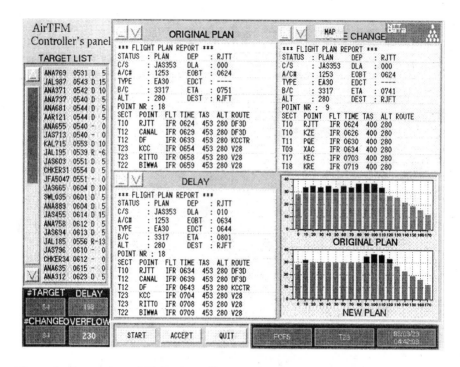

Figure 8: A snapshot of flight plan editor on AirTFM.

Target planes are sorted by priority and listed in the left box, with a scroll bar. Choosing a plane from the list displays its original flight plan in the top middle area. An alternative flight plan with a minimum delay is shown in the bottom middle area. Another alternative plan with route changes is shown at top right, but only when valid alternative routes exist for the plane. Human controllers can choose modifications from these alternative plans. In FCFS mode, the FPE automatically selects planes that need to be redirected from the top of the list, and asks humans to choose among them. In flow optimization mode, the FPE tries to find the best combination of modifications within the allotted time. Expected flow is shown as a histogram in the bottom right area of this panel, including both original flow (upper histogram) and new plan (lower histogram).

(©1997 by Elsevier Science. Reprinted by permission.)

In many applications, incremental algorithms are not practical or even possible. For example, if we need to increase the resolution of image analysis in an image understanding task, the whole process must be completely repeated and previous results cannot be used. Garvey and Lesser [4] propose Approximate Reasoning for such applications. In approximate reasoning, processing time is estimated first.

Then the processing method that best fits the deadline is chosen. As the estimation of execution time contains variance, execution must be monitored and controlled at runtime.

5.1.2 Real-Time Monitoring and Control Algorithms

Other approaches are found in the real-time search field. Korf has proposed the RTA* search algorithm, which uses a variation of minmax search, discussed in [6,7]. From the current state, RTA* looks forward to a fixed-depth horizon and applies a heuristic function to the nodes found there. The result is sent back to the current node, and the next candidate is chosen.

DYNORA II is an algorithm proposed by Hamidzadeh [5], in which the planning and execution cycle is executed as part of the search. It tries to balance planning cost and execution cost every iteration, before the new nodes are extracted. A comparison of deadline compliance between DYNORA II and RTA* is presented in [5].

5.2 Time-Dependent Heuristic Search (TDHS)

We modified the heuristic search to meet the needs of real-time problem solving. The goal of this search is to find the best solution within a previously specified time period. If the algorithm cannot find any solution by the deadline, it has no value. On the other hand, if it uses depth-first search, it might not reach a better solution in another branch of the search tree before the deadline. Thus to maximize the expected score, we need a heuristic for choosing the order in which branches are searched.

A* search is a likely candidate for the problem of determining which nodes are to be searched first. In A* search, two values are used to determine search order: the G function, which is the cost to reach a node, and the $\hat{H}$ function, which returns the estimated distance to the goal node. By choosing nodes that have minimum $G + \hat{H}$ value, the best possible node is chosen first. However, in A* search, there is no guarantee that we will get the best possible solution by a particular deadline, as the number of nodes visited depends only on these two functions; i.e., they are independent of the deadline.

We introduce the concept of processing time in the cost function of our Time-Dependent Heuristic Search (TDHS) algorithm, which is shown in Figure 9.

TDHS tries to search as widely as possible in the early stage of the search, and focuses on reaching the goal when the deadline nears. By re-evaluating the cost value of all frontier nodes in steps 17 through 19, the priority of search order changes as time passes, so that it forces the algorithm to search deeper as the deadline approaches.

```
1 :  FrontierList <- RootNode
2 :  BestSolution <- NULL
3 :  RestOfTime <- TimeLimit
4 :  PastTime <- 0
5 :  loop
6 :   CurrentNode <- TopPriority(FrontierList)
7 :   if Goalp(CurrentNode)
8 :      if Null(BestSolution)
9 :         BestSolution <- CurrentNode
10:         elseif Eval(BestSolution)<Eval(CurrentNode)
11:         BestSolution <- CurrentNode
12:   else
13:      SuccessorList <- Successors(CurrentNode)
14:      For all nodes in SuccessorList
15:         calculate h(the_node)
16:      Merge(FrontierList,SuccessorList)
17:   For all nodes in FrontierList
18:      g(the_node) <-TDG(Depth(the_node),PastTime)
19:      priority(the_node) <-g(the_node)+h(the_node)
20:   RestOfTime = RestOfTime - DeltaTime
21:   PastTime = PastTime + DeltaTime
22: until(RestOfTime <= 0)
23: return(BestSolution)
```

Figure 9: Time-dependent heuristic search (TDHS) algorithm.
TimeLimit and RootNode are provided outside of this algorithm. DeltaTime is the time required to process the loop once, starting from step 5. In step 18, the Time-Dependent G (TDG) function is re-evaluated for all the frontier nodes. By choosing a TDG function that forces the algorithm to go deeper, TDHS reaches the best possible solution within the time specified.
(©1997 by Elsevier Science. Reprinted by permission.)

5.3 Complexity Analysis of TDHS

In the domain of real-time search problems, it is often meaningless to consider the entire search space, so we usually stop before searching the entire space. Therefore, we will discuss the time complexity of extracting one node instead. This is the time complexity of one iteration, from step 5 to step 23, of the TDHS algorithm.

Given the length of the frontier list l, and the number of successor nodes b, the time complexity is presented below. Note that steps 6 to 13 are ignored as they operate in constant time.

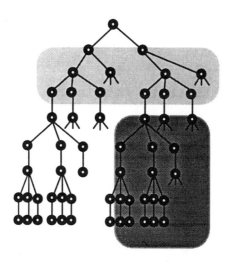

Figure 10: Search order of TDHS.
> By choosing a proper TDG function, the TDHS algorithm can search shallower levels at the beginning stages of the search, in order to avoid missing good candidates. It then tries to go deeper in the later stages of the search to reach goal nodes.

For the generation of successor nodes in steps 13 to 15, time complexity $T1$ is given by

$$T_1 = K_1 \cdot b = O(b) \tag{1}$$

For merging the frontier list in step 16

$$T_2 = K_2 \cdot b \cdot l = O(b \cdot l) \tag{2}$$

For re-evaluating and sorting the frontier list in steps 17 through 23

$$T_3 = K_3 \cdot (b+l) + K_4 \cdot (b+l) \cdot \log(b+l)$$
$$= O\big((b+l) \cdot \log(b+l)\big) \tag{3}$$

Therefore, the total complexity of one iteration is given by the following equation:

$$T_{THDS} = T_1 + T_2 + T_3$$
$$= O(b) + O(b \cdot l) + O((b+l) \cdot \log(b+l))$$
$$= O(l \cdot \log(l))$$

as $b << l$

(4)

$T3$ is not needed in the usual A* algorithm

$$T_{A*} = T_1 + T_2$$
$$= O(b) + O(b \cdot l)$$
$$= O(b \cdot l)$$

(5)

Because of the cost of re-evaluating the frontier list, the complexity of TDHS increases from $O(b \cdot l)$ to $O(l \cdot \log(l))$.

5.4 Time-Dependent Cost Function

As described in the last section, a time-dependent cost function that forces the search to progress as time passes must be chosen in order to meet the deadline. One simple linear, time-dependent cost function is defined by the following equation.

$$TDG(Time, depth) = \left| \frac{Time - TimeLimit}{TimeLimit} - \frac{depth - maxDepth}{maxDepth} \right|$$

where, $TimeLimit$ = Total Processing Time Allocated

and $maxDepth$ = Estimated Depth to Find Solution

(6)

Note that this function is normalized from 0 to 1 by dividing by TimeLimit and the estimated maximum depth of the search tree. In order to make this function work with our heuristic function, we need to assign a weight to the cost function. The profile of this function is shown as a 3D plot in Figure 11. In this function, the lowest cost is obtained by going deeper as the process progresses.

We used TDHS for the calculation of flow optimization in our test, described in the following section. The depth of the search tree is not always known before the search. In such cases, the estimated depth of the search tree can be used instead of *MaxDepth* in Equation (6). It is not always a good idea to use TDHS in search problems having an unknown depth, as reaching a goal node is not guaranteed, even with depth-first search. Another simple candidate for the time-dependent cost function is described by the following equation:

$$TDG(Time, depth) = \begin{cases} G1 & \dots if\left(Time \le \dfrac{TimeLimit}{2} \right) \\ G2 & \dots if\left(Time > \dfrac{TimeLimit}{2} \right) \end{cases}$$

where, $TimeLimit$ = Total Processing Time Allocated

and $maxDepth$ = Estimated Depth to Find Solution

(7)

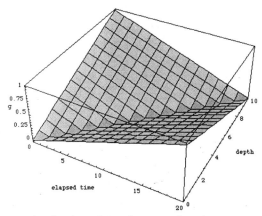

Figure 11:   An example of a time-dependent cost function.
The case of TimeLimit = 20 and MaxDepth = 10 is shown. Note
that the cost (TDG in the previous equation) value shown on the Z
axis is normalized.
(©1997 by Elsevier Science. Reprinted by permission.)

This function simply switches the cost function after half of the allocated time has
elapsed. For example, if we use search *depth* as *G1* and *(- H - depth)* for *G2*, then
TDHS will follow the usual A* search algorithm in the first half, then switch to
depth-first search in the second half.

The benefit of using this function is the reduced complexity of the algorithm. In the
case of Equation (6), we must re-evaluate the cost of each node every time we
choose the next candidate. In the time-dependent cost function of Equation (7), re-
evaluation of all of the frontier nodes is required only once, at the half time of the
search. The complexity of TDHS search using Equation (7) is the same order as
that for A* search.

5.5 Experimental Results on TDHS

5.5.1 The Traveling Salesperson Problem

To evaluate the TDHS algorithm, we chose the Traveling Salesperson Problem
(TSP). In TSP, the geological locations of cities (or the distances between cities)

are given to the problem solver. The goal is to find the order in which to visit all the cities so that the distance traveled is minimized. In this chapter, we define the evaluation metric for the search to be the shortest distance found in the time allotted. If the minimum distance is not found by the deadline, the current shortest path will be used to determine the algorithm's score. If the problem solver has not found any solution by the deadline, its score for that trial is zero.

The Traveling Salesperson Problem was chosen because it can be solved by simple search and the depth of the search tree is known. If the first city is not given, the depth of tree d will be n, where n is the total number of cities to visit. The branching factor of the search tree is $(n - i+1)$ for the ith city. Although various heuristics have been proposed for this problem, we have used two simple ones: 1) $H1 = td / ad$, and 2) $H2 = td / ad / 2$, where td is total distance traveled and ad is average distance traveled per step. These two heuristics are intended to balance the current total and average distances; if the current distance to a node is longer than average, searching its subtree will be postponed until later.

To evaluate the algorithm, Depth-first Search (DFS), Breadth-first Search (BFS), and Random Search (RANDOM) are compared to TDHS. For the TSP experiments, one hundred sets of six randomly generated locations are chosen in a 100 x 100 field. Larger-scale problems were also run, with results similar to those for the small-scale experiments. We present only the small-scale cases in this chapter. One case of a 6-city TSP is illustrated in Figure 12.

5.5.2   Base Performance of Heuristics

Figure 13 shows the results of the basic search algorithms (DFS, BFS, A* with H1 and H2, and random search) over a 6-node TSP. The X-axis shows the number of nodes generated, which corresponds to search time, assuming a constant time for visiting one node. The Y-axis shows the relative performance, calculated by the formula: $score = 300 - DistanceTraveled$. In this graph, a score of 0 means that no answer was found by the specified deadline. For example, BFS doesn't reach a goal node until almost all nodes have been searched, and thus gets no score until after about 1250 nodes have been expanded.

Both DFS and RANDOM reach a goal node in the very early stages of search but take a long time to improve their scores because they lack heuristics. On the other hand, A* search using H1 and H2 reaches the goal node later, but finds a better score than DFS or RANDOM.

From the viewpoint of real-time search algorithms, DFS and RANDOM are better than A* in cases where the search deadline is less than approximately 1000 nodes. When the deadline is longer, A* search works better than DFS and RANDOM. BFS consistently performs worse than any of the other algorithms. Comparing the two A* searches using different heuristic functions, H1 reaches a better solution earlier than H2, thus H1 is a better algorithm than H2.

5.5.3   Results of TDHS on TSP

Figure 14 shows the result of TDHS search using the heuristic function H1 and the cost function described in Equation (7) for deadlines of D = 300, 500, 1000, 1500, and 2000 nodes. For comparison, the scores of DFS, BFS, and A* using H1 are also shown.

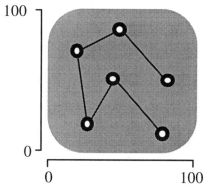

Figure 12: An example of the TSP (size = 6).
For this case, the average traveling distance per step (*ad*) was empirically determined to be approximately 52.

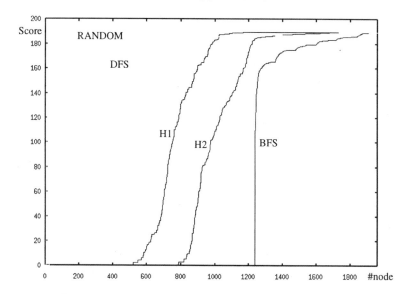

Figure 13: Results for basic search techniques.
The result of DFS, BFS, RANDOM, and A* using H1 and H2 search over the 6-city TSP are shown (average of 100 trials).

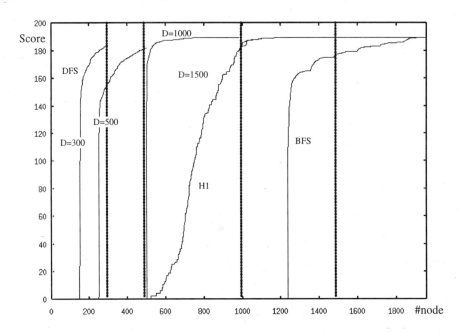

Figure 14: Performance of TDHS with heuristic function H1.

TDHS gets a higher score at each deadline than either basic A* search or DFS. For example, given a deadline of 500 nodes, the score of TDHS is about 181, while DFS gets 175 and A* has a score of 0 (no answer is found by the deadline). Similar results are obtained for other deadlines. Given enough time, such as D = 1500 or 2000 nodes, the score of TDHS equals that for A*. Therefore, even with a simple time-dependent cost function like Equation (7), TDHS works as well as or better than A* and DFS.

Figure 15 shows the results of the same experiment as Figure 14 but using the heuristic function H2. Again, TDHS gets a higher score at each deadline than either the basic A* search or DFS, supporting the results of the previous experiment. Here, however, the scores are lower because H2 is a less efficient heuristic function than H1.

The results of using TDHS with a random heuristic function are shown in Figure 16. In this case, the performance of TDHS is almost the same as both the basic RANDOM and DFS search, thus there is no benefit in using TDHS with this heuristic. With good heuristic functions, TDHS can make the best use of a given time period. Without a good heuristic, it is equivalent to the usual basic search methods.

6 Summary

We presented a model of the Air Traffic Flow Management problem. Real-time problem solving strategies were discussed and one candidate algorithm was proposed. An architecture design and implementation of an ATFM problem test system were also shown. We evaluated the real-time algorithm based on its quality of solution.

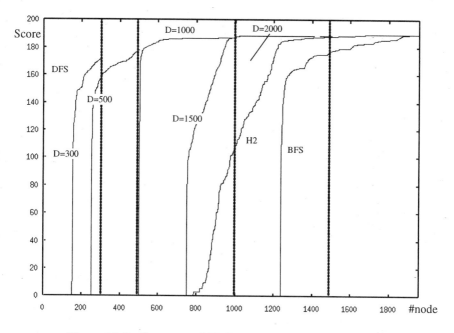

Figure 15: Performance of TDHS with heuristic function H2.

The current test cannot handle many kinds of events, such as closing down airports and the subsequent need to redirect all in-bound planes elsewhere. Such capabilities are required if we have to include more realistic scenarios in our evaluation of expert system–human interactions. We are planning to extend AirTFM to make use of a multi-agent system, to handle more dynamic situations.

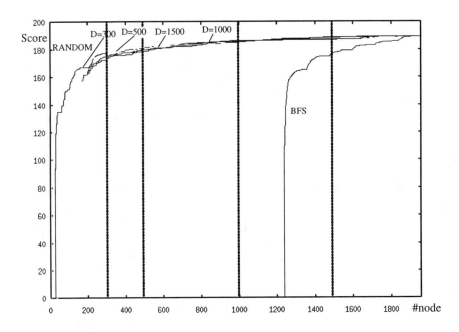

Figure 16: Performance of TDHS with a random heuristic function.

References

[1] Boddy, M. and Dean, T. (1988), An Analysis of Time-Dependent Planning, *Proceedings of the Sixth National Conference on Artificial Intelligence*, pp. 49-54.

[2] Boddy, M. and Dean, T. (1989), Solving Time-Dependent Planning Problems, *Eleventh International Joint Conference on Artificial Intelligence*, pp. 979-984.

[3] Cohen, P.R., Greenberg, M.L. Hart D.M. and Howe A.E. (1989), *Real-Time Problem Solving in the Phoenix Environment*, Department of Computer Science Technical Report, University of Massachusetts, Amherst, MA.

[4] Garvey, A. and Lesser, V. (1991), *Design-to-time Real-Time Scheduling*, Department of Computer Science Technical Report. 91-72, October, 1991, University of Massachusetts, Amherst. MA.

[5] Hamidzadeh, B. and Shekhar, S. (1992), Can Real-Time Search Algorithms Meet Deadlines? *Proceedings of the Tenth National Conference on Artificial Intelligence*, pp. 486-491.

[6] Korf, R. E. (1987), RealTime Heuristic Search First Results, *Proceedings of the the Fifth National Conference on Artificial Intelligence*.

[7] Korf, R. E. (1988), RealTime Heuristic Search New Results, *Proceedings of the Sixth National Conference on Artificial Intelligence*.

[8] Kuwata, Y., Hart, D.M. and Cohen, P.R. (1992), Steering Execution of a Large-Scale Planning and Scheduling System, *Workshop Notes for AAAI-92 SIGMAN-Workshop on Knowledge-Based Production Planning*, Scheduling & Control, pp. 62-72.

[9] Kuwata, Y., Hart, D.M. and Cohen, P.R. (1993), Steering Traffic Networks, *Workshop Notes for AAAI-93 Workshop "AI in Intelligent Vehicle Highway Systems"*, pp. 51-56.

[10] Ljungberg M. and Lucas A. (1992), *The OASIS Air Traffic Management System*, The Australian Artificial Intelligence Institute, Technical Note 28, August, 1992, La Trobe Street, Melbourne 3000, Victoria, Australia.

[11] Lucas A., Ljungberg M., Evertsz R., Tidhar G. Goldie R. and Maisano P. (1994), *New Techniques for Air Traffic Management for Single and Multiple Airports,* The Australian Artificial Intelligence Institute, Technical Note 51, April, 1994, La Trobe Street, Melbourne 3000, Victoria, Australia.

[12] Ljungberg, M. and Lucas, A. (1992), *The OASIS Air Traffic Management System,* Technical Note 28, The Australian Artificial Intelligence Institute, August.

[13] http://www.aaii.oz.au/proj/oasis/oasis.html

[14] http://www.bs.dlr.de/ff/fl/24/compas/compas.html

[15] http://pioneer.arc.nasa.gov/af/afa/ctas/

[16] Tuan, P.L., Procter, H.S., and Couluris, G.J. (1976), *Advanced Productivity Analysis Methods for Traffic Control Operations*, SRI international Technical Report FAA-RD-76-164, Menlo Park, CA, U.S.A..

[17] Wild, B, (1992), Sapporo: Towards an Intelligent Integrated Traffic Management System, *Proceedings of the International Conference on Artificial Intelligence Applications in Transportation Engineering*, San Buenaventura, CA, U.S.A., pp. 19-37.

INDEX

417